ACTA PHYSICA AUSTRIACA / SUPPLEMENTUM VI

PARTICLE PHYSICS

PROCEEDINGS OF THE
VIII. INTERNATIONALE UNIVERSITÄTSWOCHEN
FÜR KERNPHYSIK 1969 DER KARL-FRANZENS-UNIVERSITÄT
GRAZ, AT SCHLADMING (STEIERMARK, AUSTRIA)
24th FEBRUARY—8th MARCH 1969

SPONSORED BY
BUNDESMINISTERIUM FÜR UNTERRICHT
THE INTERNATIONAL ATOMIC ENERGY AGENCY
STEIERMÄRKISCHE LANDESREGIERUNG AND
KAMMER DER GEWERBLICHEN WIRTSCHAFT FÜR STEIERMARK

EDITED BY
PAUL URBAN
GRAZ

WITH 75 FIGURES

1969

SPRINGER-VERLAG / WIEN · NEW YORK

Organizing Committee:

Chairman: Prof. Dr. Paul Urban
Vorstand des Institutes für Theoretische Physik, Universität Graz

Committee Members: Prof. Dr. P. Urban
Dr. R. Baier
Dr. H. Kühnelt

Proceedings: Dr. H. J. Faustmann
Dr. P. Pesec

Secretary: M. Pail

Acta Physica Austriaca / Supplementum I
Weak Interactions and Higher Symmetries
published in 1964

Acta Physica Austriaca / Supplementum II
Quantum Electrodynamics
published in 1965

Acta Physica Austriaca / Supplementum III
Elementary Particle Theories
published in 1966

Acta Physica Austriaca / Supplementum IV
Special Problems in High Energy Physics
published in 1967

Acta Physica Austriaca / Supplementum V
Particles, Currents, Symmetries
published in 1968

ISBN 978-3-7091-7640-5 ISBN 978-3-7091-7638-2 (eBook)
DOI 10.1007/978-3-7091-7638-2

Softcover reprint of the hardcover 1st edition 1969

Library of Congress Catalog Card Number 77-97981

Titel-Nr. 9261

Contents

Contents

Ladies and Gentlemen!

It is a great pleasure for me to welcome you at our eighth Schladming winter school. As the organizer of this meeting I would like to thank all of you for accepting my invitation. About 220 participants from 19 nations, the largest number in the history of this school, and the presence of high public officials at this opening ceremony indicate that our school meets with increasing approval among specialists as well as in the public.

I have the special honor to welcome the representatives of the Ministry of Education and of the provincial government of Styria, as well as of the International Atomic Energy Agency and of the Chamber of Commerce. I would like to ask most heartily all our sponsors to accept my warmest thanks for their financial support, which made it possible to organize these "Universitätswochen". It is a great pleasure for me to have among our guests of honour the Rector of the University of Graz, Prof. Dr. O. Burkard. Last but not least I welcome Director Laurich, Mayor of our host city Schladming, who has helped us in a dedicated way in organizing this symposium, and, as we have seen already, has taken great pains to carefully prepare the surrounding slopes to satisfy also the skiing-fans among us. In this connection I would like to ask the township the favor of taking care for a constantly fair weather during our stay here.

Finally I welcome all participants of this meeting from this country and abroad and I hope you will derive great benefit from the lectures and discussions with regard to your own scientific problems. It is a great privilege for me to have suceeded in gaining the cooperation of highly outstanding experts and I want to express my thanks that they have undertaken the trouble of coming to us, which will be rewarded by the benefit their lectures will render to us.

Let me now turn to our scientific program. As was the case last year our program this year, too, deals with problems of elementary particle physics and in particular with the problems connected with strong interactions. The absence of a unified theory of strong interactions has led to the establishment of various theories and models in order to get some new information and to explain the experimental results available.

One of these attempts is the Regge Pole Theory, which during the last years has made considerable progress. By combining this theory with dispersion-theoretic statements we arrive at so-called finite energy sum rules, which lead to the concept of duality in strong interactions. Some other approaches giving insight into problems of strong interactions are provided by effective Lagrangian and Current Algebras. Among other topics we will hear something on experimental tests of CPT, CP, T and C and on problems in connection with $e^+ e^-$—scattering in storage rings.

Once more I wish to thank all of you for your coming, and in opening this meeting on "Particle Physics" I wish it every success.

Paul Urban

In memoriam Professor Gunnar Källén

About four months ago, on October 13, 1968, Professor GUNNAR KÄLLÉN of the University of Lund, Sweden, died in an airplane accident near Hannover. Undoubtedly Europe lost one of its most prominent theoretical physicists; moreover, ever since the foundation of our Schladming winterschool, KÄLLÉN was one of its most regular participants and repeatedly showed his great interest

in this meeting by lectures, critical summaries and by engaging himself in vivid discussions. His intimate relationship with our school and his continuous friendship with the author of this obituary are a special reason for honoring in this place Professor Källén and his outstanding scientific successes.

Let me first briefly state the significant dates in his remarkable scientific career:

Born on February 13, 1926 at Kristianstad, KÄLLÉN studied physics in Sweden and completed his doctorate at the University of Lund in 1950. Starting as an assistant professor at Lund from 1950 to 1952, he got to the Theoretical Study Division of CERN at Copenhagen (1952—1957), worked at NORDITA (1957—1958) until he was offered a professorship for theoretical physics at the

University of Lund. Moreover, KÄLLÉN made several journeys for the purpose of research and took part in numerous scientific conferences in about 15 countries, including the USA and the USSR.

Let me now mention some details about his scientific work: Initially he studied electrical engineering but soon he changed over to physics, especially to the problems of quantum electrodynamics; in this field he achieved most important results in the years of 1949 to 1955. His principal aim was the treatment of the theory of renormalization using, unlike other authors, consequently the Heisenberg picture instead of the interaction picture and the relations now known as Yang-Feldman equations. Considering spectral representations for two- and three-point functions he succeeded in separating the renormalization constants of quantum electrodynamics and in expressing them as integrals over certain weight functions; thus he could precisely formulate and try to solve the problem of the value of renormalization constants. Indeed, other authors are in doubt about his famous proof that at least one of the renormalization constants has to be infinite, but so far no definite answer to this question could be found.

KÄLLÉN's authority at that time in the field of quantum electrodynamics is well illustrated by the fact that it was he who was requested to write the article on this topic in the "Handbuch der Physik".

In connection with his work on quantum electrodynamics he began to study closely the analyticity properties of three- and four-point functions and obtained a number of important results, partially cooperating with WIGHTMAN and TOLL.

We must not forget a treatment of the Lee model, which KÄLLÉN did together with PAULI and where they discovered and discussed the possibility of "ghost-states".

KÄLLÉN was not only interested in the development of the general theory, he also treated many difficult concrete problems, such as in his works on vacuum polarization of higher order.

During the last years KÄLLÉN performed fundamental work in the field of radiative corrections to weak decays. Let me please talk in some detail about the article "Radiative Corrections to β-decay and Nucleon Form Factors", published in Nuclear Physics 1967, because it was KÄLLÉN himself who gave a lecture on the basic thoughts in a preliminary version at our Schladming meeting 1966 for the very first time:

Taking into account strong interactions, KÄLLÉN tried to find a physically reasonable solution to the convergence problem for radiative corrections of order α in Fermi decays. As you know, radiative corrections in β-decay and also in other semileptonic or leptonic hadron decays depend on a cutoff if one uses a "bare particle calculation" (i.e. without strong interactions). (Essentially the lifetime reads:

$$\frac{1}{\tau} = \frac{1}{\tau_0}\left(1 + \alpha f(M_F, M_{GT}, G_A/G_V) \ln \frac{\Lambda}{m_p} + \text{finite terms}\right),$$

Λ being the cutoff.)

KÄLLÉN took up the idea proposed by BERMAN and SIRLIN in 1962, namely to take into account strong interactions by the introduction of suitable form factors. But there is one crucial difference to BERMAN and SIRLIN:

KÄLLÉN uses on-mass-shell form factors, that means quantities which are experimentally measurable in principle and can therefore be used as phenomenological parameters of the theory. Of course, I can treat the formalism, which permits the introduction of such form factors in a natural way, only generally.

Essentially one decomposes the β-decay matrix element of the current-current Lagrangian into a product of matrix elements of the electron field operator and the weak hadron current by summing over intermediate states which can bring about radiative corrections of order α. As is the case in dispersion theoretical considerations states of higher mass are neglected. The matrix elements of the hadrons are calculated first for point particles and are then modified corresponding to strong interactions by the introduction of only a few form factors; for example, the matrix element $(p, \gamma \mid j^V \mid n)$ is considered as

$$(p, \gamma \mid j^V \mid n) = \mathfrak{F}\,(p, \gamma \mid j^V \mid n)_{\text{point particles}}$$

with a form factor $\mathfrak{F}$, which can be measured in principle by bremsstrahlung processes of nucleons in external fields by virtue of CVC hypothesis. Clearly we also get the wellknown electromagnetic form factors of the nucleons in the matrix element $(p \mid j^{el} \mid p')$ occurring in the development of the electron operator in order α, and also in $(p \mid j^V \mid n)$ on the account of CVC.

By suggesting for the unknown form factor a behavior similar to the usual ones one obtains higher powers of the photon momentum in the denominator, infinite integrals do not occur anymore.

A further restriction, which nevertheless is reasonable and justified is the following: Since the electromagnetic electron operator renormalization is totally independent of strong interactions and yields an infinite integral in the usual Feynman gauge, that means it is gauge dependent, only those gauges are considered which guarantee a finite result for the renormalization; in summing up all corrections affecting the lifetime, the gauge dependent contributions cancel out; KÄLLÉN called this fact "restricted gauge invariance".

There are two important features in this method: first we get finite radiative corrections (although no exact numerical results can be expected because of the approximative character of the formalism), and secondly, an estimate of the cutoff is possible. This estimate shows that the assumption $\Lambda \approx m_p$ in the bare particle calculation was a very good approximation.

KÄLLÉN's result—finite radiative corrections by means of strong interaction—is in striking disagreement with works of other authors who included the modern concept of current algebra in the calculation of radiative corrections in β-decays.

The essential point in this alternative method is the contraction of the photon in the matrix element $(\pi^0, \gamma \mid j^V \mid \pi^-)$, hereby obtaining a retarded commutator of two currents. This commutator can be calculated in case of equal times by the relations of current algebra. The result is a logarithmically divergent factor in the vector coupling constant in spite of strong interactions, as long as the vector part of the weak interaction is considered. Because of the model-dependence of the axial vector part the authors of the current algebra method get finite radiative corrections by constructing a composite model of hadrons in such a way that the resulting axial vector divergence compensates the logarithmic vector divergence.

Though KÄLLÉN did not solve the problem of the influence of strong interactions on the convergence of radiative corrections in weak decays (by means of his form factor method) it turned out to be an important controversial question in this way, still lacking a final satisfactory solution.

KÄLLÉN was one of the first who used reduction formalism, dispersion calculations and spectral representations in all his works, methods which became standard tools in modern physics. Surely KÄLLÉN's works have contributed much to the fact that field theory is applied in elementary particle physics more than ever.

Furthermore KÄLLÉN has earned considerable merits in the field of elementary particle physics as the author of an excellent book in which many problems of strong and weak interactions are treated. Here, just as in his conference lectures, KÄLLÉN proved his outstanding pedagogical talent. In his book he has shown excellently how much about mathematical methods and detailed calculations should be presented, enough to clear up the connection between theory and experiment, but not so extensively that the presentation could be spoiled.

Ladies and gentlemen, I hope that to a certain extent I have been successful in doing justice to the personality of GUNNAR KÄLLÉN and his position in science. You will certainly agree if I emphasize again that his early death undeniably has left a gap among the most outstanding theoretical physicists of Europe.

PAUL URBAN

MASSLESS PARTICLES AND ANALYTIC CONTINUATION IN MASS†

By

A.O. BARUT††

International Centre for Theoretical Physics
Trieste

I. INTRODUCTION

The fundamental importance of the massless particles in physics is seen from the fact that almost all known interactions are dominated by a massless particle (with an extrapolation to soft mesons ($m\to o$) in the case of strong interactions) (Table 1).

Interaction	"Massless" Particle	
Gravitational	graviton	$j = 2$
Electromagnetic	photon	$j = 1$
Weak (leptonic)	neutrino	$j = \frac{1}{2}$
Strong	pions(kaons)	$j = 0$

Table 1 - Massless Particles and Fundamental Interactions.

An exception seems to be the non-leptonic weak interactions of strange baryons: whether any massless virtual

† Lecture given at the VIII. Internationale Universitätswochen für Kernphysik, Schladming, February 24-March 8,1969.

††On leave from the University of Colorado,Boulder,Colorado.

states enter into these processes is not clear.

The strength of the interaction decreases with increasing spin for bosons : $j = 0$ (strong), $j = 1$(em), $j = 2$ (gravitation), $j = 3$ (no) and for fermions $j = \frac{1}{2}$ (weak), $j = \frac{3}{2}$ (no).

II. GAUGE CONDITIONS

It is wellknown that the limit from a massive particle to a massless particle is a discontinuous one: there are $(2j+1)$ states of polarization for $m>o$, but only one state of polarization for $m = 0$, no matter what the spin j is. This is a direct consequence of the properties of the unitary irreducible representations, labelled by $[m,j]$, of the restricted Poincaré group. One could treat therefore the properties of the scattering amplitudes for massive and massless particles quite independently taking into account from the beginning the condition $m=o$. There is however a device which allows one precisely to perform the discontinuous limit $m\to o$. It is important to use this limiting process, because we can then study the analyticity of the amplitudes or coupling constants as a function of the masses, in particular in the neighborhood of the points $m=o$. If the various elementary particles are intimately related to each other one would expect that the amplitudes for different processes should be related to each other by analytic continuation in the quantum numbers, in particular in mass and spin. There arise even the further, more difficult process of analytic continuation of an amplitude with N particles to another one with N-1 (or more generally to one with M) particles, which involves the vanishing of all quantum numbers and of momentum p_μ of a particle. This is indeed quite a stretch of imagination as to what the analytic continuation should be able to do, and goes much beyond anything we know to be true. But I hope to show that an analytic continuation in mass spin an other internal quantum numbers brings rather interesting results.

The device that allows us to go to the limit $m \to o$ in the amplitudes is called <u>gauge invariance</u> : One constructs the amplitudes involving spin j-particles, when $m \neq o$. These amplitudes involve irreducible spin j polarization tensors, like $\varepsilon_\mu(k)$ for j=1, and $\rho_{\mu\nu}(k)$ für j=2, etc. satisfying $\varepsilon_\mu(k)\, k = o$, $\rho_{\mu\nu} k^\mu = \rho_{\mu\nu} k^\mu = o$. One then observes that for $k^2 = o$, ε_μ or $\rho_{\mu\nu}$ is not uniquely defined: Together with ε_μ , $\varepsilon_\mu + \lambda\, k_\mu$ is also orthogonal to k_μ. Hence one obtains the <u>subsidiary conditions</u> shown in Table 2 . The occurrence of the subsidiary conditions is the prize that one has to pay in order to have a limiting process in m as $m \to o$.

Spin	Irreducible Polarization Spinors	Spinorial Amplitude M *	Gauge Condition
j=2	$\rho_{\mu\nu}(k)$ [112] $\varepsilon_\mu\ \varepsilon_\nu$	$M_n = T_{\mu\nu} \rho_n^{\mu\nu}(k)$	$T_{\mu\nu} k^\mu \varepsilon^\nu = T_{\mu\nu} \varepsilon^\mu k^\nu =$ $= T_{\mu\nu} k^\mu k^\nu = O$
j=1	$\varepsilon_n^\mu(k)$	$M_n = T_\mu\ \varepsilon_n^\mu(k)$	$T_\mu k^\mu = O$
$j=\frac{1}{2}$	$\xi_{\dot\alpha}(k)$	$M_{\dot\alpha} = T\ \ \xi_{\dot\alpha}(k)$	$M_\alpha\ k \cdot \tilde\sigma = O$ ** $T\ \xi^\alpha = O$
j=O	1	M	$\lim\limits_{k_\mu \to o} M = O$ ***

Table 2 - Subsidiary Gauge Conditions on the Amplitudes in the Limit $m \to o$.

* We use two-component spinors and their tensor products to write down the spinorial amplitudes M, for them the basis functions are entire functions (polynomials) in the momenta.

III. CONSEQUENCES OF THE GAUGE CONDITIONS

1. Discontinuities of the Coupling Constants as a Function of the Masses

We consider the vertex amplitudes involving massless particles. The "coupling constant" is a function of the square of the masses of the three particles. The gauge condition implies a discontinuous behaviour for this function. Consider the simple example of the vertex of two spin-0 particles of momenta p_1, p_2 coupled to a spin-1 particle of polarization vector $\varepsilon_\mu(k)$. The spinorial amplitude is $M = g(p_1-p_2)\cdot\varepsilon(k)$. The gauge condition gives $(k=p_1+p_2)$: $g(p_1^2-p_2^2) = 0$, or

$$(m_1^2-m_2^2)\; g(m_1^2,m_2^2,0) = 0 \tag{1}$$

Thus the coupling exists only if $m_1^2 m_2^2$. Similarly, for the vertex of two spin $-1/2$ particles coupled to a spin-1 particle, the spinorial amplitude contains two form factors: $M = f_1Y_1 + f_2Y_2$, where Y_1 and Y_2 are basis functions.

Footnotes continued - In the four-component formalism the so-called R-amplitudes have a simple form [S=I+R], but the basis functions have singularities of the type $\sqrt{p^\mu\sigma_\mu}$. The spinorial amplitudes M transform according to the representations of the homogeneous Lorentz group. For the form of M, R and their relation see A. O. Barut, The Theory of the Scattering Matrix, Macmillan, New York, 1967.

**For $k^2=0$ we have the decomposition $(k^\mu\sigma_\mu)_{\alpha\dot{\alpha}}=(k.\sigma)_{\alpha\dot{\alpha}}= =\xi_{\dot{\alpha}}(k)\xi_{\dot{\alpha}}(k)$ such that $\xi(k)$ satisfies $(k.\tilde{\sigma})\xi(k)=0$, $D^{0\,1/2}(A)\times\xi(\Lambda^{-1}k) = \xi(k)$, and is explicitly given by

$$\xi_{\dot{\alpha}} = \xi^*_\alpha \qquad \xi_\alpha = \begin{pmatrix}\sqrt{k^0+k^3}\\ \dfrac{k^1+ik^2}{\sqrt{k^0+k^3}}\end{pmatrix} \qquad \xi^\alpha = \begin{pmatrix}\sqrt{k^0-k^3}\\ \dfrac{k^1-ik^2}{\sqrt{k^0+k^3}}\end{pmatrix} \;;\; \xi_\alpha\xi^\alpha = 0$$

***The generalization of this condition valid at arbitrary k_μ is discussed in Ch. II .

The term Y_2 is automatically gauge invariant, and the gauge condition gives

$$(m_1-m_2)f_1(m_1^2,m_2^2,0) = 0 ; \qquad (2)$$

again f_1 is zero always except the point $m_1=m_2$. Similar conditions are obtained in other cases (see also Sec. IV).

2. Coupling of Massless Particles with $j\geq 3$ and $j\geq 3/2$

a) For bosons there is a proof that massless particles with $j\geq 3$ cannot couple to other particles. The argument goes as follows: For a process of n spinless (for simplicity) massive particles of momenta p_i (i=1,2, ...,n) and one massless particle of momentum k and spin j, we can write the amplitude as

$$M = \sum_{i=1}^{n} A_i \, p_i^{\mu} \, p_i^{\nu} \ldots \rho_{\mu\nu\ldots} \quad , \qquad (3)$$

where $\rho_{\mu\nu\ldots}$ is the symmetric polarization tensor of the massless particle*. There is in fact a single invariant amplitude for the process considered; Eq. (3) represents a particular choice of the basis function and we shall reduce using the gauge conditions, the n scalar functions eventually to 1. The gauge conditions for (1) imply the following set of j equations:

$$\sum_{i=1}^{n} A_i (p_i\cdot k)(p_i\cdot\varepsilon)^{j-1} = 0$$

$$\sum_{i=1}^{n} A_i (p_i\cdot k)^2 (p_i\cdot\varepsilon)^{j-2} = 0$$

$$\sum_{i=1}^{n} A_i (p_i\cdot k)^j = 0 \qquad (4)$$

* The irreducible tensors $\rho_{\mu\nu\ldots}$ are made of $\varepsilon_\mu(k)$ by Clebsch-Gordan coefficients as shown in Table 2 for j=2 .

We take all momenta as ingoing, $\sum p_i+k = 0$, hence we have the subsidiary conditions

$$\sum_{i=1}^{n} (p_i \cdot k) = 0 \quad , \quad \sum_{i=1}^{n} (p_i \cdot \varepsilon) = 0 \ . \tag{5}$$

We define now the new form factors F_i by $F_i = A_i(p_i \cdot k)$. The reason for this is that F_i in the limit $k \to o$ correspond to the coupling constants of the massless particle to other particles, for then the amplitude in terms of F_i

$$M = \sum_i F_i \frac{p_i^\mu p_i^\nu \cdots \rho_{\mu\nu\ldots}}{(p_i \cdot k)} \tag{6}$$

correspond to a sum of pole terms in the limit $k \to o$, $F_i \to \text{const.}[-2(p_i \cdot k) = (p_i - k)^2 - m_i^2]$. The gauge conditions (1) now take the symmetric form

$$\sum F_i \ (p_i \cdot \varepsilon)^{j-1} = 0$$

$$\sum F_i \ (p_i \cdot k)(p_i \cdot \varepsilon)^{j-2} = 0$$

$$\sum F_i \ (p_i \cdot k)^{j-1} = 0 \quad , \tag{4'}$$

with the subsidiary conditions (5). Because there is a single invariant amplitude and because of the symmetry of the problem we can put $F_i = F(s,t,\ldots)\eta_i$ and obtain [1]

$j = 1$: $\sum_i \eta_i = 0$

$j = 2$: $\eta_i = \eta$ (independent of i)

$j \geq 3$: $\eta_i = 0$ for all i .

b) We consider now the coupling of massless fermions with $j \geq 3/2$ [2]. The simplest vertex of a massless $j=3/2$ particle of momentum k is to a spin $-1/2$ particle of momentum k' and a spin-0 particle of momentum p, and is described by the polarization tensor

$$\rho_{\mu\nu r\dot{\alpha}}(k) = [1\ \tfrac{1}{2}\ \tfrac{3}{2}]^{n\alpha}_{r}\ \varepsilon_{\mu_n}(k)\,\sigma_{\nu\alpha\dot{\alpha}} \quad , \tag{7}$$

where the bracket is a Clebsch-Gordan coefficient and $r,n,\alpha,\dot{\alpha}$ the spinor indices. The vertex amplitude is

$$M = T^{\mu\nu}\ \rho_{\mu\nu}(k) \tag{8}$$

with the gauge conditions

$$M\ k.\tilde{\sigma} = 0 \tag{9}$$

$$T_{\mu\nu}k^{\mu} = 0 \tag{10}$$

Hence we find, because $k.\varepsilon = 0$, $k=p+k'$

$$T_{\mu\nu} = g(p-k')_{\mu}(p+k')_{\nu}$$

$$T_{\mu\nu}k^{\mu} = g(p^2-k'^2)(p+k')_{\nu} = 0$$

Thus we obtain a condition like eq. (1): $g(m^2,m'^2,0)(m^2-m'^2)=$ $=0$. Because the masses of the spin 0 and spin-1/2 particles are not equal, the coupling constant of the massless j=3/2-particle vanishes. Now we consider in addition n spinless particles of momenta $p_1 \dots p_n$.
The gauge condition (9) and $k.\varepsilon=0$ implies that the amplitude has the form

$$M = \sum_i A_i p^i_{\mu} k_{\nu} \rho^{\mu\nu}(k) \tag{11}$$

and we have the gauge-condition

$$[1\ \tfrac{1}{2}\ \tfrac{3}{2}]k.\sigma \sum_i A_i(p_i.k) = 0 \ ,$$

which is like eq. (4). The trivial solution : $A_i = A$ for all i, gives in fact M=0, because of the subsidiary condition (5). The nontrivial solution

$$M = \sum_i F_i \frac{p^i_\mu k_\nu}{(p_i . k)} \rho_{\mu\nu} \ , \ \sum_i F_i = 0 \tag{12}$$

however, introduces new poles in the amplitudes which cannot correspond to particles (poles in the spin 3/2 - spin 0 channels!). Consequently there seem to be severe limitations on the existence of massless $j \geq 3/2$ particles.

IV. $j=\frac{1}{2}$, m=0 GAUGE CONDITIONS AND S-MATRIX THEORY OF WEAK INTERACTIONS

The S-matrix theory of the massless j=1 and j=2 particles (i.e. electromagnetic and gravitational interactions) was discussed in my 1965 Schladming lectures [3] . In this Section I shall discuss the case j=1/2, m=0 (neutrino) and consequently the S-matrix theory of leptonic weak interactions.

The full impact of the powerful gauge condition for massless spinor particles (see Table 2) does not seem to have been realized before.

We first discuss some simple cases.

a) For the process $0+ \frac{1}{2} \to 0+ \frac{1}{2}$ (e.g. $\pi\nu\to\pi\nu$), the four independent amplitudes reduce to one if the j=1/2 particles are massless; the spinorial M-function is of the form

$$M = A\, k_3.\sigma \quad n.\tilde{\sigma}\, k_1.\sigma \tag{13}$$

where k_1 and k_3 are the momenta of the spinor particles and $n=k_2+k_4$.

b) The vertex $\frac{1}{2}\,\frac{1}{2} \to 0$ with one massless fermion (e.g. $\pi\to e\nu$) is described by the single amplitude $M=gk\sigma$ (k=momentum of the massless fermion).

c) The vertex $\frac{1}{2}\frac{1}{2} \to 0$ with two massless fermions (e.g. $\nu\nu\pi$) vanishes identically.

We shall discuss next the more interesting cases of leptonic weak interactions [4].

For the scattering of two spin 1/2 particles, $\frac{1}{2}+\frac{1}{2} \to \frac{1}{2}+\frac{1}{2}$ there are 16 independent amplitudes and the spinorial amplitude can be written as

$$M = \sum_{i,j=1}^{4} C_{ij}\, Y_i \otimes \bar{Y}_j \tag{14}$$

where

$$\begin{aligned} Y_1 &= (k_1+k_3)\cdot\sigma \quad , & \bar{Y}_1 &= (k_2+k_4)\cdot\sigma \\ Y_2 &= (k_1-k_3)\cdot\sigma \quad , & \bar{Y}_2 &= (k_2-k_4)\cdot\sigma \\ Y_3 &= n\cdot\sigma+k_3\cdot\sigma n\cdot\tilde{\sigma}k_1\cdot\sigma \quad , & \bar{Y}_3 &= n'\cdot\sigma+k_4\cdot\sigma n'\cdot\tilde{\sigma}k_2\cdot\sigma \\ Y_4 &= n\cdot\sigma-k_3\cdot\sigma n\cdot\tilde{\sigma}k_1\cdot\sigma \quad , & \bar{Y}_4 &= n'\cdot\sigma-k_4\cdot\sigma n'\cdot\tilde{\sigma}k_2\cdot\sigma \end{aligned} \tag{15}$$

If only one of the particles is massless (say k_2) we impose the gauge condition

$$M\, k_2\cdot\tilde{\sigma} = 0 \tag{16}$$

If two particles are massless (say k_2 and k_3) we have in addition to (16) the condition

$$k_3\cdot\tilde{\sigma}\, M = 0 \tag{17}$$

which has to be satisfied separately. The consequence of these conditions is shown in Table 3.

With the choice of basis given in (14) and (15), the conditions (16) and (17) and T-invariance gives the single amplitude

$$M = \frac{G}{\sqrt{2}}\,(k_3\cdot\sigma n\cdot\tilde{\sigma}k_1\cdot\sigma) \otimes (k_4\cdot\sigma n'\cdot\tilde{\sigma}k_2\cdot\sigma)$$

or choosing $n^\mu n'^\nu = g^{\mu\nu}$,

$$M = \frac{G(s,t)}{\sqrt{2}} \, (k_3 \cdot \sigma\tilde{\sigma}^{\mu} k_1 \cdot \sigma) \otimes (k_4 \cdot \sigma\tilde{\sigma}_{\mu} k_2 \cdot \sigma) \qquad (18)$$

which is the usual (V-A) amplitude. A different choice of the basis merely changes the function G for which we shall write a unitarity equation.

Type of reaction	Number of Amplitudes	
	with T	without T
all massive particles	10	16
one massless particle	3	8
two massless particle	1	4

Table 3 - Reduction of the Number of Amplitudes with T-invariance and gauge Condition.

For the processes of the type of β-decay, the three amplitudes can be written as V, A and the weak magnetism [4]. Thus, the neutrino gauge conditions (16), (17) are rather restrictive. To my knowledge the reduction of the number of amplitudes to one and 3 for the μ-decay and β-decay, respectively, by this gauge condition, has not been recognized before.

The above result for the μ-decay has an important consequence, I believe, on the theory of weak interactions. The question whether a conventional field theory of weak interactions exists has been much discussed recently. In some of these discussion the four-fermion interaction in the V-A form is taken to be a fundamental interaction, which means that G(s,t) in (18) is taken to be a constant. The resulting field theory is non-renormalizable, and may lead to the conclusion that the weak interactions become strong at high energies. In our view, because the most general form of the amplitude is (18), the four-fermion interaction is a phenomenological one and G(s,t) cannot be a constant. In fact one can develop now a systematic S-matrix theory of leptonic processes, similar to that of electromagnetic processes [3]. From crossing, unitarity and the assumption of unsubtracted dispersion relations one obtains in fact an integral equation for G which shows that G(s,t) behaves asymptotically like 1/s and 1/t [4]. Thus it seems possible to determine the weak interaction amplitudes to a good approximation.

V. THE CASE j=0, m=0

The gauge condition for zero spin, zero mass particles indicated in Table 2 is of a different nature than the other gauge conditions. It holds at one point, $k_\mu=0$, whereas the other gauge conditions hold for all values of k_μ. The j=0 gauge condition in this form has been given in field theory first by Nishijima [5] about ten years ago. It is based on the invariance transformation $\Phi\to\Phi+c$ of the field equations for massless scalar particles, and holds for a class of interactions [6].

Within the framework of the Poincaré group there is no room for a new gauge condition for j=0, m=0, In that sense this gauge condition is of a dynamical nature. We shall be interested in the consequences of this condition - I have given elsewhere the derivation and the generalization of this gauge condition based on the invariance under a larger group that includes dilatations [7].

First some simple consequences:

a) The $\frac{1}{2}\to\frac{1}{2}\,0$ vertex (e.g. $\Lambda\to N\pi$) is described by two coupling constants

$$M = g_1\left(\frac{p_1}{m_1} + \frac{p_2}{m_2}\right) \cdot \sigma + g_2\left(\frac{p_1}{m_1} + \frac{p_2}{m_2}\right) \cdot \sigma$$

where $g_i(m_1^2 m_2^2 \mu^2)$ and p_1 and p_2 are the momenta of the fermions. Let $k_\mu = (p_2 - p_1)_\mu$ be the momentum of the spin-zero particle. The gauge condition

$$\lim_{k_\mu \to o} M = 0 \qquad (19)$$

gives

$$g_1(m_1^2 m_2^2 0) + \frac{m_1 - m_2}{m_1 + m_2} g_2(m_1^2 m_2^2 0) = 0 \qquad (20)$$

In particular, if parity is conserved, we have

$$(m_1 - m_2)\ g_2(m_1^2 m_2^2 0) = 0$$

$$(m_1 \neq 0, m_2 \neq 0) \quad . \qquad (20')$$

b) Consider the decay of a spin zero particle into two photons (e.g. $\pi^0 \to \gamma\gamma$). Depending on parity the amplitude is either

$$M = A\ \varepsilon_{\mu\nu\lambda\rho}\ k_{1\lambda}\ k_{2\rho}\ \varepsilon^\mu(k_1)\varepsilon^\nu(k_2)$$

or

$$M = B[k_{2\mu}k_{1\nu} + Cg_{\mu\nu}]\varepsilon^\mu(k_1)\varepsilon^\nu(k_2)$$

In the first case the photon gauge is automatically satisfied, the spin-0 gauge in the limit $k_1 \to -k_2$ is also satisfied. In the second case: If we use photon-gauge first then $C = -\frac{1}{(k_1 \cdot k_2)}$, and the j=0 gauge condition implies $B(\mu^2 = o) = 0$.

Thus we obtain conditions like (1) and (2), but with a difference that g^i are now evaluated off the mass-shell, i.e. not only outside the physical region, but at analytically continued points as well. Usually the analyticity is formulated on the hypersurfaces formed by the mass-shell conditions $p_i^2=m_i^2$. Relations like (20) make sense in a framework in which we are also using an analytic continuation in mass, i.e. pass from one process to another.

In the next Section we discuss this process of analytic continuation in mass for two-body amplitudes.

VI. ANALYTIC CONTINUATION TO m=O .

The analytic continuation of the scattering amplitude in the Poincaré invariants m and j of the external legs goes beyond the usual analyticity considerations in the energy and angular momentum variables. It relates in fact one physical process to another one involving particles with different m and j. In theories using the representations of the larger groups (e.g. infinite multiplets) such analytic continuations occur. Here we want to discuss the effect of the gauge conditions when the amplitude is continued to the point $m^2=o$ for one or more particles.

If $j\neq 0$ and we continue to $m^2=0$, we must obtain the same results as those obtained for physical massless particles discussed in Sections III - IV. For example the coupling "constant" $f_1(m_1^2,m_2^2,m_\rho^2)$ of the $NN^*\rho$ - vertex satisfies (see eq. (2))

$$(m_1-m_2)\ f_1(m_1^2,m_2^2,0) = 0 \tag{21}$$

Thus, we know in principle the values of the majority of amplitudes

$$A(m_1^2m_2^2m_3^2m_4^2,\ldots,s,t,u,\ldots)$$

at $m_i^2=0$. For example, of the 10 amplitudes for the process $NN\to NN$ all but one vanish at $m_1^2=m_3^2=0$, $m_2^2\neq 0$, $m_4^2\neq 0$.

At j=0, we impose the gauge condition in the limit

$$k_\mu \to 0$$

i.e. when all four components of the momenta go to zero. Hence the limiting process must be considered more carefully.

The general method consists in writing the most general spinorial amplitude

$$M = \sum_i A_i Y_i \, ,$$

continue analytically the basis functions Y_i to the point $k_\mu \to o$, single out the pole terms in A_i which contribute at the limiting point and evaluate the limits of the form

$$\frac{Y_i}{s-m^2}$$

which are of the form

$$\frac{0}{0} \, .$$

We discuss two important cases:

1. $\pi N \to \pi N^*_{j=1/2}$ - Amplitude [8]

There are two independent spin amplitudes and two isospin amplitudes:

$$M = A\left(\frac{k_3}{m_3} + \frac{k_1}{m_1}\right).\sigma + B\frac{1}{2}\left[(k_2+k_4).\sigma + \frac{k_3.\sigma}{m_3}(k_2+k_4).\tilde{\sigma}\,\frac{k_1.\sigma}{m_1}\right] \tag{21}$$

where

$$A = A^+ \delta_{\beta\alpha} \delta_{ba} + A^- \frac{1}{2}[\tau_\beta, \tau_\alpha]_{ba} \, ,$$

similarly for B. Here k_1 and k_3 are the momenta of N and N^*, respectively, with masses m_1 and m_3 and k_2 and k_4 are the momenta of the pions. We separate the pole terms in A and B:

$$A^{\pm} = \frac{1}{4}(m_1-m_3)g'(k_4^2)\left[\frac{g(k_2^2)}{s-m_1^2} \mp \frac{g''(k_2^2)}{u-m_3^2}\right] - \frac{1}{4}(m_1-m_3)g''(k_2^2) \times$$

$$\times \left[\frac{g''(k_4^2)}{s-m_3^2} \mp \frac{g(k_4^2)}{u-m_1^2}\right] + A^{\pm R} \qquad (22)$$

$$B^{\pm} = -\frac{1}{2}g'(k_4^2)\left[\frac{g(k_2^2)}{s-m_1^2} \pm \frac{g''(k_2^2)}{u-m_3^2}\right] - \frac{1}{2}g'(k_2^2)\left[\frac{g''(k_4^2)}{s-m_3^2} \mp \frac{g(k_4^2)}{u-m_1^2}\right] + B^{\pm R}$$

where $A^{\pm R}$, $B^{\pm R}$ are the remainder of the amplitudes when poles are separated. Here $g \equiv g_{NN\pi}$, $g' \equiv g_{NN^*\pi}$ and $g'' \equiv$ $\equiv g_{N^*N^*\pi}$.

We insert (22) into (21) and use the identity

$$\frac{1}{2}\left[(k_2+k_4)\cdot\sigma + \frac{k_3\cdot\sigma}{m_3}(k_2+k_4)\cdot\tilde{\sigma}\,\frac{k_1\cdot\sigma}{m_1}\right] =$$

$$= \frac{m_1-m_3}{2}\left[\frac{k_3}{m_3} + \frac{k_1}{m_1}\right]\cdot\sigma + k_2\cdot\sigma + \frac{k_3\cdot\sigma}{m_3}\,k_2\cdot\tilde{\sigma}\,\frac{k_1\cdot\sigma}{m_1}$$

and evaluate the limits $k_2=\varepsilon k_1, \varepsilon \to o$ (for pole terms $1/(s-m_1^2)$) or $k_2=\eta k_3, \eta\to o$ (for pole term $1/(u-m_3^2)$). At the limiting point $k_2\to o$ we then use the gauge condition (19). The final result - taking also the condition (20') into account - is

$$\left(A^{\pm R} + \frac{m_1-m_3}{2}B^{\pm R}\right)\Bigg|_{k_2\to o} = \frac{1}{4m_1m_3}g'(\mu^2) \times$$

$$\times [m_3 g(o) \pm m_1 g''(o)] \qquad (23a)$$

A similar result is obtained for the limit $k_4\to o$

$$\left(A^{\pm R} - \frac{m_1-m_3}{2}B^{\pm R}\right)\Bigg|_{k_4\to o} = \frac{1}{4m_1m_3}g'(\mu^2) \times$$

$$\times[\pm m_3 g(o) + m_1 g''(o)] \qquad (23b)$$

In the equal mass case eqs. (23) reduce to the so-called Adler's "self-consistency" conditions: $A^{-R} = 0$, $A^{+R} = \frac{1}{2m} g(o) g(\mu^2)$ obtained using PCAC [9] . Furthermore, because of the antisymmetry of the M^--amplitude we have the relation

$$m_3 g(o) = m_1 g''(o) \tag{24}$$

Consequently

$$\left(A^{+R} + \frac{m_1 - m_3}{2} B^+\right)\Big|_{k_2 \to o} = \frac{g'(\mu^2) g(o)}{2m_1} =$$

$$= \frac{g'(\mu^2) g''(o)}{2m_3} \tag{25}$$

2. Electro- and Photo-Production of N^* [10]

The same method has been applied to the process $N(k_1) + \gamma(k_2) \to N^*(k_3) + \pi(k_4)$. Again we evaluate the Born terms and the limit of

$$M = \sum_i A_i Y_i$$

There are six amplitudes in the case of electro-production. Using also the condition of current conservation, we obtain from the pion-gauge condition - taking into account (20') and (24) -

$$A_1^{+R}\Big|_{k_4 \to o} = \frac{g(o) f_2'^{V}(\lambda^2)}{m_1}$$

$$A_1^{-R}\Big|_{k_4 \to o} = 0$$

$$\lambda^2 A_6^{\pm R} = 0 \tag{26}$$

where λ^2 is the photon-mass (off-shell). In the equal mass-case the results become

$$A_1^{+R} = \frac{g(o) f_2^V(\lambda^2)}{m} \; , \quad A_1^{-R} = 0 \; , \quad \lambda^2 A_6^{+R} = 0 \; ,$$

$$\lambda^2 A_6^{-R} = \frac{g(o) f_1^V(\lambda^2)}{m} \tag{27}$$

which, if compared with the PCAC-result [11] gives us the equality of the vector and axial-vector form factors:

$$\frac{g_A(k^2)}{g_A(o)} = \frac{f_1^V(k^2)}{f_1^V(o)} \tag{28}$$

VII. CONCLUSIONS

The limit from $m \neq o$ to $m=o$ is a discontinuous one. The gauge conditions allow one to perform this continuous limit and give conditions on the coupling constants and amplitudes at the points $m=o$. It is useful to start with the $m \neq o$ amplitudes, to introduce the notion of analytic continuation in mass, and then obtain the values of some of these amplitudes at $m=o$ by the gauge conditions. For physical massless particles this process gives the values of the amplitudes in the physical region, for other particles outside the physical region. The so-called "self-consistency" conditions previously obtained from PCAC are derived by such a process of analytic continuation in mass, and some new results on form factors are obtained. Furthermore, we have discussed some implications of the rather powerful spin 1/2 gauge conditions for leptonic weak decays which do not seem to have been realized before.

REFERENCES

1. S. WEINBERG, Phys. Rev. 135, B1049 (1964). For a more general treatment see A. O. BARUT, Phys. Letters 10, 356 (1964) and footnote to Table 1
2. A. O. BARUT, (to be published).
3. A. O. BARUT, In Quantumelectrodynamics, ed. by P. Urban (Springer-Verlag, 1965).
4. A. O. BARUT and A. BARAN, Trieste ICTP-report (in preparation).
5. K. NISHIJIMA, Nuovo Cim. 11, 698 (1959); J. HAMILTON, Nucl. Phys. B1, 449 (1967); M. MARTINIS, Nuovo Cim. 56, 935 (1968).
6. For a critical discussion see G. KRAMER and W. F. PALMER, Argonne preprint (Dec. 1968).
7. A. O. BARUT, IC/69/12 (Trieste preprint)
8. A. O. BARUT and B. C. ÜNAL,Lettera Nuovo Cim. 1, 145 (1969).
9. S. L. ADLER, Phys. Rev. 137, B1022 (1965).
10. A. O. BARUT and C. B. ÜNAL, A. Ü. F. F. preprint 69/1 (to be published).
11. e.g. S. L. ADLER and F. J. GILMAN, Phys. Rev. 152, 1460 (1966) and references therein.

EFFECTIVE LAGRANGIANS AND SU(3) × SU(3) SYMMETRY BREAKING†

By

S. GASIOROWICZ
University of Minnesota, Minneapolis,
and
Deutsches Elektronen Synchrotron - (DESY)
Hamburg

At last year's Schladming Winter School, Professor Gürsey discussed some aspects of the effective Lagrangian technique, which he helped to develop [1]. In the following lectures I shall apply this technique to the study of chiral SU(3)×SU(3). As background to a detailed discussion of a model Lagrangian, I will (i) review the Lagrangian formalism, (ii) discuss the transformation properties of fields, (iii) show how field algebra and symmetry breaking that leads to PCAC are implemented. In the discussion of the model, which was worked out in collaboration with D. A. Geffen, I will lay stress on the leptonic decay constants of the spin 0 and spin 1 mesons, as far as comparison with experiment is concerned. For the rest, the stress will be on the technical aspects of constructing effective Lagrangians. Baryons will not be discussed, except of a brief mention in sec. (II) [2].

---

† Lecture given at the VIII. Internationale Universitätswochen für Kernphysik, Schladming, Febr. 24 - March 8,1969.

I. THE LAGRANGIAN FORMALISM

The equations of motion described by a Lagrangian density function of certa... fields* $\Phi_A(x)$ and their derivatives, $L(\Phi_A(x)\ ,\ \partial_\mu \Phi_A(x))$, are of the form

$$\partial_\mu \frac{\partial L}{\partial\ (\partial_\mu \Phi_A(x))} = \frac{\partial L}{\partial \Phi_A(x)} \tag{1}$$

when the Lagrangian density is of second order in the derivatives. The canonical momentum is defined by

$$\pi_A(x) = \frac{\partial L}{\partial(\partial_o \Phi_A(x))} \tag{2}$$

and we shall later assume the canonical commutation relations

$$[\pi_A(x), \Phi_B(y)]_{x_o=y_o} = -i\delta_{AB}\ \delta(\underset{\sim}{x}-\underset{\sim}{y}) \tag{3}$$

Let us now turn to the definition of currents associated with a symmetry. If the fields are changed by an infinitesimal gauge transformation

$$\Phi_A(x) \rightarrow \Phi_A(x) + i\ C_{ABC}\ \alpha_B\ \Phi_C(x) \tag{4}$$

then the change in the Lagrangian density is

$$\delta L = \frac{\partial\ L}{\partial \Phi_A}\ \delta\Phi_A + \frac{\partial L}{\partial(\partial_\mu \Phi_A)}\ \partial_\mu \delta\Phi_A = \partial_\mu(\frac{\partial L}{\partial(\partial_\mu \Phi_A)}\delta\Phi_A) +$$

$$+\ \delta\Phi_A(\frac{\partial\ L}{\partial \Phi_A} - \partial_\mu \frac{\partial L}{\partial(\partial_\mu \Phi_A)}) =$$

$$= \partial_\mu(i\ \alpha_B\ \frac{\partial L}{\partial(\partial_\mu \Phi_A)}\ C_{ABC}\ \Phi_C) \tag{5}$$

---

* The labels refer to the internal symmetry states. The labels that refer to the space-time components of fields describing particles with spin, will be suppressed when there is no danger of confusion.

where the equations of motion have been used. If the gauge function α_B is constant, $\delta L = 0$ implies the conservation of the current defined by

$$j_\mu^B(x) = -i \frac{\partial L}{\partial(\partial^\mu \Phi_A)} C_{ABC} \Phi_C(x) \tag{6}$$

The proportionality constant -i is determined so that the unitary transformation corresponding to (4) is

$$U = \exp\{-i\alpha_B Q_B\} \tag{7}$$

where

$$Q_B \equiv \int d^3x j_o^B(x) = -iC_{ABC} \int d^3x \pi_A(x) \Phi_C(x) \tag{8}$$

If α_B is not constant we obtain from (5)

$$\delta L = -j_\mu^B(x) \partial^\mu \alpha_B(x) - \alpha_B(x) \partial^\mu j_\mu^B(x) \tag{9}$$

i.e.

$$\begin{aligned} j_\mu^B(x) &= -\frac{\partial(\delta L)}{\partial(\partial^\mu \alpha_B)} \\ \partial^\mu j_\mu^B(x) &= -\frac{\partial(\delta L)}{\partial\, \alpha_B} \end{aligned} \tag{10}$$

The coefficients C_{ABC} can be determined from the commutation relations for the "charges" Q_B required by the group property of the unitary transformations U. If the Lie algebra is characterized by

$$[Q_A, Q_B] = i\, f_{ABC}\, Q_C \tag{11}$$

then, using the equal time commutation relations (3) and the form (8) for the charges, we obtain

$$[Q_A, Q_B] = i\, f_{ABC}(-iC_{MCN} \int d^3x\, \pi_M(x) \Phi_N(x))$$

provided that

$$C_{MAN}\, C_{NBQ} - C_{MBN}\, C_{NAQ} = i\, f_{ABC}\, C_{MCQ} \tag{12}$$

i.e. provided that the C_{MAN} are the MN matrix element of a matrix representation (C_A) of the "charge" Q_A. For example, if the Q_A are SU(3) generators then

$$C_{MAN} = \frac{1}{2}(\lambda_A)_{MN}$$

if the fields Φ_M form a triplet, and

$$C_{MAN} = (F_A)_{MN} = - i \ f_{AMN}$$

if they form an octet.

Given a Lagrangian density L , the scattering matrix can formally be obtained by writing L as a sum of "free" Lagrangian densities + an interaction term,and

$$S = 1 + \sum_{n=1}^{\infty} \frac{i^n}{n!} \int d\ x_1 \ \dots \int dx_n \ T(L_{Int}(x_1) \ \dots \ L_{Int}(x_n)) \tag{13}$$

The terms in this expansion may be put in one-to-one correspondence with Feynman graphs. There are two properties of this formal solution that are of interest to us:

1. The scattering matrix is unaltered when the fields Φ_A are transformed into other fields according to [3]

$$\Phi_A = \chi_A \ g(\chi) \qquad g(0) = 1 \tag{14}$$

2. The scattering matrix may be expanded in a formal series in a parameter λ. The leading term contains no closed loop graphs, the second term one closed loop, etc [4]. A very compact proof due to Coleman, Wess and Zumino is the following:

Define

$$L(\Phi,\lambda) \equiv \frac{1}{\lambda^2} \ L(\lambda \ \Phi) \tag{15}$$

A given connected graph has a definite power of λ associated with it. If E is the number of external lines,

I the number of internal lines, L the number of loops, V the number of vertices and N_i (i=1,2,...,V) the number of lines attached to the i-th vertex, then the power of λ associated with the graph is

$$P = \sum_{i=1}^{V} (N_i - 2)$$

However $\sum_{i=1}^{V} N_i = E+2I$, so that

$$P = E + 2I - 2V$$

Now the number of loops is given by $L = I - V+1$ so that

$$P = E + 2L - 2$$

Thus for a given process (fixed E), terms with different numbers of loops carry different powers of λ. The three graphs with L=0 represent the lowest order term in an expansion. It follows that the invariance properties of the Lagrangian are maintained by the tree graph approximation (which is the only one we know how to handle at this time). One can also show that the tree graph approximation to the S matrix is invariant under the point transformation (14) .

II. TRANSFORMATION PROPERTIES OF FIELDS

We consider the algebra of SU(3)×SU(3), i.e. of the sixteen* generators ("charges") Q_+^a, Q_-^a (a=1,2,3,...,8) that satisfy the commutation relations

---

* We can trivially extend the theory to U(3)×U(3) by introducing two additional charges $Q_\pm^a$ and defining $f_{oij}=0$, since the additional charges commute with all the others. This extension is motivated by the existence of a nonet of vector mesons which we shall put in correspondence with a nonet of currents.

$$[Q_{\pm}^{a}, Q_{\pm}^{b}] = i\, f_{abc}\, Q_{\pm}^{c}$$
$$[Q_{+}^{a}, Q_{-}^{b}] = 0 \tag{17}$$

The generators are related by the parity operator

$$P\, Q_{+}^{a}\, P^{-1} = Q_{-}^{a} \tag{18}$$

The f_{abc} are the totally antisymmetric real SU(3) structure constants. We may define

$$Q^{a} = Q_{+}^{a} + Q_{-}^{a}$$
$$Q_{5}^{a} = Q_{+}^{a} - Q_{-}^{a} \tag{19}$$

The Q^{a} are the usual SU(3) generators, and we shall call the Q_{5}^{a} the "axial" SU(3) generators.

The transformation law for a field transforming under SU(3)×SU(3) according to $(m,\bar{n})$ is*

$$\exp\{-i\underset{\sim}{\alpha}\cdot\underset{\sim}{Q}_{+}\}\ \exp\{-i\underset{\sim}{\beta}\cdot\underset{\sim}{Q}_{-}\}\ M_{A\dot{B}}\ \exp\{i\underset{\sim}{\beta}\cdot\underset{\sim}{Q}_{-}\}\exp\{i\underset{\sim}{\alpha}\cdot\underset{\sim}{Q}_{+}\}$$
$$= (\exp\{+i\underset{\sim}{\alpha}\cdot\underset{\sim}{F}\})_{AC}\ M_{B\dot{D}}(\exp\{-i\underset{\sim}{\beta}\cdot\underset{\sim}{F}'\})_{\dot{D}\dot{B}} \tag{20}$$

where the $\underset{\sim}{F}$ and $\underset{\sim}{F}'$ are matrix representations of the SU(3) generators in the $\underset{\sim}{m}$ and $\underset{\sim}{n}$ representations respectively**. The matrices satisfy the commutation relations

$$[F_a, F_b] = i\, f_{abc}\, F_c \tag{21}$$

Let us consider some examples. A field transforming as $(\underset{\sim}{8},\underset{\sim}{1})$, for example is

---

* We label the SU(3) representations by their multiplicity, e.g. $(\underset{\sim}{1},\underset{\sim}{3},\underset{\sim}{\bar{3}},\underset{\sim}{8},\underset{\sim}{10},\underset{\sim}{\overline{10}},\ldots)$. This should cause no confusion in the cases under consideration.

** If the field transforms under $(\underset{\sim}{m},\underset{\sim}{n})$ then the $\underset{\sim}{F}'$ must be taken to be matrices in the $\bar{\underset{\sim}{n}}$ representation.

$$\exp\{-i\alpha.Q_+ -i\beta.Q_-\}M_A \exp\{i\alpha.Q_+ +i\beta.Q_-\} = (\exp\{i\alpha.F\})_{AB}M_B \tag{22}$$

For infinitesimal transformations this may be written in the form

$$\delta M_A = i(\alpha.F)_{AB}\, M_B = + f_{ABC}\, \alpha_C M_B \tag{23}$$

$$\delta' M_A = f_{ABC}\, \alpha_C\, M_B$$

Here we have introduced the notation δ for variations under infinitesimal SU(3) transformations and δ' for variations under the "axial" SU(3) transformations* . The first result is obtained by setting $\beta=\alpha$, the second by setting $\beta=-\alpha$ in (22). A field transforming under $(1,8)$ has

$$\delta M_{\dot{A}} = f_{\dot{A}\dot{B}C}\, \alpha_C\, M_{\dot{B}} \tag{24}$$

$$\delta' M_{\dot{A}} = f_{\dot{A}\dot{B}C}\, \alpha_C\, M_{\dot{B}}$$

We shall be dealing with fields that have well defined transformation properties under parity, i.e. with fields that transform under $(8.1) \oplus (1,8)$ consisting of sums and differences of the above fields**

$$\delta V_A = - f_{ABC}\, \alpha_B\, V_C$$

$$\delta A_A = - f_{ABC}\, \alpha_B\, A_C$$

---

* In other words

$$\exp\{-i\alpha.Q\}\Phi\exp\{i\alpha.Q\} \simeq \Phi - i[\alpha.Q,\Phi] \equiv \Phi + \delta\Phi$$

$$\exp\{-i\alpha.Q_5\}\Phi\exp\{i\alpha.Q_5\} \simeq \Phi - i[\alpha.Q_5,\Phi] \equiv \Phi + \delta'\Phi$$

** We shall actually treat nine vector and axial mesons, so that the range of values is $A = 0,1,2,\ldots,8$.

$$\delta' V_A = - f_{ABC}\,\alpha_B\, A_C$$

$$\delta' A_A = -f_{ABC}\,\alpha_B\, V_C \qquad (25)$$

In our discussion of the model we shall assume that the fields for the vector and axial mesons transform according to (25). For the pseudoscalar and scalar mesons we shall assume that the representation is $(\underset{\sim}{3},\underset{\sim}{\bar{3}}) \oplus (\underset{\sim}{\bar{3}},\underset{\sim}{3})$. For a field transforming according to $(\underset{\sim}{3},\underset{\sim}{\bar{3}})$ we have

$$\exp\{-i\alpha.Q_+ - i\beta.Q_-\} M_{A\dot{B}} \exp\{i\alpha.Q_+ + i\beta.Q_-\} =$$

$$= (\exp\{i\alpha.\lambda/2\})_{AC}\, M_{C\dot{D}} (\exp\{-i\beta.\lambda/2\})_{\dot{D}\dot{B}} \qquad (26)$$

i.e.

$$\delta \underline{M} = \frac{i}{2}[\alpha.\lambda, \underline{M}]$$

$$\delta' \underline{M} = \frac{i}{2}\{\alpha.\lambda, \underline{M}\} \qquad (27)$$

where $\underline{M}$ stands for the 3×3 matrix $M_{A\dot{B}}$.

For a field transforming according to $(\underset{\sim}{\bar{3}},\underset{\sim}{3})$ we note that the matrices $-\lambda_i^*$ satisfy the same commutation relations as the λ_i but are not equivalent to them, and thus $-\lambda^*/2 \equiv -\lambda^T/2$ yield a $\underset{\sim}{\bar{3}}$ representation of the generators.

Hence

$$\exp\{-i\alpha.Q_+ - i\beta.Q_-\}\, \bar{M}_{\dot{A}B} \exp\{i\alpha.Q_+ + i\beta.Q_-\} =$$

$$= (\exp\{-i\alpha.\lambda^T/2\})_{\dot{A}\dot{C}}\, \bar{M}_{\dot{C}D} (\exp\{i\beta.\lambda^T/2\})_{DB} \qquad (28)$$

i.e.

$$\delta\, \bar{M}^T = \frac{i}{2}[\alpha.\lambda, \bar{M}^T]$$

$$\delta' \bar{M}^T = -\frac{i}{2}\{\alpha.\lambda, \bar{M}^T\} \qquad (29)$$

The fields are related by

(footnote continued) The fact that the ninth mesons (A=0) do not transform under SU(3) is expressed by $f_{0AB}=0$.

$$P\,M\,P^{-1} = \bar{M}^{T} \quad M^{+} \tag{30}$$

If we write

$$\underline{M} = \underline{\sigma} + i\,\underline{\Phi} \tag{31}$$

then the 3×3 matrix fields obey the transformation laws*

$$\delta\underline{\Phi} = \frac{i}{\sqrt{2}}\,[\underline{\alpha},\underline{\Phi}]$$
$$\delta\underline{\sigma} = \frac{i}{\sqrt{2}}\,[\underline{\alpha},\underline{\sigma}]$$
$$\delta'\underline{\Phi} = \frac{1}{\sqrt{2}}\,\{\underline{\alpha},\underline{\sigma}\}$$
$$\delta'\underline{\sigma} = -\frac{1}{\sqrt{2}}\,\{\underline{\alpha},\underline{\Phi}\} \tag{32}$$

Incidentally it follows from (26) that

$$M' = \exp\{i\lambda.\alpha/2\}\,M\,\exp\{i\lambda.\alpha/2\}$$
$$(MM^{+})' = \exp\{i\lambda.\alpha/2\}\,MM^{+}\exp\{-i\lambda.\alpha/2\} \tag{33}$$

so that invariants can be constructed as follows

$$(\mathrm{tr}(MM^{+}))^{n}$$
$$(\mathrm{tr}(MM^{+}MM^{+}))^{m} \tag{34}$$

Also

$$(\det M)' = \det(\exp\{i\alpha.\lambda\}) \cdot \det M = \det M$$

if $\alpha_o = 0$, so that $\det(\underline{\sigma}+i\underline{\Phi})$ + c.c. is chiral invariant and parity conserving. This may be rewritten in the form

$$I = (\mathrm{tr}\underline{\sigma})^{3} - 3\mathrm{tr}\underline{\sigma}\;\mathrm{tr}\underline{\sigma}^{2} - 3\mathrm{tr}\underline{\sigma}(\mathrm{tr}\underline{\Phi})^{2} + 3\mathrm{tr}\underline{\sigma}\;\mathrm{tr}\underline{\Phi}^{2} + $$
$$+ 6\mathrm{tr}\underline{\Phi}\;\mathrm{tr}\underline{\sigma\Phi} + 2\mathrm{tr}\underline{\sigma}^{3} - 6\mathrm{tr}\underline{\sigma\Phi}^{2} \tag{36}$$

---

* We write $\underline{\alpha} = \frac{1}{\sqrt{2}} \sum_{i=o}^{8} \alpha_i \lambda_i$ and similarly identify σ_i and Φ_i in the 3×3 matrices $\underline{\sigma}$ and $\underline{\Phi}$.

We shall not discuss baryon sin our model. For completeness, however, let me briefly discuss how one goes about constructing effective Lagrangians with baryons. For a spin 1/2 octet,

$$\delta\psi_A = - i\psi_B (\alpha F)_{BA} = - f_{ABC}\, \alpha_B\, \psi_C$$
$$\delta'\psi_A = - f_{ABC}\, \alpha_B\, \gamma_5 \psi_C \tag{37}$$

If we introduce

$$\underline{\psi} = \frac{1}{\sqrt{2}} \sum \lambda_A\, \psi_A$$

$$\underline{\psi}_{\pm} = (1 \pm \gamma_5)\, \underline{\psi}$$
$$\underline{\bar{\psi}}_{\pm} = \underline{\bar{\psi}}(1 \mp \gamma_5) \tag{38}$$

then

$$\delta \underline{\psi}_{\pm} = \frac{i}{\sqrt{2}} [\underline{\alpha}, \underline{\psi}_{\pm}]$$

$$\delta \underline{\bar{\psi}}_{\pm} = \frac{i}{\sqrt{2}} [\underline{\alpha}, \underline{\bar{\psi}}_{\pm}]$$

$$\delta' \underline{\psi}_{\pm} = \frac{i}{\sqrt{2}} [\underline{\alpha}, \underline{\psi}_{\pm}] \tag{39}$$

$$\delta' \underline{\bar{\psi}}_{\pm} = \frac{i}{\sqrt{2}} [\underline{\alpha}, \underline{\bar{\psi}}_{\pm}]$$

Terms like $\bar{\psi}_{\pm}\psi_{\pm}$ vanish, but terms like $\bar{\psi}_{\pm}\gamma_{\mu}\psi_{\pm}, \bar{\psi}_{\pm}\not{\partial}\psi_{\pm}\ldots$ do not. Mass terms cannot appear* . One can get around this difficulty by constructing baryon fields that transform like $(\underset{\sim}{3},\underset{\sim}{\bar{3}}) \oplus (\underset{\sim}{\bar{3}},\underset{\sim}{3})$; it is easy to check that

$$\underline{\psi}_{+}\, \underline{M}\ ,\ \underline{M}\, \underline{\psi}_{-}\ ,\ \underline{\bar{\psi}}_{+}\, \underline{M}\ ,\ \ \underline{M}\, \underline{\bar{\psi}}_{-} \sim \underline{M}$$

$$\underline{M}^{+}\ \underline{\psi}_{+},\ \ \underline{\psi}_{-}\underline{M}^{+},\ \ \underline{M}^{+}\ \underline{\bar{\psi}}_{+},\ \ \underline{\bar{\psi}}_{-}\ \underline{M}^{+} \sim \underline{M}^{+} \tag{40}$$

---

* This is the well known result that $\bar{\psi}\psi$ is not γ_5 invariant.

Mass terms will appear when in a combination like $tr(\bar{\psi}_+ M\psi_- M^+ + \bar{\psi}_- M^+ \psi_+ M)$ the scalar field acquires a constant component (nonvanishing vacuum expectation value).

Once the transformation properties of fields are given, chiral invariants are easy to construct. For example a chiral invariant Lagrangian involving baryons and spin 0 mesons might be of the form

$$
\begin{aligned}
L = {} & \frac{i}{4}\, tr(\bar{\psi}_+ \not\partial \psi_+ + \bar{\psi}_- \not\partial \psi_-) + \frac{1}{2} tr(\partial_\mu M \partial^\mu M^+) + \\
& + \frac{1}{4}\, a\, tr\,(\bar{\psi}_+ M \psi_- M^+ + \bar{\psi}_- M^+ \psi_+ M) - \frac{1}{2}\mu^2 tr M M^+ + \\
& + \frac{1}{4}\, i b_1 tr(\bar{\psi}_+ \gamma_\mu \psi_+ \partial^\mu M M^+ + \dots) + \\
& + \frac{1}{4}\, i b_2 tr(\bar{\psi}_+ \partial^\mu M M^+ \gamma_\mu \psi_+ + \dots) + \qquad (41) \\
& + f(tr M M^+,\ tr\, M M^+ M M^+,\ \det M + \det M^+) + \dots
\end{aligned}
$$

Using techniques that will be described in sec. (IV) one may use this Lagrangian (e.g. in the SU(2)×SU(2) limit for simplicity) to derive the Goldberger-Treiman relation and the Adler-Weisberger relation. The meson-meson term f is irrelevant, a is proportional to the nucleon mass and b_2 enters into the definition of G_A/G_V (b_1 does not enter in the SU(2)×SU(2) limit).

III. FIELD ALGEBRA, SYMMETRY BREAKING AND PCAC

A direct way of implementing the experimentally well established notion of vector meson dominance of the currents [5] is to build into the Lagrangian the requirement that the currents be proportional to the vector meson field operators. This means that the vector fields must enter in a way such that $\delta L \propto \partial^{\mu} \alpha V_{\mu}$.

Given the transformation laws for the spin 1 mesons

$$\delta \underline{V}_{\mu} = \frac{i}{\sqrt{2}} [\underline{\alpha}, \underline{V}_{\mu}]$$
$$\delta A_{\mu} = \frac{i}{\sqrt{2}} [\underline{\alpha}, \underline{A}_{\mu}] \quad , \ldots \tag{42}$$

it is clear that a conventional "kinetic energy" term like $tr(\partial_{\mu}\underline{V}_{\mu})^2$ will give rise to a current of the form* $\underline{j}_{\mu} = c\frac{i}{\sqrt{2}}[\underline{V}_{\nu}, \partial_{\mu}\underline{V}^{\nu}]$. It turns out that to obtain the field current proportionality (from now on, we will call this Field Algebra), it is necessary to make two modifications:

(i) This is to modify the transformation laws of the spin 1 mesons in the following manner**

$$\delta \underline{V}_{\mu} = \frac{1}{\gamma_o} \partial_{\mu} \underline{\alpha} + \frac{i}{\sqrt{2}} [\underline{\alpha}, \underline{V}_{\mu}]$$
$$\delta \underline{A}_{\mu} = \frac{i}{\sqrt{2}} [\underline{\alpha}, \underline{A}_{\mu}]$$
$$\delta' \underline{V}_{\mu} = \frac{i}{\sqrt{2}} [\underline{\alpha}, \underline{A}_{\mu}]$$
$$\delta' \underline{A}_{\mu} = \frac{1}{\gamma_o} \partial_{\mu} \underline{\alpha} + \frac{i}{\sqrt{2}} [\underline{\alpha}, \underline{V}_{\mu}] \tag{43}$$

(ii) With these transformation laws, first suggested by Yang and Mills [6] it is easy to check that the field quantities defined by

$$\underline{F}_{\mu\nu} = \partial_{\mu}\underline{V}_{\nu} - \partial_{\nu}\underline{V}_{\mu} - \frac{i\gamma_o}{\sqrt{2}}[\underline{V}_{\mu}, \underline{V}_{\nu}] - \frac{i\gamma_o}{\sqrt{2}} [\underline{A}_{\mu}, \underline{A}_{\nu}]$$
$$\underline{G}_{\mu\nu} = \partial_{\mu}\underline{A}_{\nu} - \partial_{\nu}\underline{A}_{\mu} - \frac{i\gamma_o}{\sqrt{2}}[\underline{V}_{\mu}, \underline{A}_{\nu}] + \frac{i\gamma_o}{\sqrt{2}} [\underline{V}_{\nu}, \underline{A}_{\mu}] \tag{44}$$

transform according to

---

* In the derivation we use $\frac{1}{2} \sum_{i=o}^{8} (\lambda_i)_{ab} (\lambda_i)_{cd} = \delta_{ad}\delta_{bc}$

** Only for vector fields is such a modification of the transformation law compatible with the Jacobi identity.

$$\delta\underline{F}_{\mu\nu} = \frac{i}{\sqrt{2}} [\underline{\alpha},\underline{F}_{\mu\nu}]$$

$$\delta\underline{G}_{\mu\nu} = \frac{i}{\sqrt{2}} [\underline{\alpha},\underline{G}_{\mu\nu}]$$

$$\delta'\underline{F}_{\mu\nu} = \frac{i}{\sqrt{2}} [\underline{\alpha},\underline{G}_{\mu\nu}]$$

$$\delta'\underline{G}_{\mu\nu} = \frac{i}{\sqrt{2}} [\underline{\alpha},\underline{F}_{\mu\nu}] \qquad (45)$$

Thus a term like

$$-\frac{1}{4} \text{tr} (\underline{F}_{\mu\nu}\underline{F}^{\mu\nu} + \underline{G}_{\mu\nu}\underline{G}^{\mu\nu}) \qquad (46)$$

is invariant under space-time dependent gauge transformations, and no current is defined* . If, however, the chiral invariant mass term

$$\frac{1}{2} m_o^2 \text{tr}(\underline{V}_\mu \underline{V}^\mu + \underline{A}_\mu \underline{A}^\mu) \qquad (47)$$

is added to the kinetic energy term, then

$$\delta (\quad) = \frac{m_o^2}{\gamma_o} \text{tr} (\underline{V}_\mu \partial^\mu \underline{\alpha} + \ldots)$$

$$\delta' (\quad) = \frac{m_o^2}{\gamma_o} \text{tr} (\underline{A}_\mu \partial^\mu \underline{\alpha} + \ldots)$$

i.e.

$$\underline{j}_\mu = -\frac{m_o^2}{\gamma_o} \underline{V}_\mu$$

$$\underline{j}_{5\mu} = -\frac{m_o^2}{\gamma_o} \underline{A}_\mu \qquad (48)$$

When other fields are included, we must, in analogy with electromagnetic couplings of matter, replace derivatives of fields by "covariant derivatives" which transform as the fields. For example for spin 0 mesons we require that

---

*Just as the Maxwell Lagrangian does not define an electric current.

$$\delta(\Delta_\mu \underline{\Phi}) = \frac{i}{\sqrt{2}} [\underline{\alpha}, \Delta_\mu \underline{\Phi}]$$

$$\delta(\Delta_\mu \underline{\sigma}) = \frac{i}{\sqrt{2}} [\underline{\alpha}, \Delta_\mu \underline{\sigma}]$$

$$\delta'(\Delta_\mu \underline{\Phi}) = \frac{1}{\sqrt{2}} \{\underline{\alpha}, \Delta_\mu \underline{\sigma}\}$$

$$\delta'(\Delta_\mu \underline{\sigma}) = - \frac{1}{\sqrt{2}} \{\underline{\alpha}, \Delta_\mu \underline{\Phi}\} \quad (49)$$

Such covariant derivatives are

$$\Delta_\mu \underline{\Phi} = \partial_\mu \underline{\Phi} - \frac{i\gamma_o}{\sqrt{2}} [\underline{V}_\mu, \underline{\Phi}] - \frac{\gamma_o}{\sqrt{2}} \{\underline{A}_\mu, \underline{\sigma}\}$$

$$\Delta_\mu \underline{\sigma} = \partial_\mu \underline{\sigma} - \frac{i\gamma_o}{\sqrt{2}} [\underline{V}_\mu, \underline{\sigma}] + \frac{\gamma_o}{\sqrt{2}} \{\underline{A}_\mu, \underline{\Phi}\} \quad (50)$$

and their presence in a Lagrangian does not change the field algebra structure.

With all of the tools for the construction of chiral invariant Lagrangians at our disposal, we must now turn to the question of symmetry breaking terms. In order to restrict the number of additional parameters that enter the Lagrangian we shall require "pure" PCAC, i.e. we shall demand strict proportionality between the divergences of the axial currents and the pseudoscalar field operators. This means that $\delta' L \propto tr(\underline{\alpha f \Phi})$. A glance at eq. (32) shows that if the only symmetry-breaking term has the form

$$L^{SB} = tr(\underline{f}\, \underline{\sigma}) \quad (51)$$

then

$$\delta' L_{SB} = -tr(\underline{f}\{\frac{1}{\sqrt{2}} \underline{\alpha}, \underline{\Phi}\}) = - \frac{1}{\sqrt{2}} tr(\underline{\alpha}\{\underline{f}, \underline{\Phi}\})$$

i.e.

$$\partial^\mu \underline{j}_{5\mu} = \frac{1}{\sqrt{2}} \{\underline{f}, \underline{\Phi}\} \quad (52)$$

and

$$\delta L_{SB} = \frac{i}{\sqrt{2}}\, tr(\underline{\alpha}[\underline{\sigma},\underline{f}])$$

i.e.

$$\partial^{\mu}\underline{j}_{\mu} = -\frac{i}{2}[\underline{\sigma},\underline{f}] \tag{53}$$

To arrange for chiral symmetry breaking and SU(3) breaking with i-spin and hypercharge conservation maintained, we take f_o and f_8 nonvanishing. This means that

$$\partial^{\mu} j^{(k)}_{5\mu} = (f_o d_{ok\ell} + f_8 d_{8k\ell})\Phi_{\ell} \tag{54}$$

and

$$\partial^{\mu} j^{(k)}_{\mu} = f_8\, f_{8k\ell}\, \sigma_{\ell} \tag{55}$$

Thus none of the axial currents are conserved, and the strangeness-changing vector currents are not conserved. We shall discuss the relation of the parameters f_o and f_8 to observed quantities in the next section.Before doing so we will briefly discuss symmetry breaking for the vector mesons and its implications for field algebra. It will be sufficient to discuss vector mesons alone.

With octet symmetry breaking, the simple SU(3) Yang-Mills Lagrangian is modified to read

$$L = -\frac{1}{4} f^{i}_{\mu\nu}(\delta_{ij}+\sqrt{3}\xi d_{8ij}) f^{\mu\nu}_{j} + \frac{1}{2} m_o^2 V^{i}_{\mu}(\delta_{ij}+\sqrt{3}\xi d_{8ij}) V^{\mu}_{j} \tag{56}$$

Note that terms with f_{8ij} are excluded by symmetry. Although we need only deal with an octet of vector mesons (recall our assignment $(\underset{\sim}{8},\underset{\sim}{1}) \oplus (\underset{\sim}{1},\underset{\sim}{8})$), it is no trouble to include the ninth vector meson. In that case, we let the sums go from 0 to 8, and use

$$d_{oij} = \sqrt{2/3}\; d_{ij}$$

If we define the matrix D_8 by

$$(D_8)_{ij} = d_{8ij} \tag{57}$$

and write the f's and V's as nine-component vectors, then

$$L = -\frac{1}{4}\tilde{f}_{\mu\nu}(1+\sqrt{3}\,\xi D_8)f^{\mu\nu} + \frac{1}{2}m_0^2\tilde{V}_\mu(1+\sqrt{3}\,\xi' D_8)V^\mu \tag{58}$$

where

$$\sqrt{3}\,D_8 = \begin{pmatrix} 1 & & \\ & -1/2 & \\ & & -1 \end{pmatrix}_{\text{octet}} \quad \text{or} \quad \begin{pmatrix} 1 & & & \\ & -1/2 & & \\ & & -1 & \sqrt{2} \\ & & \sqrt{2} & 0 \end{pmatrix}_{\text{nonet}} \tag{59}$$

The boxes refer to rows and columns labeled 1,2,3; 4,5, 6,7 and 8 or 8,0 respectively. In the latter case the (8,0) submatrix can be diagonalized

$$U\begin{pmatrix} -1 & \sqrt{2} \\ \sqrt{2} & 0 \end{pmatrix}U^{-1} = \begin{pmatrix} -2 & 0 \\ 0 & 1 \end{pmatrix}$$

With appropriate mixing* the Lagrangian reads

$$L = -\frac{1}{4}(1+\xi)\sum_{i=0}^{3}(f^i_{\mu\nu})^2 + \frac{1}{2}m_0^2(1+\xi')\sum_{i=0}^{3}(V^i_\mu)^2 -$$

$$-\frac{1}{4}(1-\frac{\xi}{2})\sum_{i=4}^{7}(f^i_{\mu\nu})^2 + \frac{1}{2}m_0^2(1-\frac{\xi'}{2})\sum_{i=4}^{7}(V^i_\mu)^2 -$$

$$-\frac{1}{4}(1-2\xi)(f^8_{\mu\nu})^2 + \frac{1}{2}m_0^2(1-2\xi')(V^8_\mu)^2 \tag{60}$$

To discuss the current commutation relations, we need the commutation relations among the fields, and to use the canonical commutation relations we must renormalize the fields so that the kinetic energy terms have unit coefficients. We thus introduce

---

* The mixing angle is given by $\tan^2\theta = \frac{1}{2}$; we refer to this angle as the nonet mixing angle

$$\hat{V}^i_\mu = \begin{pmatrix} (1+\xi)^{1/2} \\ (1-\frac{1}{2}\xi)^{1/2} \\ (1-2\xi)^{1/2} \end{pmatrix} V^i_\mu \qquad \begin{matrix} i = 0,1,2,3 \\ i = 4,\ldots,7 \\ i = 8 \end{matrix}$$

The current comes from the mass terms. It has the form

$$j^i_\mu = -\frac{m_o^2}{\gamma_o}\begin{pmatrix} 1+\xi' \\ 1-\frac{1}{2}\xi' \\ 1-2\xi' \end{pmatrix} V^i_\mu = -\frac{m_o^2}{\gamma_o}\begin{pmatrix} (1+\xi')(1+\xi)^{-1/2} \\ (1-\frac{1}{2}\xi')(1-\frac{1}{2}\xi)^{-1/2} \\ (1-2\xi')(1-2\xi)^{-1/2} \end{pmatrix} \hat{V}^i_\mu$$

(62)

Also, from (2) it follows that

$$\hat{\pi}^i_k = \hat{f}^i_{ko} \tag{63}$$

Since we now have

$$L = -\frac{1}{4}\sum_i (\hat{f}^i_{\mu\nu}) + \frac{1}{2}m_o^2\,\frac{1+\xi'}{1+\xi}\sum_{i=o}^{3}(\hat{V}^i_\mu)^2 + \ldots$$

it follows that the equations of motion are

$$-\partial^\mu \hat{f}^i_{\mu\nu} = m_o^2\begin{pmatrix} (1+\xi')\;(1+\xi)^{-1} \\ (1-\frac{1}{2}\xi')(1-\frac{1}{2}\xi)^{-1} \\ (1-2\xi')(1-2\xi)^{-1} \end{pmatrix} \hat{V}^i_\nu + \ldots \tag{64}$$

so that

$$\hat{V}^i_o = -\frac{1}{m_o^2}\begin{pmatrix} (1+\xi)(1+\xi')^{-1} \\ \vdots \\ \vdots \end{pmatrix} \underset{\sim}{\nabla}\cdot\underset{\sim}{\hat{\pi}} + \ldots \tag{65}$$

Hence the current commutators

$$[j^k_o, j^\ell_m] = \left(\frac{m_o^2}{\gamma_o}\right)^2\begin{pmatrix} (1+\xi')^2(1+\xi)^{-1} \\ \vdots \\ \cdot \end{pmatrix} [\hat{V}^k_o, \hat{V}^\ell_m]$$

have Schwinger terms [7] of the form

$$\frac{im_o^2}{\gamma_o}\begin{pmatrix}1+\ \xi' \\ 1-\ \frac{1}{2}\xi' \\ 1-\ 2\xi'\end{pmatrix} \partial_m \delta(\underset{\sim}{x}-\underset{\sim}{y}) \tag{66}$$

Thus if there is symmetry breaking in the mass terms, the Schwinger terms are <u>not</u> invariant under SU(3), and one cannot obtain a generalization of the Weinberg sum rules. Furthermore, an examination of the remaining terms in the commutator shows that with $\xi' \neq 0$, even the field algebra is no longer maintained [8]. We thus set $\xi' = 0$. One consequence of this assumption is that (as can be seen from (64)) that

$$m_\omega^2 = m_\rho^2 = \frac{m_o^2}{1+\xi}\ ;\quad m_{K^*}^2 = \frac{m_o^2}{1-\frac{1}{2}\xi}\ ;\ m_\phi^2 = \frac{m_o^2}{1-2\xi} \tag{67}$$

which is the Gell-Mann-Okubo mass formula for reciprocal masses squared, with the standard nonet mixing angle. The ω-ρ mass degeneracy is independent of our choice of ξ and ξ'. It can be removed by splitting off the ninth vector meson (which does not transform under SU(3) anyway because $f_{0ij} = 0$) by adding a term like

$$-\frac{1}{4}\alpha(f_{\mu\nu}^o)^2 + \frac{1}{2}m_o^2\beta(V_\mu^o)^2$$

to the Lagrangian. The nonet symmetry does, however, seem to be a very good approximation.

The chiral extension of such a kinetic energy mixing term is

$$\mathrm{tr}(\underline{F}_{\mu\nu}\lambda_8\underline{F}^{\mu\nu} + \underline{G}_{\mu\nu}\lambda_8\underline{G}^{\mu\nu}) \tag{68}$$

We shall see that in spite of appearances, there is no analog of the ω-ρ degeneracy for the axial mesons. Terms like those in (68) violate PCAC for the K and η mesons, and in the following discussion we shall modify the Lagrangian in such a way that the mixing (68) emerges as a result of the introduction of L_{SB} in (51) ; in this way we can preserve PCAC.

IV. A MODEL FOR SPIN 0 AND SPIN 1 MESONS*

I will now describe the construction of a model which describes the interaction of spin 0 and spin 1 particles. In contrast to the lectures of Gürsey last year [9], I shall not say much about nonlinear realizations, in which some or all of the scalar fields are constructed out of other (pseudoscalar) fields. The main reason for this is that there is now some evidence for the existence of scalar particles [10], and their masses seem to lie in the same region as those of the vector and axial mesons. In a discussion of low energy pion interactions, these masses are very large, and it is reasonable to treat them as infinitely large; a careful limiting procedure leads to the nonlinear realizations [11].

Let us now list the kind of terms that we want to include:

1.

$$-\frac{1}{4}\mathrm{tr}(\underline{F}_{\mu\nu}\underline{F}^{\mu\nu} + \underline{G}_{\mu\nu}\underline{G}^{\mu\nu}) + \frac{1}{2}\, m_0^2 \mathrm{tr}\, (\underline{V}_\mu \underline{V}^\mu + \underline{A}_\mu\, \underline{A}^\mu) \qquad (69)$$

Here $\underline{F}_{\mu\nu}$, $\underline{G}_{\mu\nu}$ are the Yang-Mills field introduced in (44); the mass terms lead to field algebra. Note that the bare masses of the vector and axial mesons are all the same.

2. We write the corresponding "free" terms for the mesons:

---

* A very similar model was studied by P. K. Mitter and J. L. Swank, Nuclear Physics B8, 205 (1968).

$$\frac{1}{2}\mathrm{tr}(\Delta_\mu \underline{\sigma}\Delta^\mu \underline{\sigma} + \Delta_\mu \underline{\Phi}\Delta^\mu \underline{\Phi}) - \frac{1}{2}\mu^2 \mathrm{tr}(\underline{\sigma}^2 + \underline{\Phi}^2) \qquad (70)$$

The covariant derivatives are defined in eq. (50); the scalar and pseudoscalar bare masses are all the same.

3. We next include meson-meson scattering terms. There are no criteria to guide us in the selection of these terms. We have chosen

$$\frac{1}{4}\lambda(\mathrm{tr}(\underline{\sigma}^2 + \underline{\Phi}^2))^2 + \frac{1}{8}\lambda'\mathrm{tr}((MM^+)^2+(M^+M)^2) +$$
$$+ \frac{1}{4}\eta_1 \mathrm{tr}(\Delta_\mu M \Delta_\mu M^+ MM^+ + \Delta_\mu M^+ \Delta^\mu MM^+ M) +$$
$$+ \frac{1}{4}\eta_2 \mathrm{tr}(\Delta_\mu \underline{M}\underline{M}^+ \Delta^\mu \underline{M}\underline{M}^+ + \Delta_\mu \underline{M}^+\underline{M}\Delta^\mu \underline{M}^+\underline{M}) \qquad (71)$$

The particular form does not affect most of the conclusions that I will discuss in these lectures. The gradient terms seem to be necessary to help with the mass fits.

4. Another spin 1 - spin 0 interaction term which appears to be necessary to fit the A_1 decay is

$$\frac{i\gamma_0 \delta}{2m_0^2}\mathrm{tr}((\underline{F}_{\mu\nu} + \underline{G}_{\mu\nu})\Delta^\mu \underline{M}\Delta^\nu \underline{M}^+ + (\underline{F}_{\mu\nu} - \underline{G}_{\mu\nu})\Delta^\mu \underline{M}^+\Delta^\nu \underline{M}) \qquad (72)$$

This term contributes to the D-wave decay matrix element for $A_1 \rightarrow \rho + \pi$. The parameter δ is determined by the A_1 widths or by the partial widths for different helicity states. Present data are not good enough to determine δ: they do suggest its presence.

5. There are more terms that we will want to add to the Lagrangian. Rather than doing this at this point, we shall proceed with the Lagrangian that we already have, and then see very clearly why the additional terms have to come in.

We certainly need to break chiral symmetry and SU(3), and we do this by introducing

$$L_{SB} = \mathrm{tr}(\underline{f}\ \underline{\sigma}) \qquad (73)$$

This, as already discussed, implies PCAC. In a Feynman graph expansion of the scattering matrix, terms generated by (73) lead to so-called "tadpole graphs" in which a scalar meson (with T=0 and Y=0) just disappears into the vacuum. The graphs are there because the scalar fields σ_o and σ_8 can have nonvanishing vacuum expectation values, without violating anything except chiral symmetry and SU(3). To get rid of the tadpole graphs, we introduce new fields $\underline{\sigma}'$, that have vanishing vacuum expectation values, by means of

$$\underline{\sigma} = \underline{\sigma}' + \underline{\Sigma} \qquad (74)$$

where

$$\underline{\Sigma} = \begin{pmatrix} a & 0 & 0 \\ 0 & a & 0 \\ 0 & 0 & b \end{pmatrix} \qquad (75)$$

When the substitution (74) is made in the Lagrangian that we already have, we find that there still are terms linear in $\underline{\sigma}'$. These can be eliminated by a proper choice of $\underline{\Sigma}$. The relation between $\underline{\Sigma}$ and $\underline{f}$, written in the form $\underline{f}=u\,\underline{\Sigma}+v$ is unaffected by the presence of the vector meson terms, and at this stage reads

$$u = \mu_1^2 - \lambda'(a^2 + ab + b^2)$$

$$v = \lambda'\, a\, (a+b)\, b$$

where

$$\mu_1^2 = \mu^2 - \lambda(2a^2+b^2) \,. \qquad (76)$$

The reason for writing this down in detail is that one can see the possibility of a "spontaneously broken symmetry" with a solution u=v=0, which does <u>not</u> correspond to a=b=0. We shall see that the solutions corresponding to "spontaneous breaking" lead to massless particles. For that we need expressions for the masses of the particles.

6. The masses of the particles cannot simply be read off from the Lagrangian ((70) + terms quadratic in the fields after the substitution (74) is made), because symmetry breaking introduces a coupling between the pseudoscalar particles and the axial vector mesons* (also the scalar K and the no longer "conserved" strangeness changing vector meson K^*). The mesonic kinetic energy terms aree also changed by the presence of the η_1 and η_2 terms in (71) but for purposes of exposition we shall leave them out. The coupling referred to comes about in the covariant derivatives in eq. (50). $\Delta_\mu \underline{\Phi}$ will now contain a term <u>linear</u> in $\underline{A}_\mu$, viz $-\gamma_0/\sqrt{2}\ \{\underline{A}_\mu, \underline{\Sigma}\}$; similarly $\Delta_\mu \underline{\sigma}$ contains a term linear in $\underline{V}_\mu$, viz. $- i\gamma_0/\sqrt{2}\ \times$ $\times[\underline{V}_\mu, \underline{\Sigma}]$. Thus the kinetic energy terms for the mesons contain terms of the form $-\gamma_0/\sqrt{2}\ \mathrm{tr}(\partial^\mu \underline{\Phi}\{\underline{A}_\mu, \underline{\Sigma}\})$, for example, which lead to transitions of the type shown in the diagram below

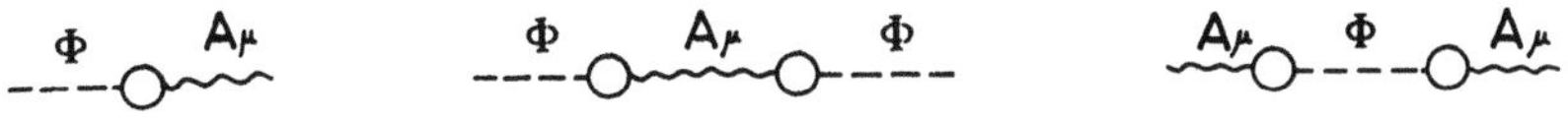

and which "mixing" we need to get rid of. We define a new axial meson field $\underline{A}_\mu$ by **

$$\underline{A}_\mu = \underline{A}_\mu - \widetilde{D_\mu \underline{\Phi}} \tag{77}$$

---

* In the symmetry limit no such coupling is possible since the spin 0 mesons belong to $(\underset{\sim}{3},\underset{\sim}{\bar{3}}) \oplus (\underset{\sim}{\bar{3}},\underset{\sim}{3})$ and the spin 1 mesons to $(\underset{\sim}{8},\underset{\sim}{1}) \oplus (\underset{\sim}{1},\underset{\sim}{8})$.

** This mixing was noted by Wessaand Zumino and by Lee and Nieh in [1] and discussed last year by Gürsey for SU(2)×SU(2). Identically,

$$D_\mu \underline{\Phi} \equiv \partial_\mu \underline{\Phi} - \frac{i\gamma_0}{\sqrt{2}}[\underline{V}_\mu, \underline{\Phi}] .$$

where $\widetilde{D_\mu \underline{\Phi}}$ is defined as follows:
If $D_\mu \underline{\Phi}$ is symbolically denoted by

$$D_\mu \underline{\Phi} = \begin{pmatrix} \cdot & \cdot & \cdot \\ \cdot & \cdot & \cdot \\ \cdot & \cdot & \cdot \end{pmatrix}$$

then,

$$\widetilde{D_\mu \underline{\Phi}} = \begin{pmatrix} \alpha\cdot & \alpha\cdot & \beta\cdot \\ \alpha\cdot & \alpha\cdot & \beta\cdot \\ \beta\cdot & \beta\cdot & \gamma\cdot \end{pmatrix}$$

i.e. the 11 matrix element of $D_\mu \underline{\Phi}$ is multiplied by α, etc.- The coefficients α,β and γ are determined by the requirement that there be no $A^i_\mu \partial^\mu \Phi^j$ terms in the Lagrangian. Such terms come from the kinetic energy terms for mesons and from the mass term for the axial mesons. Similarly we introduce a new vector meson field

$$\underline{U}_\mu = \underline{V}_\mu - i\underline{\gamma}\left[\Delta_\mu \underline{\sigma}', \underline{\Sigma}\right] \tag{78}$$

where

$$\Delta_\mu \underline{\sigma}' = \Delta_\mu \underline{\sigma} + \frac{i\gamma_o}{\sqrt{2}} \underline{V}_\mu, \underline{\Sigma}$$

Note that $\underline{\Sigma}$ only contains the diagonal matrices λ_o, λ_8 and $[\lambda_8, \lambda_j] = 0$ unless $j = 4,5,6,7$ so that only the K^* parts of U differ from V.

After some algebra one finds that, after the introduction of standard nonet mixing (to get rid of $\Phi_o\Phi_8$ terms)

$$\Phi_\eta = \sqrt{2/3}\ \Phi_o + \sqrt{1/3}\ \Phi_8$$

$$\Phi_X = \sqrt{1/3}\ \Phi_o - \sqrt{2/3}\ \Phi_8 \tag{79}$$

the 0^- kinetic energy terms, for example, are

$$\frac{1}{2}((D_\mu \Phi_\pi)^2 + (D_\mu \Phi_\eta)^2) \frac{m_o^2}{m_o^2 + 2\gamma_o^2 a^2} + \frac{1}{2}(D_\mu \Phi_X)^2 \frac{m_o^2}{m_o^2 + 2\gamma_o^2 b^2} +$$

$$+ \frac{1}{2}(D_\mu \Phi_K)^2 \frac{m_o^2}{m_o^2 + \gamma_o^2 (a+b)^2/2} \tag{80}$$

and the mass terms are

$$-\frac{1}{2}(\mu_1^2-\lambda' a^2)(\phi_\pi^2+\phi_\eta^2)-\frac{1}{2}(\mu_1^2-\lambda'(a^2-ab+b^2))\phi_K^2-\frac{1}{2}(\mu_1^2-\lambda' b^2)\phi_X^2 \tag{81}$$

The crucial point here, independent of the particular choice of meson-meson terms in (71) is that the pion and η masses turn out to be degenerate. This characteristic nonet degeneracy is very attractive for the vector mesons ($m_\omega \sim m_\rho$) but clearly disagrees with experiment for the pseudoscalar mesons. If we want to fit the masses, we must add

$$\frac{1}{3}\mu g(\det M + \det M^+) \tag{82}$$

which splits the π and η masses, but also breaks PCAC for the ninth pseudoscalar meson* . Incidentally, before introducing this complication note that in (76) setting u=v=0 but not a=b=0 implies that at least one of the pseudoscalar masses vanishes. The same holds true when (82) is added. This changes eq. (76) but also the expression for the pseudoscalar masses**. We do not have time to get into a more detailed discussion of this illustration of the Goldstone theorem (see Bardeen and Lee, ref.[11], for example).

7. In order to identify the fields with the appropriate particles, we must set the coefficients of the kinetic energy terms equal to unity. This is done by defining a new set of renormalized operators

$$\hat{\phi}_\pi = Z_\pi^{-1/2}\,\phi_\pi$$
$$\hat{\phi}_K = Z_K^{-1/2}\,\phi_K \tag{83}$$

* The expression (82) is chiral invariant, but not under U(3)×U(3) .

** With a=0, the K mass vanishes, with $\lambda' b=2\mu g$ the pion mass vanishes. The latter is the more interesting case.

$$\hat{\sigma}_\pi = Z_{\sigma\pi}^{-1/2} \sigma_\pi$$

$$\hat{\sigma}_K = Z_{\sigma K}^{-1/2} \sigma_K$$

....

with $Z_\pi = (1+2\gamma_o^2 a^2/m_o^2)$ etc. and working with these fields. They modify the mesonic decay constants. After some algebra we rewritte (54) and (55) in the form

$$\partial^\mu j_{5\mu}^\pi = m_\pi^2 f_\pi \hat{\phi}_\pi$$

$$\partial^\mu j_{5\mu}^K = m_K^2 f_K \hat{\phi}_K$$

$$\partial^\mu j_\mu^K = m_{\sigma K}^2 f_{\sigma K} \hat{\sigma}_K \quad (84)$$

and deduce

$$f_\pi = \sqrt{2}\, a\, Z_\pi^{-1/2}$$

$$f_K = \frac{a+b}{\sqrt{2}} Z_K^{-1/2}$$

$$f_{5K} = \frac{a-b}{\sqrt{2}} Z_{\sigma K}^{-1/2} \quad (85)$$

Note the following: if we take the K_{e3} form factor at zero momentum transfer squared*

$$f_+(0) = \frac{f_\pi^2+f_K^2-f_{\sigma K}^2}{2f_K f_\pi} \quad (86)$$

we see that in the absence of the renormalization effects that arise from the coupling to vector mesons and axial mesons, i.e. when all the Z's are 1, then $f_+(0) = 1$. The experimental requirement that**

---

* This result was first derived by Glashow and Weinberg (Phys. Rev. Letters 20, 224 (1968)) using assumptions equivalent to those that go into our effective Lagrangian, including the assumption that f_+ when extrapolated off the π and K mass shells does not depend on the extrapolated values of the masses, which is also true of the model.

**One assumes the equality of the axial and vector Cabibbo angles and uses data from $\pi_{\mu 2}$, $K_{\mu 2}$ and K_{e3} .

$$\frac{f_\pi f_+(0)}{f_K} = \frac{1}{2}\left(1+\frac{f_\pi^2-f_{\sigma K}^2}{f_K^2}\right) = 0.84 \tag{87}$$

places very stringent constraints on the parameters that appear in the Lagrangian. This value to good to 5 % : the experimental errors are small (∿ 2 %) but the radiative corrections have not been calculated.

8. Let us now turn to a spin 1 mesons. It is clear that the vector masses remain degenerate, except that the K^* is shifted because of mixing with the scalar K field (the K^* field appears in $\Delta_\mu \underline{\sigma}$; compare the discussion after eq. (78)). To remove this degeneracy and maintain field algebra we could introduce a term like that shown in (68). Such a term destroys PCAC for K mesons, and it is more appealing to lift the degeneracy by means of a chiral invariant term that looks like (68) as a result of symmetry breaking. Such a term is

$$\frac{1}{8}\,\xi\,\mathrm{tr}[(\underline{F}_{\mu\nu}+\underline{G}_{\mu\nu})^2\underline{M}\underline{M}^+ + (\underline{F}_{\mu\nu}-\underline{G}_{\mu\nu})^2\,\underline{M}^+\underline{M}\,] \tag{88}$$

One may add

$$-\frac{1}{12}\,\xi_\nu\,(\mathrm{tr}\,\underline{F}_{\mu\nu})^2$$

to break the ω-ρ degeneracy, but we have not done so*. The axial mesons are automatically split away from the vector mesons by the mixing with the pseudoscalar mesons. For example one obtains

$$1-\frac{m_\rho^2}{m_A^2} = \frac{\gamma_\rho^2 f_\pi^2}{m_\rho^2} \tag{89}$$

which shows that the KSFR relation [13]

---

* There is uncertainty about the T=0 axial mesons, and we don't know whether a corresponding axial term is needed.

$$2\gamma_\rho^2 f_\pi^2/m_\rho^2 = 1$$

and the Weinberg formula $m_{A_1} = \sqrt{2}\, m_\rho$ are not independent results, and they do not follow from chiral symmetry, PCAC and field algebra alone. The only other result that follows if the η_1 and η_2 terms are left out is the expression ($m_{A_1}^2 \simeq 2m_\rho^2$)

$$\left(\frac{m_{KA}}{m_{K^*}}\right)^2 = \frac{2-(f_{\sigma K}/f_\pi)^2}{2-(f_K/f_\pi)^2} \tag{90}$$

9. Consider the vector mesons: the presence of (88) implies that the vector (and axial) kinetic energy terms have coefficients different from one. We thus again renormalize the fields

$$\hat{V}_\pi = Z_{v\pi}^{-1/2}\; U_\pi$$
$$\hat{V}_\eta = Z_{v\eta}^{-1/2}\; U_\eta$$

where

$$Z_{v\pi}^{-1} = 1 - \xi a^2$$

$$Z_{vk}^{-1} = 1 - \frac{1}{2}(a^2+b^2)\,\xi$$

. . . .

The masses are then changed to

$$m_{v\pi}^2 = m_o^2\, Z_{v\pi}$$

$$m_{v\eta}^2 = m_o^2\, Z_{v\eta}$$

$$m_{vK}^2 = m_o^2\, Z_{vK}\, \Gamma_{\sigma K} \tag{91}$$

where

$$\Gamma_{\sigma K} = 1 + \frac{1}{2}(b-a)^2\gamma_o{}^2/m_o^2$$

The factor $\Gamma_{\sigma K}$ accounts for the "mixing" with the scalar K meson. It differs from unity as soon as $\langle\sigma_8\rangle \neq 0$.

Let us now discuss the spin 1 to vacuum matrix elements of the vector currents. The latter are given by the field algebra expressions

$$j_\mu^i = -\frac{m_o^2}{\gamma_o} V_\mu^i \qquad i=0,1,2,\ldots,8 \qquad (92)$$

where the V's are the original unrenormalized vector fields. In terms of the renormalized physical vector fields we get

$$j_\mu^i = -\frac{m_o^2}{\gamma_o} Z_{v\pi}^{1/2} \hat{V}_\mu^i \qquad i=1,2,3(\rho)$$

$$j_\mu^i = -\frac{m_o^2}{\gamma_o} Z_{vK}^{1/2} \hat{V}_\mu^i + \ldots. \qquad i=4,\ldots,7(K^*)$$

$$j_\mu^8 = -\frac{m_o^2}{\gamma_o} [Z_{v\eta}^{1/2}\cos\Theta\, \hat{V}_\mu^\eta + Z_{vX}^{1/2} \sin\Theta\, \hat{V}_\mu^X]$$

$$j_\mu^o = -\frac{m_o^2}{\gamma_o} [-Z_{v\eta}^{1/2} \sin\Theta \hat{V}_\mu^\eta + Z_{vX}^{1/2} \cos\Theta\, \hat{V}_\mu^X] \qquad (93)$$

where $\tan^2\Theta = \frac{1}{2}$ for the ω-ρ degenerate model. With the constants F_ρ, F_ω, ... defined by

$$\langle 0|j_\mu^i|\rho,\lambda\rangle = \varepsilon_\mu^{(\lambda)} F_\rho$$

$$\langle 0|j_\mu^8|\phi,\lambda\rangle = \varepsilon_\mu^{(\lambda)} F_\phi$$

we get

$$F_\rho = -Z_{v\pi}^{1/2} m_o^2/\gamma_o \quad ; \quad F_{K*} = -Z_{vK}^{1/2} m_o^2/\gamma_o \qquad (94)$$

$$F_\omega = -Z_{vX}^{1/2} m_o^2 \sin\Theta/\gamma_o \; ; \; F_\phi = -Z_{v\eta}^{1/2} m_o^2 \cos\Theta/\gamma_o$$

We assume that the hadronic electromagnetic current is of the form suggested by the Gell-Mann-Nishijima form, $j_\mu = j_\mu^{(3)} + \frac{1}{\sqrt{3}} j_\mu^{(8)}$ so that we need not consider $j_\mu^{(o)}$ in a comparison with the leptonic decay rates of these mesons (e.g. $\rho \to e^+ + e^-$) with experiment. After eliminating the Z's in terms of the physical masses (eq.(91)) and defining γ_ρ by

$$F_\rho = - m_\rho^2/\gamma_\rho$$

so that

$$\frac{m_o}{\gamma_o} = \frac{m_\rho}{\gamma_\rho}$$

we get

$$\frac{F_\phi}{F_\rho} = \frac{m_\phi}{m_\rho} \cos \Theta ; \quad \frac{F_\omega}{F_\rho} = \frac{m_\omega}{m_\rho} \sin \Theta \tag{95}$$

The"generalized mixing angle" depends on ξ and on ξ_V if the ω-ρ mass difference is included. With the experimental masses we find that $\Theta \simeq 40.3^o$ which is to be compared with 35.3^o, the ideal nonet (degenerate ω,ρ) mixing angle. The recent Orsay and DESY determinations of this angle from leptonic decays yield $40^o \pm 8^o$.

Our assumptions about the effective Lagrangian fit rather closely with those made by Oakes and Sakurai [14] who however took $f^2_{\sigma K} = 0$. Their result for the angle reads

$$\frac{\cos^2\Theta}{m^2_\phi} + \frac{\sin^2\Theta}{m^2_\omega} = \frac{4}{3m^2_{K^*}} - \frac{1}{3m^2_\rho} \tag{9 }$$

whereas our results can be recast in the form

$$\frac{\cos^2\Theta}{m^2{}_\phi} + \frac{\sin^2\Theta}{m^2_\omega} = \frac{4\Gamma_{\sigma k}}{3m^2_{K^*}} - \frac{1}{3m^2_\rho} \tag{97}$$

Thus the experimental angle determines $\Gamma_{\sigma K}$ * .
If we write

$$\Gamma_{\sigma K} = 1 + E \simeq 1.08$$

---

* If we add a $\frac{1}{2}\, \delta m^2 V^o_\mu\, V^{\mu o}$ term, we get an additional parameter and more freedom to increase $f_{\sigma K}$

we can also determine $f_{\sigma K}$ by equating the vector Schwinger terms for i=1 and i=4. The result may be written in the form

$$\frac{f_{\sigma K}^2}{f_\pi^2} = \left(\frac{m_\rho^2}{2f_\pi^2\gamma_\rho^2}\right)\frac{2\varepsilon}{1+\varepsilon} \simeq 0.16 \qquad (98)$$

This value, 0.16, is much smaller than that obtained by Glashow and Weinberg (who obtain 0.34 by fixing the axial meson masses $m_{A_1}^2 = 2m_\rho^2$, $m_{KA}^2 = 2m_{K^*}^2$ and using eq. (90)). One can show that a smaller value of ε implies a larger value for the scalar K meson mass. Requiring that all the renormalization constants be positive leads to the result

$$\left(\frac{m_{\sigma K}}{m_K}\right)^2 \gtrsim 0.23\,\frac{1+\varepsilon}{\varepsilon} \qquad (99)$$

We thus see that if we want to get a scalar K mass in the vicinity of 1 GeV we must give up trying to fit the vector meson masses (or mixing angle) exactly. If we take $m_\rho = m_\omega = 770$ MeV ($\theta \simeq 35^\circ$) we obtain $\varepsilon \simeq 0.05$. With $(f_{\sigma K}/f_\pi)^2$ of the order of 0.10, we find that to satisfy (87) we require $(f_K/f_\pi)^2 \simeq 1.32$. This in turn, using eq. (90) implies a rather large axial K mass ($\sim$ 1500 MeV); this is one reason for keeping the η_1, η_2 terms.

To summarize

(i) The fitting of parameters to the masses is not trivial; for example gradient-coupling meson-meson terms are required to avoid too large an axial K meson mass. Also the ω-ρ mass difference strongly affects the value of the scalar K meson mass. Given this strong sensitivity, it is not clear whether, given the uncertainties in some of the axial and scalar masses, it makes much sense to tie down the parameters at this time.

(ii) The model shows that the larger values of the mixing angle ($\theta \sim 35 - 40^{o}$) does not necessarily imply that the Oakes - Sakurai [14] assumptions are wrong. If the contribution of the scalar K meson is included, the O-S angle fits experiment much better.

(iii) The model does not naturally fit what appears to be the physical situation, i.e. an octet of 0^- mesons and a nonet of vector mesons, since the transformation properties usually assigned are $(\underset{\sim}{3},\underset{\sim}{\bar{3}}) \oplus (\underset{\sim}{\bar{3}},\underset{\sim}{3})$ and $(\underset{\sim}{8},\underset{\sim}{1}) \oplus (\underset{\sim}{1},\underset{\sim}{8})$ respectively. This may imply the need for configuration mixing, or it may suggest a fundamental limitation on the usefulness of the tree approximation. A tentative suggestion that one cannot really do without closed loops comes from the fact that it is impossible, within the rules outlined above, to construct a term that under the influence of symmetry breaking leads to the coupling ω-ρ-π .

REFERENCES

1. The pioneering papers in this field are
J. SCHWINGER, Ann. Phys. N. Y. 2, 407 (1957); F. GÜRSEY, Nuovo Cimento 16, 230 (1960); Ann. Phys. N. Y. 12, 91 (1961); G. KRAMER, H. ROLLNIK and B. STECH, Z. Physik 154, 564 (1959); M. GELL-MANN and M. LÉVY, Nuovo Cimento 16, 705 (1960)
of which the last is closed to current interests. Recent developments were stimulated by
S. WEINBERG, Phys. Rev. Letters 18, 188 (1967) and carried on by
J. SCHWINGER, Physics Letters 24B, 473 (1967); J. WESS and B. ZUMINO, Phys. Rev. 163, 1727 (1967); J. CRONIN, Phys. Rev. 161,1483 (1967); W. A. BARDEEN and

B. W. LEE, Nuclear Physics and Particle Physics (ed. B. Margolis and C. Lam), W. A. Benjamin, New York 1968; B. W. LEE and H. T. NIEH, Phys. Rev. 166, 1507 (1968), R. ARNOWITT, M. H. FRIEDMAN and P. NATH, Phys. Rev. Letters 19, 1085 (1967).

2. A discussion closed to ours in spirit may be found in B. W. LEE, Phys. Rev. 170, 1359 (1968); For a brief review see the review article by S. GASIOROWICZ and D. A. GEFFEN (to be published).
3. J. S. R. CHISHOLM, Nuclear Physics 26, 469 (1961); S. KAMEFUCHI, L. O'RAIFEARTAIGH and A. SALAM, Nuclear Physics 28, 529 (1961).
4. Y. NAMBU, Phys. Letters 26B, 626 (1968); S. COLEMAN, J. WESS and B. ZUMINO (to be published).
5. J. J. SAKURAI, Ann. Phys. N. Y. 11, 1 (1960); M. GELL-MANN and F. ZACHARIASEN, Phys. Rev. 124, 953 (1961).
6. C. N. YANG and R. L. MILLS, Phys. Rev. 96, 191 (1954).
7. For a discussion of Schwinger terms see G. KÄLLÉN, Acta Physica Austriaca, Suppl. V (1968).
8. This observation was also made by P. K. MITTER and L. J.SWANK (to be published), and I. KIMMEL, Phys. Rev. Letters 21, 177 (1968).
9. F. GÜRSEY, Acta Physica Austriaca, Suppl. V (1968).
10. See the rapporteur talk by H. HARARI, Proceedings of the XIV International Conference on High Energy Physics held in Vienna, 1968.
11. S. WEINBERG, Phys. Rev. Letters 18, 188 (1967); W. A. BARDEEN and B. W. LEE, to be published. A nonlinear realization of SU(3)×SU(3) was discussed at this School by K. DIETZ.
12. P. K. MITTER and L. J. SWANK, to be published; S. GASIOROWICZ and D. A. GEFFEN, unpublished.
13. K. KAWARABAYASHI and M. SUZUKI, Phys. Rev. Letters, 16, 255 (1966).

FAYYAZUDDIN and RIAZUDDIN, Phys. Rev. 147, 107 (1966).

14. R. J. OAKES and J. J. SAKURAI, Phys. Rev. Letters 19, 1266 (1967).

EQUAL TIME COMMUTATORS, SUM RULES AND LOW ENERGY PION PHYSICS†

By

G. FURLAN

Istituto di Fisica Teorica
dell' Università, Trieste, Italy

INTRODUCTION

A fruitful advance in our understanding elementary particle physics has been achieved through current algebras [1]. This approach is characterized first of all by the choice of currents as fundamental quantities in describing the weak and electromagnetic interactions of hadrons. Then a very attractive suggestion follows in order to establish a bridge between the observable hadron currents and symmetry operators of the theory: according to Gell-Mann we identify the charges (associated to those vector and axial vector currents) with the generators of the SU(3) × SU(3) group *.

In this way the equal time commutation relations between charges generate the algebra of the corresponding group and we assume they are exact, independently whether the generators are conserved or not. Thus we have two SU(3) octets of currents, the vector and axial vector

\+ The introduction of currents and divergences in the framework of the Lagrangian formalism is discussed in the parallel lectures of Prof. S. Gasiorowicz.

† Lecture given at the VIII. Internationale Universitätswochen f. Kernphysik, Schladming, Feb. 24,-March 8, 1969.

ones, the corresponding 16 charges $Q^\alpha(t) = \int j_o^\alpha(\vec{x},t)d\vec{x}$ and the following commutation relations hold:

$$\text{(A)} \qquad [Q_\beta^V, Q_\alpha^V] = if_{\beta\alpha\gamma}\, Q_\gamma^V \; ; \quad [Q_\beta^V, V_\alpha^\mu] = if_{\beta\alpha\gamma} V_\gamma^\mu$$

$$[Q_\beta^V, Q_\alpha^A] = if_{\beta\alpha\gamma}\, Q_\gamma^A \; ; \quad [Q_\beta^V, A_\alpha^\mu] = if_{\beta\alpha\gamma} A_\gamma^\mu$$

$$[Q_\beta^A, Q_\alpha^A] = if_{\beta\alpha\gamma}\, Q_\gamma^V \; ; \quad [Q_\beta^A, A_\alpha^\mu] = if_{\beta\alpha\gamma} V_\gamma^\mu \; .$$

More specific hypotheses concern the commutators between local densities like

$$\text{(B)} \qquad [j_o^\beta(\vec{x},o),\, j_o^\alpha(o)] = if_{\beta\alpha\gamma} j_o^\gamma(o)\; \delta(\vec{x})$$

while for (time,space) commutators additional (Schwinger) terms are required

$$\text{(C)} \qquad [j_o^\beta(\vec{x},o),\, j_k^\alpha(o)] =$$

$$= if_{\beta\alpha\gamma} j_k^\gamma(o)\,\delta(\vec{x}) + S_{km}^{\beta\alpha}\, \partial_m\, \delta(\vec{x})$$

which makes (C) much more model dependent than (B). We will not discuss the case of space-space commutators *. Eqs. (A), (B), (C) represent our fundamental assumptions (in order of decreasing reliability) and the aim is to deduce some consequences from them.

Among the many things we learned from current algebra there is a systematic way of establishing low energy theorems for massless pions of zero momentum [1]. These relations are exact. The only disadvantage, of course, is they hold in an unphysical point (the threshold for production of zero mass pions).

Once we assume (this is the PCAC or smoothness assumption) that there is no appreciable variation in continuing those informations to physical pions, the result is a theoretical calculation of the low energy parameters of

* They occur if we consider the SU(6) algebra, for instance.

pion physics like s-wave (π-target) scattering lengths or $\ell = 0$ electroproduction multipoles or form factors for pionic decays of strange particles.

When the process is simple kinematically we obtain spectacular results like the Tomozawa-Weinberg formulae [2] for $a_{1/2}$, $a_{3/2}$ in πN *. For more complicated cases, like electroproduction, where more amplitudes are required, the current algebra prediction is not unambiguous nor sufficient to determine the complete amplitude. For processes with more pions, like $K \to 3\pi$ or $\eta \to 3\pi$, many problems still persist.

Thus we became convinced that the mass extrapolation from soft to physical pions, connected with this exploitation of current algebra, is an important problem with a twofold aspect. First of all we need a discussion on the nature and importance of the $O(\mu^2)$ terms which correct the zero order (in the pion mass) low energy theorem. Secondly we must specify the physical point where one can compare the current algebra predictions with experiments. In fact different points of the physical region coincide when the pion momentum $q_\mu \to 0$ and the choice of a best extrapolation line depends on additional arguments, which supplement the original commutators.

The answer to these questions is known [3]: in particular it is clearly understood now, that the simplest approach to the problem is the saturation of the $SU(2)\times SU(2)$ equal time commutators in a suitable reference frame. In so doing the low energy theorems appear as the dominant contribution of a sum rule obtained by saturating the axial <u>charge</u> commutators between states of zero or nearly zero momentum. The other terms of the sum rule represent the $O(\mu^2)$ corrections. All results refer to

---

* Actually the reason of the success is a bit more involved, as we will discuss later.

physical pions produced at rest.

However one can convince himself that current algebra is not enough, in many cases, to determine the complete low energy amplitude and new informations are required. We will find it more adherent to the whole approach to supplement these assumptions in the form of higher equal time commutators (for instance of field algebra), but clearly this introduces a model dependent element. Moreover we can ask how to enlarge our set of starting assumptions (read: commutators) in order to describe processes involving physical pions of non-vanishing three momentum. You can easily imagine that this will be achieved by introducing the equal time commutators between current densities.

In order to cover this material I will divide my lectures into two parts. The first part will be devoted to a unified treatment of the dispersive sum rules and low energy limits arising from the saturation of a charge commutator in a general reference frame. The second one will deal with the generalization of this approach to pions of any momentum.

I. EQUAL TIME COMMUTATORS AND REFERENCE FRAMES

1. The necessary insight into the heart of the matter is obtained by discussing the general aspects of the saturation of an equal time commutator between axial charges* . According to the SU(2) × SU(2) algebra

$$\langle p_2|[Q^A_\beta, Q^A_\alpha]|p_1\rangle = i\varepsilon_{\beta\alpha\gamma}\langle p_2|Q^V_\gamma|p_1\rangle \qquad (1)$$

where $|p_2\rangle$, $|p_1\rangle$ are one nucleon states. After insertion

* Since we are interested in pion processes, but an analogous discussion can be done in any case.

of completeness one has the sum rule

$$\sum_n \langle p_2 | Q^A_\beta | n \rangle \langle n | Q^A_\alpha | p_1 \rangle \text{ - crossed term } = \tag{2}$$

$$= i\varepsilon_{\beta\alpha\gamma} \langle p_2 | Q^V_\gamma | p_1 \rangle = (2\pi)^3 \delta(\vec{p}_2 - \vec{p}_1) 2Ei\varepsilon_{\beta\alpha\gamma} \frac{\tau_\gamma}{2}$$

The first thing to do is to exhibit in a complete manner the structure of the various contributions to the sum rule. Since the procedure is quite general and it will be used often in the next developments, I prefer to do it in the more general case of a sum rule of the form

$$\rho^{BA} = \sum_n \langle p_2 | B | n \rangle \langle n | A | p_1 \rangle F(\vec{p}_n, \vec{p}_2, \vec{q}_2; E_n, E_2, q_{20}) -$$

$$- \begin{pmatrix} B \leftrightarrow A \\ \\ \vec{q}_2 \leftrightarrow -\vec{q}_1 \end{pmatrix} \tag{3}$$

where B, A are local operators and F is a function of the momenta of the intermediate states (p_n) and of the external particles (p_2, p_1) and operators (q_2, q_1). In our particular example (2), $B, A = A^\beta_o, A^\alpha_o$ and $F = (2\pi)^6 \delta(\vec{p}_2 - \vec{p}_1) \delta(\vec{p}_2 - \vec{p}_n)$ (A_o is the axial current density associated to the definition of $Q^A = \int A_o(\vec{x}, 0) d\vec{x}$) .

It is convenient to rewrite ρ^{BA} in the compact form

$$\rho^{BA} = \langle p_2 | (BF(P)A - c.t.) | p_1 \rangle \tag{4}$$

being $P = (\vec{P}, H)$ the total four momentum operator.

A meaningful picture is obtained now by extracting, via the reduction formula, the nucleon from the asymptotic one particle states:

$$\langle p_2 | B = \langle 0 | O_B + \langle 0 | Ba(\vec{p}_2)$$

$$O_B = \int e^{ipx} dx (i\gamma\partial - m) \; T\psi(x) B(o) \tag{5}$$

Thus (4) becomes

$$\rho^{BA} = \langle 0|\{O_B F(P) O_A + Ba(\vec{p}_2) F(P) O_A + O_B F(P) a^+(\vec{p}_1) A + $$
$$+ Ba(\vec{p}_2) F(P) a^+(\vec{p}_1) A - \text{c.t.}\}|0\rangle \quad (6)$$

and in order to perform an convenient reordering of the operators we exploit the commutation relations (remember we are working with asymptotically free states)

$$[a(\vec{p}_1), a^+(\vec{p}_2)]_+ = 2E(2\pi)^3 \delta(\vec{p}_1 - \vec{p}_2)$$
$$[a(\vec{p}), P_\mu]_- = p_\mu \, a(\vec{p})$$
$$[P_\mu, a^+(\vec{p})]_- = p_\mu \, a^+(\vec{p}) \quad (7)$$

so that

$$a(p)\, F(P) = F(P+p)\, a(p)$$
$$F(P) a^+(p) = a^+(p)\, F(P+p) \quad (8)$$

Then we can shift "a" to the right and "a^+" to the left with the result

$$\rho^{BA} = \langle 0|\{O_B F(P) O_A + BF(P+p_2) a(p_2) O_A + O_B a^+(p_1) F(P+p_1) A -$$
$$- Ba^+(p_1) F(P+p_1+p_2)\, a(p_2) A +$$
$$+ (2\pi)^3 2E\delta(\vec{p}_2 - \vec{p}_1) BF(P) A - \text{c.t.}\}|0\rangle$$
$$(9)$$

By going back to the completeness sum (and referring explicitly to our example) we can separate four different contributions to ρ^{BA} :

$$\rho = \rho_I + \rho_{II} + \rho_{III} + \rho_{IV}$$

$$\rho_I = \sum_n \overline{\langle p_2|B|n\rangle} F(p_n) \overline{\langle n|A|p_1\rangle} + \sum_n \overline{\langle p|Q^A_\beta|n\rangle}\,\overline{\langle n|Q^A_\alpha|p\rangle} - \text{c.t.} \tag{10}$$

$$\rho_{II} = \sum_n \langle O|B|n\rangle F(p_n+p_2) \overline{\langle n,p_2|A|p_1\rangle} + \sum_n \overline{\langle p_2|B|n,p_1\rangle} F(p_n+p_1) \langle n|A|O\rangle - \text{c.t.} + \sum_n \langle O|Q^A_\beta|n\rangle \overline{\langle n,p|Q^A_\alpha|p\rangle} + \sum_n \overline{\langle p|Q^A_\beta|n,p\rangle} \langle n|Q^A_\alpha|O\rangle - \text{c.t.} \tag{11}$$

$$\rho_{III} = -\sum_n \langle O|B|n,p_1\rangle F(p_n+p_1+p_2) \langle n,p_2|A|O\rangle + -\sum_n \langle O|Q^A_\beta|n,p\rangle \langle n,p|Q^A_\alpha|O\rangle - \text{c.t.} \tag{12}$$

$$\rho_{IV} = \sum_n \langle O|B|n\rangle F(p_n) \langle n|A|O\rangle \langle p_2|p_1\rangle + \sum_n \langle p_2|p_1\rangle \langle O|Q^A_\beta|n\rangle \langle n|Q^A_\alpha|O\rangle \quad . \tag{13}$$

$\overline{\langle p|Q|a\rangle} \equiv \langle O|O_Q|a\rangle$ is the connected part of the matrix element $\langle b|Q|a\rangle$.

The interpretation of these terms is rather easy (Fig. 1):

ρ_I represents the usual connected contributions due to intermediate states of baryonic number one, propagating in the s, $\bar{s}$ channels.

ρ_{II} corresponds to "mass singularities" coming from states with the same quantum numbers as Q^A_β, Q^A_α ($J^P=O^-$, $I^G=1^+$) : one pion, three pions ... .

ρ_{III} contains the so-called Z graphs in which the intermediate state $|n\rangle$ has baryonic number -1: the simplest one consists of a single antinucleon.

ρ_{IV} finally is due to the disconnected diagram in which the nucleon travels freely and we shall understand this contribution as subtracted away once for all.

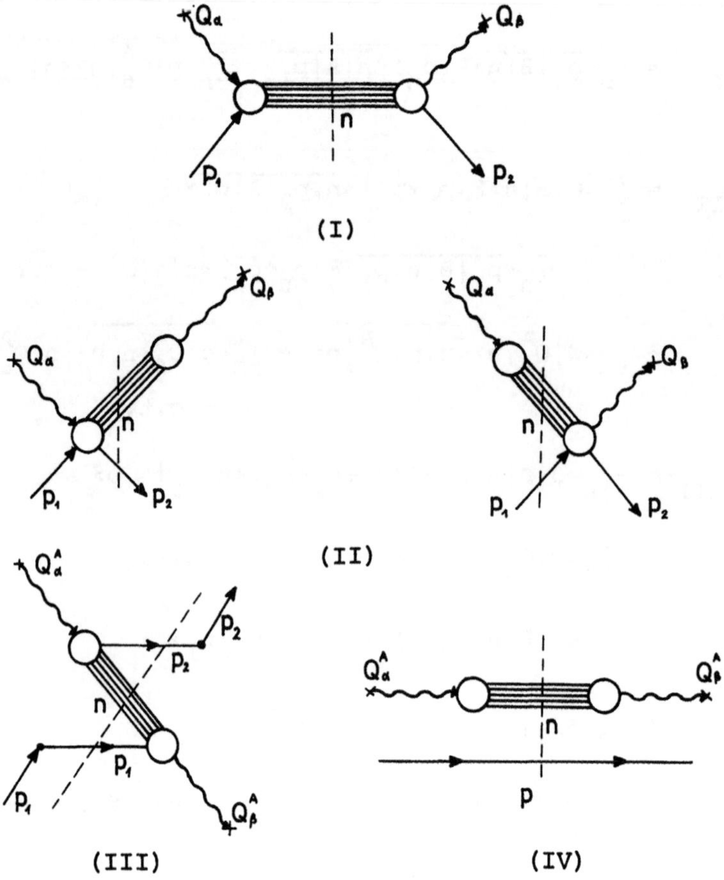

Fig. 1

2. The real point is now the frame dependence of the sum rule. From a general point of view one can understand this feature by remembering that the "equal time" notion is not a Lorentz invariant one and therefore one should expect the explicit evaluation of the commutator to depend on the particular instants you consider the two charges. Put in another way, since the frame can be specified by the state of motion of the external particles, we have an infinite set of sum rules, one for each value of the momentum of the external nucleons. The practical effect of choosing a particular

frame will be a different manner of evaluating the completeness sum, by enhancing some classes of contributions with respect to others.

The exceptional case is represented by time independent charges: in this case no frame dependence appears since it is irrelevant to evaluate the commutator at a particular time.

These considerations can be emphasized by looking at the commutation relations of a charge with the generators of the Poincaré group:

$$[\vec{P}, Q_\alpha] = 0 \qquad i[H,Q_\alpha] = \int D_\alpha(\vec{x},o)\,d\vec{x}$$

$$[\vec{M}, Q_\alpha] = 0 \qquad [\vec{K}, Q_\alpha] = i\int \vec{x} D_\alpha(\vec{x},o)\,d\vec{x}$$

$$D_\alpha = \partial_\mu A^\alpha_\mu \tag{14}$$

Therefore the frame dependence will be more or less strong according to the "quasi" conserved character of the charge, namely according to the strength of the "leakage" matrix elements $\langle p|D_\alpha|p'\rangle$.

We will make more quantitative this discussion in the next section. In order to gain a first hint I discuss here more closely the case of strictly conserved charges. Since $\dot{Q} = i[H,Q] = 0$, this means that Q connects only states of equal energy (and three momentum)* . Let us try to recognize how this can be realized for the three classes of contributions, ρ_I, ρ_{II} , ρ_{III} of eqs. (10),(11),(12). Since there is no frame dependence we can choose to do this in the rest frame $\vec{p} = \vec{0}$: as a consequence, for kinematical reasons, only $1/2^-$ states

---

* $0 = \langle E_2|[H,Q]|E_1\rangle = (E_2-E_1)\langle E_2|Q|E_1\rangle$ i.e.

if $\langle E_2|Q|E_1\rangle \neq 0$ then $E_2 = E_1$.

contribute to ρ_I and $1/2^+$ antibaryons to ρ_{III}.

Then in ρ_I, $\langle N(\frac{1}{2}^+)|Q^A|N'(\frac{1}{2}^-)\rangle \neq 0$ only if M=M' namely if there is a multiplet of opposite parity, with respect to N, and same mass. This mechanism to realize the symmetry requires the existence of a degenerate parity doublet. If this is not the case $\rho_I=0$ (when $\dot{Q} = 0$).

Analogously in ρ_{III}, $\langle 0|Q^A|N\bar{N}\rangle \neq 0$ only when $M+\bar{M} = 0$ i.e. for vanishing nucleon mass. While this is a rather unrealistic situation for baryons, it is not so for leptons where the axial current is rigorously conserved for the zero mass neutrino.

Finally in order to have $\rho_{III} \neq 0$ we need a zero mass pion* (the famous Goldstone boson).

It is worth to point out the different predictions one obtains in the three limits of symmetry. In cases (I) (M=M') or (III) (M=0) the matrix elements $\langle N|Q^A|N'\rangle$ or $\langle N|Q^A|N\rangle$ turn out, from the sum rule, to be equal to a pure Clebsch-Gordan coefficient: therefore there is no renormalization of the axial vector coupling constant between NN' or NN. On the contrary the predictions from mechanism II concern amplitudes involving the zero mass boson.

Of course these considerations are quite general and they can be developed for any symmetry group, SU(3) for instance. If we forget about the possibility of vanishing baryon masses, being in that case the charged 0^+ objects, $\rho_I \neq 0$ requires a multiplet of baryons $1/2^+$ degenerate in mass with the nucleon (the octet) and for $\rho_{II} \neq 0$ we need a massless 0^+ boson. What does Nature tell? SU(2) × SU(2) or SU(3) charges are not exactly conserved, since no one of the above mechanisms is realized: no rigorously degenerate multiplets or massless bosons exist and the breaking of the symmetry will be proportional to the mass

* From the technical point of view one must have some care and evaluate $\langle 0|Q|\pi\rangle \sim \mu$ first with a small but not vanishing mass and then let $\mu \to 0$.

splitting M'-M and to the boson mass μ^2 .

Actually there seems to be a preferred way in Nature to break the symmetry, namely that mechanism is realized which gives the best approximation, and the corrections to the corresponding symmetry limit are the smallest. Thus a first indication can be given by the ratio $\mu^2/(M'-M)^2$. For SU(2) × SU(2) where, at least for nucleon processes*, $\mu^2/(M'-M)^2 \sim 5\%$ (and $\mu^2/4M^2 \sim 0.5\%$!), we prefer the boson mechanism (transition between $|0\rangle$ and $|\pi\rangle$), and ascribe the breaking to the finite pion mass. For SU(3) where**) $(M'-M)^2/\mu^2 \simeq 7\%$ the multiplet mechanism should be the main one (transition between $|N\rangle$ and $|\Sigma\rangle$), as one can see from the evident traces in nature.

A further suggestion can be gained: even if the charges are only partially conserved we can try to be smart and profit by the frame dependence to enhance, in each case, the mechanism which gives the best realization of the symmetry.

3. In order to make these simple considerations more concrete, it is convenient to work out a few details about the sum rule (2).

I select first of all (for technical reasons) the one nucleon term and I rewrite the remaining part of the sum rule by using the relation

$$\langle p|Q^A_\beta|p'\rangle = (2\pi)^3\delta(\vec{p}-\vec{p}\,')\frac{\langle p|D^A_\beta|p'\rangle}{i(E-E')} \qquad (15)$$

Thus eq. (2) becomes

** $\mu \simeq 1$ BeV (k mesons?), M = 9.38 MeV , $M'(\Sigma) \simeq 1190$

* $\mu \simeq 136$ MeV , $M(N\frac{1}{2}^+) \simeq 938$ MeV , $M'(N'\frac{1}{2}^-) \simeq 1570$ MeV

$$\frac{E^2-m^2}{E^2}\, g_A^2\left[\frac{\tau_\beta}{2}, \frac{\tau_\alpha}{2}\right] + \sum_{n\neq N} \frac{(2\pi)^3}{2E}\, \delta(\vec{p}-\vec{p}_n) \times$$

$$\times \left[\frac{\langle p|D^A_\beta|n\rangle\langle n|D^A_\alpha|p\rangle}{(E - E_n)^2} - \text{c.t.}\right] = 1 \, . \left[\frac{\tau_\beta}{2}, \frac{\tau_\alpha}{2}\right] \tag{16}$$

and after introducing the auxiliary variable q_o, one has

$$\frac{E^2-m^2}{E^2}\, g_A^2 + \frac{1}{\pi}\int \frac{dq_o}{q_o^2}\, D^{(-)}(q_o,E) = 1 \tag{17}$$

where

$$D^{\beta\alpha} = \pi(2\pi)^3 \sum_{n\neq N} \delta(\vec{p}-\vec{p}_n)\,(\langle p|D_\beta|n\rangle\langle n|D_\alpha|p\rangle\,\delta(E+q_o-E_n) -$$

$$- \text{c.t.}) = \frac{1}{2}[\tau_\beta\, ,\tau_\alpha]\, D^{(-)} + \delta_{\beta\alpha}\, D^{(+)} \tag{18}$$

This form of the sum rule exhibits in an explicit way the dependence on the reference frame, namely on E. Thus the one nucleon contribution is enhanced and respectively killed in the two extreme configurations $E \to \infty$, $E=m$. As far as the continuum is concerned, this dependence can be conveniently emphasized by going to Lorentz invariant variables

$$\nu = p.q = E\, q_o \qquad u = q^2 = q_o^2 = \frac{\nu^2}{E^2} \tag{19}$$

so that, when q_o varies, ν, u run on an integration line which is a parabola in the (ν,u) plane. Accordingly the continuum part of the sum rule takes the form

$$\int \frac{dq_o}{q_o^2 E}\, D^{(-)}(q_o,E) = \int \frac{d\nu}{\nu^2}\, du\, \delta\!\left(u- \frac{\nu^2}{E^2}\right) D^{(-)}(\nu,u) \tag{20}$$

and the relevant point is that the integration path depends on the reference frame; the frame dependence will manifest itself in the different way the various contributions are weighted.

In fact according to the general discussion of the previous sections we can recognize in $D^{\beta\alpha}$ the three fundamental contributions

$$D_I = \pi(2\pi)^3 \sum_{n\neq N} \delta(p+q-p_n)\overline{\langle p|D_\beta|n\rangle}\,\overline{\langle n|D_\alpha|p\rangle} - \text{c.t.}$$

$$D_{II} = \pi(2\pi)^3 \sum \delta(q-p_n)\,(\langle 0|D_\beta|n\rangle\overline{\langle n,p|D_\alpha|p\rangle} + $$
$$+\overline{\langle p|D_\beta|n,p\rangle}\langle n|D_\alpha|0\rangle) - \text{c.t.}$$

$$D_{III} = -\pi(2\pi)^3 \sum \delta(p-q+p_n)\langle 0|D_\beta|p,n\rangle\langle n,p|D_\alpha|0\rangle - \text{c.t}\ .$$

$$q \equiv (\vec{0},q_0) \qquad (21)$$

and we can locate them on the parabola by looking at its intersections with the lines (Fig. 2):

$$\left.\begin{array}{l} s = (p+q)^2 = M_n^2 \\ \bar{s} = (p-q)^2 = M_n^2 \end{array}\right\} \quad (\text{for } D_I, D_{III})$$

$$u \equiv q^2 = M_n^2 \qquad (\text{for } D_{II})$$

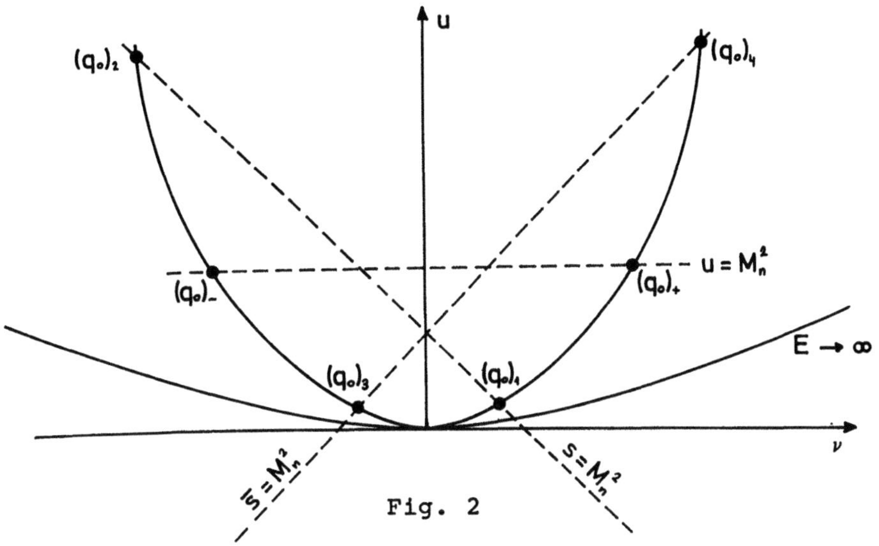

Fig. 2

We have in particular the four intersections

$$(q_o)_{1,2} = - E \pm \sqrt{\vec{p}^2 + M_n^2}$$

$$(q_o)_{3,4} = + E \mp \sqrt{\vec{p}^2 + M_n^2} \tag{22}$$

corresponding to the direct and crossed term in D_I ($(q_o)_1$, $(q_o)_3$) and to the direct and crossed Z graphs ($(q_o)_4$, $(q_o)_2$). Then in correspondence to D_{II} there are

$$(q_o)_\pm = \pm M_n \tag{23}$$

Let us discuss the dependence of the parabola on the parameter E (the frame dependence). The curvature is determined by

$$a = \frac{1}{E^2} \tag{24}$$

and the two extreme choices $E \to \infty$, $E \to m$ correspond to $a \to 0$, $a \to 1/m^2$. In the first case the parabola degenerates into the straight line u=0 and the intersections $(q_o)_{2,4}$, $(q_o)_\pm$ are pushed to infinity and $\nu_1 = - \nu_2 = E(q_o)_1 \to 1/2(M_n^2 - m^2)$. Thus we can obtain a fixed mass representation, with the usual singularities coming from D_I, while the contribution of D_{II}, D_{III} (vanishing, a subtraction constant, divergent(!)) depends on the asymptotic behaviour of the involved amplitudes, namely on dynamics.

The opposite choice gives the steepest parabola of maximum curvature and offers the advantage of strong selection rules due to angular momentum and parity conservation. Since $\vec{q} = 0$, in the rest frame $\vec{p} = 0$, $\vec{p}_n = 0$ and only states with the same angular momentum as the target and opposite parity (D^A is a 0^-), can contribute to D_I, D_{III} : only s-wave excitation is allowed.

Therefore we can conclude that the choice $E \to \infty$ is convenient to emphasize the structure of the s-channel singularities (which in particular cases, like SU(3), means the "multiplet aspect" of a broken symmetry); this will lead us to relations among the weak and electromagnetic vertices of resonances.

On the contrary the rest frame saturation will be particularly suited to study phenomena where the main contribution is due to the low mass boson; in our example the result will be a low energy limit for pion nucleon scattering.

4. Since the pion contribution is particularly relevant, it is the right moment now to select it.

In the example we are studying there is an additional difficulty due to the fact that, at first sight, the pion term gives rise to an infinite contribution since we are in the forward direction and $\langle p,\pi|D_\alpha|p\rangle$ (see eq. (21)) has a pole at $u=\mu^2$ *.

To deal with this additional pole, we resort to the higher commutator

$$[D_\beta(\vec{x},o),\ D_\alpha(o)] = \delta(\vec{x})\ C_{\beta\alpha} \tag{25}$$

from which one easily deduces the sum rule

$$\frac{1}{\pi}\int d\nu\ du\ \delta(u-\frac{\nu^2}{E^2})\ D^{(-)}(\nu,u) = C^{(-)} \tag{26}$$

* From the point of view of dispersion theory it is a double pole singularity with a spectral function $\delta'(u-\mu^2)$.
This means that the straightforward evaluation of the sum rule (17) (rather than (27)) introduces beside the physical πN amplitude, its derivative with respect to q_o namely a term $\propto \mu^2 \left|\partial T^{(-)}_{\pi N}(q_o)/\partial q_o\right|_{q_o=\mu}$, whose evaluation is not simple. Thus I prefer to follow the attitude of using higher equal time commutators and exploit the additional sum rule (26) to eliminate that model dependent contribution. The opposite point of view has been taken by B. Stech and coworkers. (D. Dillenburg, G. Jacob, Th. Maris and B. Stech, preprint by Instituto de Fisica, Universidade do Pôrto Alegre, Brasil).

(the nucleon does not contribute as $\vec{q} = \vec{0}$). We combine (26) and (20) to obtain

$$\frac{E^2-m^2}{E^2} g_A^2 + \frac{1}{\pi} \int \frac{d\nu du}{\nu^2\mu^2} \delta(u-\frac{\nu^2}{E^2})(\mu^2-u)\ D^{(-)} =$$

$$= 1 - \frac{C^{(-)}}{\mu^2 E^2} \qquad (27)$$

or

$$\frac{E^2-m^2}{E^2} g_A^2 + \frac{f_\pi}{\pi} \int \frac{d\nu}{\nu^2} du\ \delta(u-\frac{\nu^2}{E^2})\ F^{(-)}(\nu,u) =$$

$$= 1 - \frac{C^{(-)}}{\mu^2 E^2} \qquad (27')$$

where

$$F^{\beta\alpha} = \pi(2\pi)^3 \sum_{n\neq N} \delta(p+q-p_n) \langle p|D_\beta|n\rangle\langle n|\chi_\alpha|p\rangle - \text{c.t.} \qquad (28)$$

$$\chi_\alpha = (\Box + \mu^2) \quad \Phi^\alpha = (\Box + \mu^2)\ D^\alpha/\mu^2 f_\pi$$

is the pion source with the property $\langle 0|\chi|\pi\rangle = \langle\pi|\chi|0\rangle = 0$ and I used the PCAC relation $D^\alpha = \mu^2 f_\pi\ \Phi^\alpha$ which is exact on the pion pole.

Then the π contribution is obtained from F_{II} (Eq. 21) and by using $\langle 0|D_\beta|\pi_\gamma\rangle = \delta_{\beta\gamma} f_\pi \mu^2$ one gets

$$(F_{II}^{\beta\alpha})_\pi = \mu^2\pi\delta(\mu^2-u) f_\pi \langle p\pi_\beta|\chi_\alpha|p\rangle =$$

$$= \mu^2\pi f_\pi \delta(\mu^2-u)\ T_{\pi N}^{\beta\alpha} \qquad (29)$$

Therefore the sum rule becomes, taking crossing into account,

$$T_{\pi N}^{(-)}(\nu_o) = \frac{\nu_o}{f_\pi^2} - \frac{C^{(-)}}{\nu_o f_\pi^2} - \frac{\nu_o^2-\mu^2 m^2}{\nu_o} \frac{g_A^2}{f_\pi^2} - \qquad (30)$$

$$- 2\mu^2 \frac{\nu_o}{\pi} \int_0^\infty \frac{d\nu}{\nu^2} \frac{du}{\mu^2-u} \delta(u-\frac{\nu^2}{E^2})\ \tau^{(-)}(\nu,u)$$

where $T_{\pi N}^{\beta\alpha}(\nu_o)$ is the pion nucleon scattering amplitude at the <u>physical point</u> $u = \mu^2$, $\nu_o = \mu E$.

Moreover to visualize the leakage character of the higher contributions I introduced the further quantity

$$f_\pi \, \tau^{\beta\alpha} = (\mu^2-u) F^{\beta\alpha} = \tag{31}$$

$$= f_\pi \pi (2\pi)^3 \sum_n \delta(p+q-p_n) \langle p|\chi_\beta|n\rangle\langle n|\chi_\alpha|p\rangle - \text{c.t.}$$

Eq. (30) is the most general sum rule one obtains by saturating the starting commutator: we can now particularize the configuration and look at the two extreme choices $\vec{p}=0$, $|\vec{p}|\to\infty$. In the first case (30) takes the very simple form

$$T^{(-)}_{\pi N} \text{ (th.)} = \tag{32}$$

$$= \frac{\mu m}{f_\pi^2} \left(1 - \frac{C^{(-)}}{\mu^2 m^2} - \mu^2 f_\pi^2 \frac{2}{\pi} \int_0^\infty \frac{d\nu}{\nu^2} \frac{du}{\mu^2-u} \delta\left(u - \frac{\nu^2}{m^2}\right) \tau^{(-)}\right)$$

and we recognize the Weinberg formula [2], originally deduced for soft pions, plus corrections of order μ^2 (each D^α is $\sim \mu^2$ and therefore $C^{(-)}/\mu^2 \sim \mu^2$).

Thus the lesson we learn is that the rest frame saturation of the $[Q^A, Q^A]$ commutator not only reproduces in the $\mu \to 0$ limit the results of the standard theory of soft pions but also enables us to answer the question of their extrapolation to the physical region. In particular the existence of the selection rule, which enforces an extremely simple spectrum of singularities in the s-channel and therefore minimizes the corresponding correction, suggests strongly to choose the steepest parabola as the most convenient line along which to extrapolate the current algebra limit from the origin $\nu=u=0$ to the physical region. (It is the "best" sum rule from the point of view of correcting the symmetry limit).

Then the limit $E \to \infty$ corresponds to $\nu_o \to \infty$ and any conclusion depends on the dynamical model for the asymptotic behaviour of $T^{(-)}$: if one assumes a Regge pole model, $T^{(-)}(\nu_o) \sim \nu_o^{\alpha_\rho(o)}$ and since $\alpha_\rho(o) \gtrsim 1/2$, $T^{(-)}(\nu_o)/\nu_o \to 0$ for $\nu_o \to \infty$. As a consequence we obtain the Adler-Weisberger formula [4]

$$g_A^2 = 1 - \frac{2}{\pi} f_\pi^2 \int_0^\infty \frac{d\nu}{\nu^2} \tau^{(-)}(\nu,o) =$$

$$= 1 - \frac{2}{\pi} f_\pi^2 \int_0^\infty \frac{d\nu}{\nu^2} \mathrm{Im}\, T^{(-)}(\nu,o) \tag{33}$$

(The standard form is obtained by further using the Goldberger-Treiman relation $f_\pi = m\, g_A/g_{\pi N}$). The identification $\tau^{(-)}(\nu,o) = \mathrm{Im}\, T^{(-)}(\nu,o)$ is based on the approximation of using PCAC for zero mass pions.

Of course, beside the above perspicuous cases, one can use the general sum rule (30) in a variety of ways. For instance one can write for the physical amplitude $T^{(-)}(\nu)/\nu$ an unsubtracted dispersion relation*, at fixed t=0:

$$T_{\pi N}^{(-)}(\nu_o) = \frac{g_{\pi N}^2 \mu^2 \nu_o}{\nu_o^2 - \mu^4/4} + \frac{2\nu_o}{\pi} \int_0^\infty \frac{d\nu}{\nu^2 - \nu_o^2} \mathrm{Im}\, T^{(-)}(\nu) \tag{34}$$

Compare then eqs. (30) and (34) and differentiate with respect to E^2 at $E^2 = m^2$. After some simple manipulations one can get

$$f_\pi^2 = \frac{m^2 g_A^2}{g_{\pi N}^2} \left\{ 1 - \frac{2\mu^2 m^4}{g_{\pi N}^2 \pi} \int_0^\infty \frac{d\nu}{(\nu^2 - \mu^2 m^2)^2} \mathrm{Im}\, T^{(-)}(\nu) + \right. \tag{35}$$

$$\left. + \frac{2\mu^2 m^4}{g_{\pi N}^2} \frac{1}{\pi} \frac{\partial}{\partial E^2} \int_0^\infty \frac{d\nu}{\nu^2} \left(\frac{E^2}{\nu^2 - \mu^2 E^2} \right) \tau^{(-)}\left(\nu, u = \frac{\nu^2}{E^2}\right) \Big|_{E=m} \right\}^{-1}$$

which gives in principle an alternative way to calculate the corrections to the Goldberger-Treiman relation

---

* which is justified since $T^{(-)}/\nu \to 0$ as $\nu \to \infty$.

(which is exact in the limit $\mu^2 \to 0$). The final evaluation is not simple since strong cancellations can occur between the two integrals.

We can also use the previous formulae to check that in the limit of conserved charges, $\mu \to 0$, all frames are trivially equivalent. Indeed for finite E we get from (30) the unique result

$$\lim_{\substack{\nu \to 0 \\ (\mu^2=0)}} \frac{T^{(-)}(\nu)}{\nu} = \frac{1}{f_\pi^2} \tag{36}$$

The equivalence with the $E \to \infty$ frame is achieved by expressing $\lim_{\nu \to 0} T^{(-)}(\nu)/\nu$ through the dispersion relation (34). By comparing (36) and (34) the Adler-Weisberger relation (33) follows and it is exact .

5. I want to devote this section to a more complete examination of the threshold sum rule (32).

As we already remarked its interest lies in the fact that it gives for the physical pion amplitude the representation closest to the current algebra limit (36), with a particularly simple structure of the $O(\mu^2)$ corrections (only s-wave excitation).

First of all we will neglect the higher commutator $C^{(-)}$ by using extensively PCAC $D_\alpha = \mu^2 f_\pi \phi_\alpha$ plus the canonical commutator $[\phi,\phi] = 0$. Thus

$$T_{\pi N}^{(-)}(\text{th.}) = \frac{m\mu}{f_\pi^2} \left(1 - 2\,\frac{f_\pi^2 \mu^2}{\pi} \int_0^\infty \frac{d\nu}{\nu^2(\mu^2 - \nu^2/m^2)} \tau^{(-)}(\nu, u = \frac{\nu^2}{m^2})\right) =$$

$$= \frac{\mu m}{f_\pi^2} (1 - \delta T) \tag{32'}$$

Then one can classify the corrections δT_I, δT_{II}, δT_{III} according to the decomposition (21) for τ : they contain among the other things a factor $\mu^2 m^2/\bar{\nu}^2$, where $\bar{\nu}$ is the effective "ν" of the states contributing to the dispersion integrals, so that

I. δT_I contains a term $\mu^2/(m-m')^2$, where m' is the mass of a $\frac{1}{2}^-$ isobar;

II. δT_{II} contains a term μ^2/μ'^2 where μ' is the mass of a higher pion isobar;

III. δT_{III} contains a term $\mu^2/(m+m'')^2$ where m" is the mass of a $\frac{1}{2}^+$ nucleon isobar.

In particular m"=m and the contribution is $\sim \mu^2/4m^2$.

Therefore the validity of the uncorrected soft pion formulae requires

$$\text{I. } \mu^2/(m-m')^2 \ll 1 \quad , \quad \text{II . } \mu^2/\mu'^2 \ll 1 \tag{37}$$

$$\text{III. } \mu^2/m^2 \ll 1$$

Condition (II) is a well known one and it is already required by the validity of the Goldberger-Treiman relation. Conditions (I),(III) can be rather clearly understood in the light of our discussion in Section 2: they merely express the requirement that the best way to realize chiral symmetry in Nature is through the Goldstone boson mechanism, with respect to the multiplet structure (cond. I) or to the vanishing fermion mass (cond.III) mechanisms.

From the practical point of view we expect a large correction if there is a closed parity doublet with respect to the target. For πN we have a $\frac{1}{2}^-$ at 1570 MeV and the effect is small($m'-m \sim 5\mu$). On the contrary there should be a sizeable deviation from the current algebra result for $\pi\Sigma$ - scattering where we find a $\Lambda'(\frac{1}{2}^-)$ at 1405 MeV so that $m'-m \simeq 1.5\mu$ or $\pi\rho$ scattering where there is the A_1 and $m_{A_1} - m_\rho \sim 2\mu$. (Of course any quantitative conclusion depends also on the strength of the $NN^*\pi$ vertices).

Before we try an effective estimate of the corrections

it is important to stress a general property of these higher terms. Our approach is based on weak interaction quantities (like A_μ^α or D^α): by selecting the dominant q^2 singularity we obtain a pion amplitude in the physical point $q^2 = \mu^2, \nu = \nu_o$. However this goal cannot be achieved for the other contributions. Since we are integrating on the parabola both energy and mass vary and in particular q^2 is no longer fixed at the value μ^2 : as $\nu > \mu m$, $q^2 > \mu^2$ and one has to deal with the weak vertices $\langle p|D^\alpha|n\rangle$ (or strong vertices with the pion off the mass shell) in the timelike region. Thus the problem of mass extrapolation is not eliminated but it is shifted from the pion to the next contributions, which are however of a higher order in μ^2 and therefore only the less important part of the sum rule is affected by the unavoidable approximations.

In order to have a first indication on the size of the corrections we keep only the δT_I contribution. We notice that the apparent singularity at the lower limit of integration μm must be cancelled by an analogous factor in $\tau^{(-)}$, which in this point is rigorously equal to the imaginary part of the physical s-wave amplitude (vanishing at threshold). Therefore I put

$$\tau^{(-)}(\nu,u) = \kappa(\nu)\sqrt{s}\, b^{(-)}(\mu m,\mu^2) + \bar{\tau}^{(-)}(\nu,u) \tag{38}$$

where $\kappa(\nu)$ is the c.m. three momentum

$$\sqrt{s}\;\kappa(\nu) = (\nu^2-\mu^2 m^2)^{1/2} \tag{39}$$

Moreover using elastic unitarity for partial waves

$$b^{(-)}(\mu m,\mu^2) = \left[\frac{\operatorname{Im}T^{(-)}(\nu)}{\sqrt{s}\;\kappa(\nu)}\right]_{\nu=\mu m} = \frac{8\pi}{3}(a_1^2-a_3^2) \tag{40}$$

with a_1, a_3 the $I=\frac{1}{2}$, $\frac{3}{2}$ scattering lengths ($T(\text{th.})=8\pi(m+\mu)a$). Expressing everything in terms of "$a^{(-)}$", we find finally, after some approximations,

$$a^{(-)} = L + \frac{2}{3\pi} (a_1^2 - a_3^2)\,\mu + R^{(-)} \tag{41}$$

where L is the universal length

$$L = \frac{\mu}{8\pi f_\pi^2} \simeq \frac{0.09}{\mu} \tag{42}$$

and $R^{(-)}$ is the dispersion integral over $\tau^{(-)}$, dominated by the high part of the spectrum (the $\frac{1}{2}^-$ isobars) and presumably negligible owing to the strong convergence factors. If we neglect $R^{(-)}$, eq. (41) can be considered as a consistency condition*: it is amusing to remark that it represents the algebraic translation of the standard, non-linear integral equation one obtains from elastic unitarity.

6. I conclude this chapter by shortly discussing two additional points. First of all I will complete the analysis of πN scattering by studying the symmetric part of the amplitude, then some comments will be added about the relation of our results to Lorentz invariance.

a) In order to describe the symmetric part of the πN amplitude from the point of view of equal time commutators, the so-called σ term is required, that is the commutator

$$[Q_\beta^A \,,\, D_\alpha^A] = i\, S_{\beta\alpha} \tag{43}$$

where $S_{\beta\alpha}$ is a scalar, isospin symmetric operator.

---

* If we accept, at this stage, the experimental indication $a_1 + 2a_3 \simeq 0$, it follows from (41)
$\mu a^{(-)} \simeq \mu L \simeq 0.09$ while the experimental value is 0.085.

After combination with the higher commutator*

$$[\dot{D}_\beta(\vec{x}), D_\alpha(o)] = i\, C^{(+)} \delta_{\beta\alpha}\, \delta(\vec{x}) \tag{44}$$

it is easy to deduce the general forward sum rule

$$\frac{f_\pi}{\pi} \int \frac{d\nu}{\nu}\, du\, \delta(u - \frac{\nu^2}{E^2})\, F^{(+)}(\nu,u) = \sigma - \frac{C^{(+)}}{\mu^2 E^2} \tag{45}$$

where

$$\sigma\delta_{\beta\alpha} = \langle p_2 | S_{\beta\alpha} | p_1 \rangle \tag{46}$$

Selection of the π pole gives:

$$T^{(+)}(\nu_o) = \frac{\sigma}{f_\pi^2} - \frac{C^{(+)}}{\nu_o f_\pi^2} - \frac{2\mu^2}{\pi} \int_o^\infty \frac{d\nu du}{\nu(\mu^2 - u)}\, \delta(\frac{\nu^2}{E^2} - u)\, \tau^{(+)}(\nu,u) \tag{47}$$

In particular, in the rest frame the result follows

$$a^{(+)} \approx \frac{\sigma}{m\mu} L + \frac{1}{3}(a_1^2 + 2a_3^2)\mu + R^{(+)} - \frac{C^{(+)}}{\mu^3\ m^3} \tag{48}$$

In the limit $\mu \to 0$ we recover the Adler consistency condition [5] $a^{(+)} = 0$ $(\sigma \propto \mu^2)$. All the frames are equivalent in this limit, but the $E \to \infty$ one. Indeed the limit $E \to \infty$ $(\nu_o \to \infty)$ on Eq. (47) is not defined since $T^{(+)}(\nu_o) \sim \nu_o^{\alpha_+(o)}$ $\alpha_+(o) \simeq 1$ and therefore the sum rule diverges as $T^{(+)}(\nu_o) \to \infty$, $\nu_o \to \infty$. (That shows by passing the rather particular character of the $E \to \infty$ configuration).

b) The previous discussions on the frame dependence could let arise some doubts. Since any theory must be in agreement with special relativity, the commutators we used are valid in any reference frame. Therefore, even if the choice of a particular frame enables us to emphasize some specific aspects of a problem, finally the results

* Notice that the knowledge of $C^{(+)}$ commutator is actually beyond the frame-work of current and field algebra since the use of the canonical commutator $[\phi, \dot{\phi}] = i\, Z_\pi^{-1}\, \delta(\vec{x})$ actually depends on the form of the coupling.

have to be equivalent. Let me put the question as follows: since it is possible to go from a frame to another by a Lorentz transformation, there is a relation among the sum rules obtained by working on different parabolae.

The simplest way to get a relation of this sort is to differentiate (17) with respect to E (which means to perform an infinitesimal Lorentz transformation from E to δE). By using the invariant form (20) and including the nucleon contribution in the integral, one gets:

$$\int d\nu du\ \delta'(u-\frac{\nu^2}{E^2})\ D^{(-)}(\nu,u) = 0 \tag{49}$$

which can be conveniently rewritten as

$$\int d\nu du\ \delta(u-\frac{\nu^2}{E^2})\frac{\partial}{\partial u}\ D^{(-)}(\nu,u) = \int d\nu\ \frac{\partial}{\partial\lambda}\ D^{(-)}(\nu,\frac{\nu^2}{E^2}+\lambda)\Big|_{\lambda=0} = 0 \tag{50}$$

When you make explicit the various contributions to the sum rule, derivatives of $T^{(-)}$ and of the various form factors appear, which make (50) or analogous sum rules not very useful, at least at present.

Moreover the sum rule becomes trivial in the symmetry limit ($\mu\to 0$) since all contributions are of order μ * (or higher), which confirms that for conserved charges the equivalence of the various sum rules is automatic.

It is important to trace back the origin of the sum rule (50) to a more general argument, as follows: let us start from the commutator

$$[Q^A(0)\ ,\ Q^A(0)] = Q^V \tag{51}$$

and commute it with the generators H, $\vec{K}$ of the Poincaré group. Since Q^V is conserved we obtain (use eq. (14)):

---

* Practically this is due to the absence of the "ν^2" denominator, which also makes worse the convergence of the sum rule .

$$\int(1,\vec{x})\{[Q_\beta,\ D_\alpha] - [Q_\alpha,\ D_\beta]\}d\vec{x} = 0 \tag{52}$$

which merely tells that $[Q_\beta,\ D_\alpha]$ must be symmetric in (β,α). If we next Lorentz transform

$$[Q^A(0)\ ,\ D^A(0)] = S(0) \tag{53}$$

the further condition appears

$$\int d\vec{x}\ \vec{x}[D_\beta(\vec{x},o),\ D_\alpha\ (o)] = 0 \tag{54}$$

I write in the equivalent form*

$$\frac{\partial}{\partial\vec{q}}\int e^{-i\vec{q}\vec{x}}\ d\vec{x}\langle p_2|[D_\beta,D_\alpha]|p_1\rangle\Big|_{\vec{q}=0} = 0 \tag{55}$$

and after using completeness it can be easily verified that (55) leads directly to the sum rule (50).

II. LOW ENERGY PION PHYSICS AND EQUAL TIME COMMUTATORS

1. In the previous chapter we have studied the general features of saturating an axial charge commutator; in particular we have shown that very simple predictions are obtained for physical pion amplitudes in terms of equal time commutators plus a completeness sum. The results refer to pions produced at rest. The aim of this chapter is to generalize the previous treatment to pions of any momentum. In particular we will obtain a systematic indication on what are the additional commutators, beyond the framework of current algebra (like [D, D] of eq. (25)), necessary to achieve a complete knowledge of the whole amplitude [6].

---

* The general statement is that $[D_\beta,\ D_\alpha]$ cannot be more singular than a $\delta(\vec{x})$.

As a warming up exercise I want to recall a fifteen years old formulation of pion processes in terms of equal time commutators and completeness [7] . In this way we will learn the general technique to handle the problem as well as the aspects of the old fashioned pion theory we hope to improve by using current algebra.

I introduce first of all the Heisenberg pion fields $\phi^\alpha(\vec{x},t)$, $\Phi^\alpha(\vec{q},t) = \int d\vec{x}\; e^{-i\vec{q}\vec{x}}\; \Phi^\alpha(\vec{x},t)$ with the normalization

$$<0|\Phi^\alpha(\vec{q},0)|\pi^\beta> = (2\pi)^3\; \delta_{\beta\alpha}\,\delta(\vec{q}-\vec{\pi}) \qquad (56)$$

Then it is convenient to define two operators ϕ^α and $\phi^{\alpha *}$ with the property (analogous to that of free field creation and annihilation operators)

$$<\pi|\phi^\alpha|0> = 0$$
$$<0|\phi^{\alpha *}|\;\pi> = 0 \qquad (57)$$

It follows $(\omega = \sqrt{\vec{q}^2 + \mu^2})$

$$\phi^\alpha(\vec{q},t) \;= \omega\; \Phi^\alpha(\vec{q},t) + i\; \dot{\Phi}^\alpha(\vec{q},t) \qquad (58)$$

and

$$<0|\phi^\alpha(\vec{q})|\pi^\beta> = <\pi^\beta|\phi^{*\alpha}(\vec{q})|0> =$$
$$= 2\omega_q(2\pi)^3\delta(\vec{q}-\vec{\pi})\,\delta_{\beta\alpha} \qquad (59)$$

In order to reach a meaningful expression for the matrix element of ϕ^α between any states, it is important to notice that, as a consequence of the definition (58), ϕ^α obeys the differential equation

$$\dot{\phi}^\alpha(t) = \omega\dot{\Phi}^\alpha + i\ddot{\Phi}^\alpha = -i\omega\phi^\alpha + i\ddot{\Phi} + i\omega^2\Phi^\alpha =$$
$$= i\chi^\alpha - i\;\omega\;\phi^\alpha \qquad (60)$$

I used the definition $\chi^\alpha(\vec{q},t) = (\omega^2+d^2/dt^2)\phi^\alpha(\vec{q},t)$ and $\chi^\alpha(\vec{q},t)$ is the Fourier transform of the pion source

$$\chi^\alpha(\vec{x},t) = (\Box+\mu^2)\,\phi^\alpha(\vec{x},t)$$

$$\langle 0|\chi^\alpha|\pi\rangle = \langle\pi|\chi^\alpha|0\rangle = 0 \tag{61}$$

Eq. (60) can be rewritten in the integral form

$$\phi^\alpha(t) = e^{-i\omega t}\,\phi^\alpha(+\infty) + i\,e^{-i\omega t}\int_{+\infty}^{t} e^{i\omega t'}\chi^\alpha(t')dt' \tag{62}$$

from which one gets (t=0)

$$\langle b|\phi^\alpha|a\rangle = \langle b|\phi^\alpha(+\infty)|a\rangle + \frac{\langle b|\chi^\alpha|a\rangle}{E_b+\omega-E_a+i\varepsilon} \tag{63}$$

If the ingoing state $|a\rangle$ contains a single pion (for simplicity) $|a\rangle = |b',\pi\rangle$ we have further

$$\langle b|\phi|a\rangle = \frac{\langle b|\chi|a\rangle}{E_b+\omega-E_a+i\varepsilon} + \langle b|b'\rangle\langle 0|\phi_{out}|\pi\rangle \tag{64}$$

which clearly exhibits the separation into a connected plus a disconnected part.

Now to the Low representation. What I need is the equal time commutator

$$\langle p_2|[\phi^\beta(\vec{q}_2),\chi^\alpha(-\vec{q}_1)]|p_1\rangle = \tag{65}$$

$$= (2\pi)^3\delta(\vec{p}_2+\vec{q}_2-\vec{p}_1-\vec{q}_1)\langle p_2|C^{\beta\alpha}|p_1\rangle$$

and I have to assume $C^{\beta\alpha}$ as one of the inputs given by the theory. By using completeness*

$$\langle p_2|C^{\beta\alpha}|p_1\rangle = \sum_n (2\pi)^3\delta(\vec{p}_2+\vec{q}_2-\vec{p}_n)\langle p_2|\phi^\beta|n\rangle\langle n|\chi^\alpha|p_1\rangle - \text{ c. t.} \tag{66}$$

Notice that while $\omega = \sqrt{\vec{q}_2^{\,2}+\mu^2}$, i.e. ϕ^β corresponds to a

* The matrix elements are those of coordinate space operators $\langle a|\phi^\beta(\vec{x},0)|b\rangle|_{\vec{x}=0}$.

physical pion, this is not necessarily the case for χ^α: the fourth component q_{10}, associated to $\vec{q}_1$ and defined by $q_{10} = E_2 + \omega_2 - E_1$, can be different from the on-shell quantitiy

$$\omega_1 = \sqrt{\vec{q}_1^{\,2}+\mu^2} \quad .$$

One can now use the result (64) to select the disconnected term:

$$\sum_n \langle p_2|\phi^\beta|n\rangle\langle n|\chi^\alpha|p_1\rangle\delta(\vec{p}_2+\vec{q}_2-\vec{p}_n)(2\pi)^3 =$$

$$= \sum_{p,\pi} (2\pi)^3\delta(\vec{p}_2+\vec{q}-\vec{p})\langle p_2|p\rangle\langle 0|\phi^\beta|\pi\rangle\langle p_2,\pi|\chi^\alpha|p_1\rangle +$$

$$+ \sum_n \frac{\langle p_2|\chi^\beta|n\rangle\langle n|\chi^\alpha|p_1\rangle}{E_2+\omega_2-E_n+i\varepsilon}(2\pi)^3\delta(\vec{p}_2+\vec{q}-\vec{p}_n) - \text{c.t.} =$$

$$= \langle p_2,\pi^\beta(\vec{q}_2)|\chi^\alpha|p_1\rangle + \ldots \qquad (67)$$

and $\langle p_2,\pi^\beta(\vec{q}_2)|\chi^\alpha|p_1\rangle$ is the amplitude $T^{\beta\alpha}(\chi^\alpha N_1 \to \pi^\beta N_2)$ for production of a <u>physical</u> pion of momentum $\vec{q}_2$ from the "off shell" source χ^α. In particular πN - scattering is obtained if $q_{10} \to \omega_1 = \sqrt{\vec{q}_1^{\,2}+\mu^2}$.

Therefore according to our starting aim, we have obtained a representation for $T^{\beta\alpha}$ in terms of equal time commutators and intermediate state sum:

$$T^{\beta\alpha} = C^{\beta\alpha} - \sum_n (2\pi)^3\delta(\vec{p}_2+\vec{q}_2-\vec{p}_n)\frac{\langle p_2|\chi^\beta|n\rangle\langle n|\chi^\alpha|p_1\rangle}{E_2+\omega_2-E_n+i\varepsilon} -$$

$$- \text{c.t.} =$$

$$= C^{\beta\alpha} + \frac{1}{\pi}\int d\,Q_o'\,\frac{t^{\beta\alpha}(Q_o')}{Q_o'-Q_o-i\varepsilon} \qquad (68)$$

where

$$t^{\beta\alpha}(Q_o) = \pi(2\pi)^3\sum_n\delta(P+Q-p_1)\langle p_2|\chi^\beta|n\rangle\langle n|\chi^\alpha|p_1\rangle -$$

$$- \text{c.t.} \quad ; \qquad P = \frac{p_1+p_2}{2} \; ; \quad Q = \frac{q_1+q_2}{2} \qquad (69)$$

A completely analogous result holds for other pion processes, for instance photoproduction (or in general vector production). In this case the relevant amplitude is $T_{\mu}^{\beta\alpha}=i\langle p_2,\pi^{\beta}(\vec{q})|V_{\mu}^{\alpha}|p_1\rangle$ and the starting equal time commutator is

$$[\phi^{\beta}(\vec{q}),\ V_{\mu}^{\alpha}(-\vec{k})]$$

For representation (68) one can repeat remarks analogous to those already made in Sections (1), (2), (3) or:

a) $t^{\beta\alpha}$ can be decomposed into three meaningful parts t_I, t_{II}, t_{III} (connected s, $\bar{s}$ channel contributions, mass singularities, Z graphs). In particular as far as t_{II} is concerned there is no pion contribution any longer since $\langle 0|\chi|\pi\rangle = 0$. This explains the usefulness to start directly with $[\phi,\chi]$: a single pion contribution in the direct channel is automatically selected. Of course there is a prize to pay: in evaluating $[\phi,\chi]$ we need, beside $[\phi,\phi]$ higher commutators like $[\phi,\dot{\phi}]$, $[\phi,\ddot{\phi}]$, $[\dot{\phi},\ddot{\phi}]$ which are strongly model dependent. Accordingly $C^{\beta\alpha}$ introduces t dependent factors corresponding to the exchange of particles in the t channel*: their number and importance depends of course on the particular Lagrangian we use.

b) The dispersion relation (68) is performed at fixed t, i.e. the energy and the "masses" associated to $\phi^{\alpha,\beta}$ and χ^{α} vary with the Q_0 integration. This means that the intermediate state sum contains "off shell" matrix elements as $\langle p|\chi|n\rangle$, whose momentum transfer $(E_2-E_n)^2-\vec{q}_2^{\,2}$ varies with the mass of the intermediate state.

Moreover the evaluation of the completeness sum is frame dependent: if we go to Lorentz scalar variables the integration lines in the $(\nu, q_{1,2}^2)$ planes are parabolae,

* From the point of view of dispersion theory $C^{\beta\alpha}$ looks like a subtraction constant at infinity which determines the asymptotic behaviour of the real part of $T^{\beta\alpha}$. In the original papers $C^{\beta\alpha}$ was thrown away by assuming a source χ independent of $\phi,\dot{\phi}$.

whose parameters depend on P_o, $\vec{P}^2$, $\vec{P}.\vec{Q}$, $\vec{Q}^2$,... .

If one wants to recover fixed t and fixed mass dispersion relations, the $P_o \to \infty$ frame is suggested; then however the parabolae degenerate into the straight lines $q_i^2 \to -\vec{q}_i^2$ and the prediction holds for pions of imaginary mass* (as in the first simple derivation of dispersion relations via the reduction formula).

2. The aim of this Section is to improve the above description of physical pion processes by substituting to $C^{\beta\alpha}$ of Eq. (65) a set of better known commutators mainly arising from current algebra.

In so doing the connection with the low energy limits will be ensured and the final representation for $T^{\beta\alpha}$ will resemble a subtracted dispersion relation. Therefore the sum rule will be more rapidly convergent and will exhibit explicit kinematical factors (like ω or $\vec{q}$) which guarantee the dominance of the equal time contribution in the low energy region. Since pion amplitudes are the object of interest, we will exploit the intimate connection between pion and axial vector current: actually the successful and simplest relation is the PCAC one $\partial_\mu A_\mu^\alpha = \mu^2 f_\pi \ \phi^\alpha$, but commutation relations involving the divergence are already beyond the pure current algebra framework. Therefore, roughly speaking, we try to substitute to $\phi^\alpha \to A_o^\alpha$, better we look for a combination of A_o^α and ϕ^α, let us call it " a^α", such that

$$\langle\pi|a|0\rangle = \langle 0|a^*|\pi\rangle = 0 \qquad (70)$$

Then $(\langle 0|A_\mu^\alpha|\pi^\beta\rangle = i\ \pi_\mu\ \delta_{\alpha\beta}\ f_\pi)$

* To simplify the calculations go in forward direction t=o and take $\vec{p}.\vec{q}=0$; if you want to keep fixed $\nu=E\omega$ in the amplitude T, $\omega = \sqrt{\mu^2+\vec{q}^2} = \frac{\nu}{E} \to 0$ as $E\to\infty$, i.e. $\mu^2 \to -\vec{q}^2$.

$$a^{\alpha}(\vec{q},t) = \omega f_{\pi}\Phi^{\alpha}(\vec{q},t) - i\, A_0^{\alpha}(\vec{q},t) \tag{71}$$

and

$$<0|a^{\alpha}(\vec{q})|\pi^{\beta}> = (2\pi)^3\delta(\vec{q}-\vec{\pi})2\omega f_{\pi}\ \delta_{\alpha\beta} \tag{72}$$

"a^{α}" is the analogue of ϕ^{α} (Eq. (58)) and we still need an object with the properties of $\chi^{\alpha}(<0|\chi|\pi> = 0)$. Thus I introduce B_0^{α} :

$$B_{\mu}^{\alpha} = A_{\mu}^{\alpha} + f_{\pi}\,\partial_{\mu}\,\Phi^{\alpha} \tag{73}$$

and

$$\partial_{\mu}\,B_{\mu}^{\alpha} = \partial_{\mu}\,A_{\mu}^{\alpha} + f_{\pi}\Box\Phi^{\alpha} = f_{\pi}(\mu^2+\Box)\Phi^{\alpha} = f_{\pi}\chi^{\alpha} \tag{74}$$

We will also exploit the relation

$$a^{\alpha} = f_{\pi}(\phi^{\alpha}-i\ \dot{\Phi}^{\alpha}) = i\ A_0^{\alpha} = f_{\pi}\phi^{\alpha} - i\ B_0^{\alpha} \tag{75}$$

The analogue of the differential equation (60) becomes (in momentum space)

$$\dot{a}^{\alpha}(\vec{q},t) = \omega f_{\pi}\dot{\Phi}^{\alpha}(\vec{q},t)-i\dot{A}_0^{\alpha}(\vec{q},t) = \omega(B_0^{\alpha}-A_0^{\alpha})-i\dot{B}_0^{\alpha}+if_{\pi}\ddot{\Phi}^{\alpha} =$$

$$= \omega B_0^{\alpha}-i\omega a^{\alpha}+if_{\pi}\omega^2\phi^{\alpha}-i\dot{B}_0^{\alpha}+if_{\pi}\ddot{\Phi}^{\alpha} =$$

$$= \omega B_0^{\alpha}-i\omega a^{\alpha}+i(f_{\pi}\chi^{\alpha}-\dot{B}_0^{\alpha}) = \omega B_0^{\alpha}-i\omega a^{\alpha}+ib^{\alpha} \tag{76}$$

where

$$b^{\alpha}(\vec{q},t) = f_{\pi}\chi^{\alpha}(\vec{q},t) - \dot{B}_0^{\alpha}(\vec{q},t) = i\vec{q}.\vec{B}^{\alpha}(\vec{q},t) \tag{77}$$

Let us go to the determination of the physical amplitudes. As before one has to start from the equal time commutator $[\phi,\chi]$ for πN or $[\phi,V_{\mu}]$ for vectorproduction. For simplicity it is convenient to begin with the last example. Consider therefore $f_{\pi}[\phi^{\beta}(\vec{q}),V_{\mu}^{\alpha}(-\vec{k})]$ and use Eq. (75).

$$f_{\pi}[\phi^{\beta}(\vec{q}),V_{\mu}^{\alpha}(-\vec{k})] = [a^{\beta}(\vec{q}),V_{\mu}^{\alpha}] + i[B_0^{\beta}(\vec{q}),V_{\mu}^{\alpha}] =$$

$$= i[B_o^\beta, V_\mu^\alpha] + \omega_q[\Phi^\beta, V_\mu^\alpha] f_\pi - i[A_o^\beta, V_\mu^\alpha] \quad . \qquad (78)$$

The right use of this identity is the following: we must decide what commutators to saturate by inserting intermediate states and what to take as known. Of course $[\phi, V_\mu]$ has to be saturated as in Section 1, in order to get the amplitude of interest, while one should avoid saturation of $[\Phi, V_\mu]$, $[A_o, V_\mu]$ since Φ, A_o have pion pole contributions and would introduce additional, unwanted pion amplitudes. $[B_o, V_\mu]$ can be saturated, on the other hand, since B_o does not pick up pion terms. Thus, once we have decomposed $[\phi, V_\mu]$ into simpler commutators, the recipe is to saturate those commutators which do not pick up further pion terms (and they contain B_o, b, χ) while the remaining ones should be taken from the theory and it will be gratifying to check they are mainly given by current and field algebra.

In order to make evident the distinction, from now on I will collect on the l.h.s. the commutators to be saturated and leave on the r.h.s those which are assumed known.

Therefore (78) has to be understood as follows:

$$f_\pi \underbrace{[\phi^\beta(\vec{q}),\ V_\mu^\alpha(-\vec{k})] - i[B_o^\beta, V_\mu^\alpha]}_{\text{saturate}} = \omega_q f_\pi \underbrace{[\Phi^\beta, V_\mu^\alpha] - i[A_o^\beta, V_\mu^\alpha]}_{\text{assume known}} \qquad (78')$$

The Low representation follows by taking (78') between one nucleon states and inserting completeness. It is convenient to select the one nucleon contribution and to use for the other matrix elements involving B_o the relation

$$i\langle p|B_o|n\rangle = \frac{\langle p|\dot{B}_o|n\rangle}{E-E_n} = \frac{\langle p|(f_\pi\chi - i\vec{q}\vec{B})|n\rangle}{E-E_n} \qquad (79)$$

Thus one obtains

$$f_\pi T_\mu^{\beta\alpha}(VN_1 \to \pi N_2) = \langle p_2|(if_\pi\omega[\phi^\beta, V_\mu^\alpha] + [A_o^\beta, V_\mu^\alpha])|p_1\rangle -$$

$$- i\langle p_2|a^\beta|N\rangle\langle N|V_\mu^\alpha|p_1\rangle + i\sum_n (2\pi)^3\delta(\vec{p}_2+\vec{q}-\vec{p}_n) \times$$

$$\times \left\{ f_\pi \frac{\langle p_2|\chi^\beta|n\rangle\langle n|V_\mu^\alpha|p_1\rangle}{(E_2-E_n)(E_2+\omega_q-E_n)} - \frac{i\vec{q}\langle p_2|\vec{B}^\beta|n\rangle\langle n|V_\mu^\alpha|p_1\rangle}{E_2-E_n} \right\} - \text{c.t.} \tag{80}$$

which can be also put in the form

$$f_\pi T_\mu^{\beta\alpha} = \langle p_2|[A_o^\beta, V_\mu^\alpha]|p_1\rangle + T_\mu(\text{one nucleon}) + \\ + \omega R_\mu + \vec{q}.\vec{S}_\mu \tag{81}$$

Following these rules we can work out the same procedure for πN scattering by starting from

$$[\phi^\beta(\vec{q}_2),\ \chi^\alpha(-\vec{q}_1)]\ .$$

The basic identity reads:

$$f_\pi^2[\phi^\beta(\vec{q}_2), \chi^\alpha(-\vec{q}_1)] - if_\pi[B_o^\beta, \chi^\alpha] + \omega_2[B_o^\beta, B_o^\alpha] + i[b^\beta, B_o^\alpha] = \\ = [a^\beta, b^\alpha] + (d/dt + i\omega_2)[a^\beta, B_o^\alpha] \tag{82}$$

After taking (82) between one nucleon states and using completeness, it is easy to obtain the representation for $T^{\beta\alpha}(\chi^\alpha N_1 \to \pi^\beta N_2)$.

I give here the final result for πN scattering $q_{10}=\omega_1$, by choosing, for aesthetic reasons, a frame in which $E_1=E_2$, $\omega_1=\omega_2$:

$$f_\pi^2 T_{\pi N}^{\beta\alpha} = T_{\text{nucleon}}^{\beta\alpha} + \langle p_2|(\omega[A_o^\beta, A_o^\alpha] - \tfrac{1}{2}[A_o^\beta, \vec{q}_1.\vec{B}^\alpha] - \\ - \tfrac{1}{2}[\vec{q}_2.\vec{B}^\beta, A_o^\alpha])|p_1\rangle + \omega^2 R + \\ + \omega\vec{q}_2.\vec{S}^{(2)} - \omega\vec{q}_1.\vec{S}^{(1)} - q_{2i}q_{1j}U_{ij} \tag{83}$$

where the one nucleon term is

$$T^{\beta\alpha}_{\text{nucleon}} = \sum_N (2\pi)^3\delta(\vec{p}_2+\vec{q}_2-\vec{p}_N)\{\langle p_2|a^\beta|N\rangle\langle N|\chi^\alpha|p_1\rangle -$$

$$- \langle p_2|q_2\cdot B^\beta|N\rangle\langle N|B^\alpha_o|p_1\rangle\} \qquad (84)$$

and

$$R = \langle p_2|(if_\pi[\phi^\beta,A^\alpha_o]+if^2_\pi[\phi^\beta,\dot\phi^\alpha]-\omega f^2_\pi[\phi^\beta,\phi^\alpha])|p_1\rangle -$$

$$- \sum_n \delta(\vec{p}_2+\vec{q}_2-\vec{p}_n)\ \frac{\langle p_2|\chi^\beta|n\rangle\langle n|\chi^\alpha|p_1\rangle}{(E-E_n)^2(E+\omega-E_n)^2} - \text{c.t.} \qquad (85)$$

$$\vec{S}^{(2)} = \langle p_2|if_\pi[\vec{B}^\beta,\phi^\alpha]|p_1\rangle + if_\pi \sum_n \delta(\vec{p}_2+\vec{q}_2-\vec{p}_n)\ \cdot$$

$$\cdot\ \frac{\langle p_2|\vec{B}^\beta|n\rangle\langle n|\chi^\alpha|p_1\rangle}{(E-E_n)^2} - \text{c.t.} \qquad (86)$$

$$\vec{S}^{(1)} = \langle p_2|(-if_\pi[\phi^\beta,\vec{B}^\alpha])|p_1\rangle - if_\pi \sum_n \delta(\vec{p}_2+\vec{q}_2-\vec{p}_n)\ \cdot$$

$$\cdot\ \frac{\langle p_2|\chi^\beta|n\rangle\langle n|\vec{B}^\alpha|p_1\rangle}{(E-E_n)^2} - \text{c.t.} \qquad (87)$$

$$U_{ij} = f^2_\pi \sum_n \delta(\vec{p}_2+\vec{q}_2-\vec{p}_n)\ \frac{\omega+E-E_n}{(E-E_n)^2}\langle p_2|B^\beta_i|n\rangle\langle n|B^\alpha_j|p_1\rangle - \text{c.t.} \qquad (88)$$

In the general case $q_{10} \neq \omega_1$ one would obtain the amplitude for π production by the virtual π source χ^α: this process can be measured in principle in neutrino experiments by looking at the longitudinal part of the axial vector current. The transversal* contribution can be related to the matrix element $\langle p_2,\pi^\beta|\vec{B}^\alpha|p_1\rangle$ and knowledge of these two quantities should be enough to determine the axial neutrino production of pions. $\langle p_2,\pi^\beta|\vec{B}^\alpha|p_1\rangle$ can be then represented in a form analogous to eq. (81) just by $V_\mu \to \vec{B}$.

* Actually the transversal part is $A^T_\mu = A_\mu - \frac{1}{\Box}\partial_\mu(\partial_\nu A_\nu)$ which coincides with B_μ only at $\to\Box - \mu^2$.

3. The considerations of the previous sections enabled us to write a general representation for scattering and weak and electromagnetic production of pions on nucleons. (The analysis of other pion processes as $K\to\pi\ell\nu$ is completely analogous). In general two main contributions were separated: a potential like part, which corresponds to the exchange of particles in the t-channel, and a sum over intermediate states which can be considered as the resonance scattering part. Since the motivation for the last complicated manipulations was an improvement of the Low equation (68), let us discuss the advantages of the new formulation. Those concern both the nature of the commutators we need and the structure of the remaining contributions.

As far as the equal time commutators are concerned, it is easy to recognize, by looking at the general formulae (80),(83), that we require for vector production: $[A_o,V_o]$, $[A_o,\vec{V}]$, $[\Phi,\vec{V}]$, for axial production and πN scattering: $[A_o,A_o]$, $[A_o,\vec{A}]$, $[A_o,\Phi]$, $[\vec{A},\Phi]$, $[\Phi,\Phi]$, $[\Phi,\dot{\Phi}]$.

Therefore the necessary inputs are the current algebra commutators $[A_o,V_\mu]$, $[A_o,A_\mu]$, the field algebra ones $[\Phi,\vec{A}]$, $[\Phi,\vec{V}]$, $[\Phi,\Phi]$ and $[\Phi,A_o]$ (the σ-term), $[\Phi,\dot{\Phi}]$.

It can be instructive to compare the commutators we introduced at the different stages of the theory. The simplest C.A. + P.C.A.C. predictions, valid for zero mass pions, make use of only the charge commutators $[Q_A,Q_A]$, $[Q_A,A_\mu]$, $[Q_A,V_\mu]$. The continuation of these results to physical pions at rest is achieved by introducing the new quantities $[\Phi,\Phi]$, $[\Phi,Q_A]$, $[\Phi,\dot{\Phi}]$, as we learnt in the previous chapter. Finally the present generalization to pions of any momentum and the determination of the complete* amplitude require the extension

* This means that we are doing something more than simply extrapolating the pure C.A. results.

$Q_A \to A_o$ plus the field algebra (for instance) informations. These higher commutators introduce a model dependent (indeed up to date) element; although unpleasant, I consider this an unavoidable feature of any detailed treatment of physical pion processes, based on the connection between pion and axial current.

If one accepts in particular the field algebra suggestion $[\Phi,\Phi] = [\Phi,\vec{A}] = [\Phi,\vec{V}] = 0$, it is gratifying to check that, from the algebraic point of view, no really new commutators appear. Correspondingly, the "potential" part can be considered as due, for the various processes, to the exchange of the ρ, A_1, π (σ?) in the t channel.

If we further compare the generalized Low representations (80), (83) with the original one (68), it is seen that the separation between potential and resonance part is emphasized by the explicit ω, $\vec{q}_1$, $\vec{q}_2$ factors multiplying the intermediate states contribution. This resembles a subtracted dispersion relation, which actually shows up in the higher number of denominators (at least in the (χ,χ) part) and improved convergence. Indeed the $\omega, \vec{q}_1, \vec{q}_2$ factors remember the real origin of our approach which, roughly speaking, consists in an extrapolation of the current algebra limit from the point $|\vec{q}| = \omega = 0$ to farther and farther points of the physical region.

These remarks are well visualized by the πN case (83). No zero order terms in ω, $\vec{q}_1$, $\vec{q}_2$ appear for $T_{\pi N} - T_{nucleon}$ according to the Adler consistency condition; for $\vec{q}_1 = \vec{q}_2 = 0$ the term linear in ω gives the Weinberg formula if $\omega \to 0$, and its corrections if $\omega = \mu$. The quadratic terms ω^2, $q_i q_j$, ωq_j are the general corrections corresponding to physical pions of nonvanishing three momentum. We can therefore say that current algebra and the static model, or isobaric models, give two complementary descriptions of the πN amplitude (or its partial wave components). Current algebra will describe quite well, at low energies, those waves where resonances are unimportant, as the s-wave near threshold, we studied in the previous chapter. On the contrary the equal time commutator (ρ-exchange) is irrelevant for the p-wave dominated by the (33) resonance [8].

We have of course a prize to pay for this general treatment. First of all the fact of using A_o rather than Q_A makes more complicated the spectrum of singularities in the q^2 channel, since the A_1 (or other 1^+ states) appears, beside the pion. However if one works with small values of $\vec{q}^{\,2}_{1,2}$ the error in neglecting those new contributions should be of the order of $\vec{q}^{\,2}/m_A^2 \ll 1$.

Secondly since we enlarged our description by introducing the axial current to describe the pion, things become more general but also more complicated. Thus in the expression for $T_{\pi N}$, both matrix elements of χ and $\vec{B}$ appear. If one tries to use a "one pion" approximation by taking only the connected contribution from $|n\rangle \simeq |p,\pi\rangle$ the result is now a system of coupled integral equations for $\langle p,\pi|\chi|p\rangle$, $\langle p,\pi|\vec{B}|p\rangle$, while the simpler Eq. (68) would lead to the standard non-linear integral equation for $\langle p,\pi|\chi|p\rangle$ only.

Last, but not least, we meet the frame dependence, which is more crucial here than in the case of the charge commutators, since quantities like A_o do not exhibit any partial conservation property. Symmetry considerations* suggest the choice of the nucleon Breit frame $\vec{P} = \vec{0}$ ($P = p_1+p_2$) as a particularly convenient one.

A final remark about the comparison of our approach with the "hard pion" [9] or phenomenological Lagrangian [10] treatment of low energy pion processes. Practically,

---

* The reason is essentially the existence of simple properties under crossing of the two nucleons.

in these theories Feynman diagrams, rather than dispersive saturation of equal time commutators, are used to perform mass extrapolation; vector dominance is assumed everywhere, which should amount to accept field algebra in the evaluation of $[\phi,\phi]$, $[\phi, \vec{V}]$ ... . The most obvious difference is in the evaluation of a resonant contribution: in the dispersive treatment there are the connected and Z graph energy denominators with numerator form factors taken at different values of the momentum transfer. By neglecting this dependence, the two denominators combine and one is led back to a single Feynman denominator and coupling constant.

BIBLIOGRAPHY

1. An exhaustive bibliography on the subject can be found in the recent monographies:
   B. RENNER: "Current algebras and their applications" Pergamon Press (1968);
   S. L. ADLER and R. DASHEN, "Current algebras", Benjamin (1968).
2. S. WEINBERG, Phys. Rev. Lett. 17, 616 (1966).
   Y. TOMOZAWA, Nuovo Cimento 46A, 707 (1966).
3. S. FUBINI, G. FURLAN, Ann. of Physics (1968).
   V. de ALFARO, C. ROSSETTI, Invited paper at the Conference of the Italian Physical Society, 1967.
   See also: G. FURLAN, C. ROSSETTI, Lectures at the International Seminar on Elementary Particles, Varna (1968); S. FUBINI, Lectures at the Erice School (1968).
4. S. L. ADLER, Phys. Rev. 140 B, 376 (1965);
   W. I. WEISBERGER, Phys. Rev. 143, 1302 (1966).
5. S. L. ADLER, Phys. Rev. 137, B1022 (1965); 139,B1638 (1965).

6. The content of this Section is based on the paper "Equal time commutators and low energy pion physics", V. de ALFARO, S. FUBINI, G. FURLAN, C. ROSSETTI, Inst.f.Adv.Study preprint, Princeton.
7. F. LOW, Phys. Rev. 97, 1392 (1955).
8. See e.g. D. AMATI and S. FUBINI, Ann. Rev. Nucl. Science 12, 359 (1962).
9. H. SCHNITZER and S. WEINBERG, Phys. Rev. 164, 1828 (1967).
10. See for instance R. ARNOWITT and P. NATH, Lectures presented at the Summer Theoretical Physics Institute, Boulder (1968).

NONLEPTONIC HYPERON DECAYS†

By

D. FLAMM

Institut für Hochenergiephysik d. Österreichischen Akademie d. Wissenschaften
Vienna, Austria

1. INTRODUCTION

There has been quite some progress in recent years concerning nonleptonic hyperon decays. On the experimental side accurate measurements of the decay parameters have been performed from which the signs and magnitudes of all amplitudes have been determined. On the theoretical side, on the other hand, we now have a better understanding of strong interactions.

While let us say up to 1965 the absolute value of the amplitudes could not be deduced theoretically current algebra and PCAC provide a means to tackle this problem.

We shall talk about the following seven decay processes:

$$\Lambda^o \to P\pi^- : \Lambda^o_- \qquad \Sigma^+ \to N\pi^+ : \Sigma^+_+$$

$$\to N\pi^o : \Lambda^o_o \qquad \Sigma^- \to N\pi^- : \Sigma^-_-$$

$$\Sigma^+ \to P\pi^o : \Sigma^+_o \qquad \Xi^- \to \Lambda^o\pi^- : \Xi^-_-$$

$$\Xi^o \to \Lambda^o\pi^o : \Xi^o_o$$

†Lecture given at the VIII. Internationale Universitätswochen f. Kernphysik, Schladming, Feb. 24 - March 8,1969.

On the right hand side is the usual abbreviation for the decay amplitude. In it the decaying hyperon has as additional lower index the charge of the outgoing pion.

The most general form of the decay matrix element is

$$M = N\bar{u}(p_2)\{S-\gamma_5 P\}u(p_1)$$

$$N = \left(\frac{m_1 m_2}{2K^o E_1 E_2}\right)^{1/2} \tag{1.1}$$

In the rest frame of the decaying hyperon this reads

$$M = N\left(\frac{E_2+m_2}{2m_2}\right)^{1/2} \chi^{\dagger}(p_2)\left[S-\frac{\vec{\sigma}\,\vec{K}}{E_2+m_2}\,P\right]\chi(o) \tag{1.2}$$

where E_2, m_2 are the energy and the mass of the outgoing baryon, $\vec{K}$ is the momentum of the outgoing pion and $\chi(\vec{p})$ are two component spinors.

From eq. (1.1) it is clear that S is the parity violating amplitude, while P is parity conserving one. Eq. (1.2) shows that S is an s-wave amplitude; while P is p-wave. In the center of mass system other partial waves are not allowed due to angular momentum conservation.

Instead of eq. (1.2) experimentalists [1] like to use the effective Hamiltonian ($\vec{p}_2 = -\vec{K}$)

$$H_{eff} = s + p\,\vec{\sigma}\,\vec{p}_2/|\vec{p}_2| \tag{1.3}$$

The connection between the amplitudes defined in this way and the ones in eq. (1.1) is very simple

$$s = S$$

$$p = +bP$$

$$b = \frac{|\vec{K}|}{E_2+m_2} = \left(\frac{(m_1-m_2)^2-m_\pi^2}{(m_1+m_2)^2-m_\pi^2}\right)^{1/2} \sim \frac{m_1-m_2}{m_1+m_2} \tag{1.4}$$

The hyperon decay rate is then

$$\Gamma = \frac{|\vec{K}|}{8\pi m_1^2}\{(m_1+m_2)^2 - m_\pi^2\}(|s|^2+|p|^2) \tag{1.5}$$

In order to separate s and p one has to measure the asymmetry parameters. If the decaying hyperon has a polarization $\vec{P}_1$ the angular distribution of the decay baryon in the hyperon rest frame is given by

$$W(\theta) = \text{const.}(1+\alpha|\vec{P}_1|\cos\theta)$$

$$\cos\theta = \frac{\vec{P}_1\vec{p}_2}{|\vec{P}_1||\vec{p}_2|} \tag{1.6}$$

The connection of the parameter α with the amplitudes s and p is

$$\alpha = \frac{2\,\mathrm{Re}(s^*p)}{|s|^2+|p|^2} \tag{1.7}$$

Even more information can be obtained by measuring the polarization of the outgoing baryon. The expectation value of the spin operator of the outgoing baryon is [2]

$$\langle\vec{\sigma}_2\rangle = \frac{1}{1+\alpha\vec{e}_2\vec{P}_1}\{(\alpha+\vec{e}_2\vec{P}_1)\vec{e}_2+\beta\vec{e}_2 x\vec{P}_1+\gamma((\vec{e}_2 x\vec{P}_1)x\vec{e}_2)\} \tag{1.8}$$

with $\vec{e}_2 = \vec{p}_2/|\vec{p}_2|$ and

$$\beta = \frac{2\mathrm{Im}(s^*p)}{|s|^2+|p|^2} \tag{1.9}$$

$$\gamma = \frac{|s|^2-|p|^2}{|s|^2+|p|^2} \tag{1.10}$$

The polarization parameters α, β, and γ are not linearly independent, but fulfil the relation (1.11) as can easily be checked using eqs. (1.14).

$$\alpha^2 + \beta^2 + \gamma^2 = 1 \tag{1.11}$$

Thus it is customary to reexpress these three parameters in terms of two independent ones, namely, α and the angle ϕ

$$\beta = \sqrt{1-\alpha^2}\,\sin\phi$$
$$\gamma = \sqrt{1-\alpha^2}\,\cos\phi$$
$$\phi = \mathrm{arctg}\left(\frac{\beta}{\gamma}\right) \tag{1.12}$$

Next let us briefly recall what we can learn from time reversal invariance about the amplitudes. It tells us that the amplitudes s and p are relative real apart form final state interactions. The phase of s and p is related to the scattering phase shift for the decay baryon-pion system δ_s for orbital angular momentum L = 0 and δ_p for L = 1, respectively, by [3]

$$s = s_o e^{i\delta_s} = \pm|s|e^{i\delta_s}$$
$$p = p_o e^{i\delta_p} = \pm|p|e^{i\delta_p} \tag{1.13}$$

where s_o and p_o are amplitudes with the final state interaction switched off. We may thus introduce the phase $\Delta = \delta_s - \delta_p + n\pi$ (n=0, ±1); $n\pi$ corresponds to the relative sign of s_o and p_o, while Δ is the phase angle of s relative to p. For small final state interaction we expect $\Delta \sim 0$ or $\pm 180^o$.

From eqs. (1.7), (1.9) and (1.13) we have

$$\alpha = \frac{2|s||p|\cos\Delta}{|s|^2+|p|^2} = \frac{2s_o p_o \cos(\delta_p-\delta_s)}{|s|^2+|p|^2} \tag{1.14}$$

$$\beta = -\frac{2|s||p|\sin\Delta}{|s|^2+|p|^2}$$

The relative phase angle for the s and p wave amplitudes in the same decay process can be measured experimentally. The phases for different decay processes, on the other hand, depend on the phase convention, which can be chosen arbitrary. In the following we shall choose the SU(3) phase convention of De Swart [4]. The phase shifts for the pion nucleon system are known experimentally [5] (table 1).

	L=0, J=1/2		L=1, J=1/2	
$\|\vec{K}\|$ (MeV)	$\delta_1(I=\frac{1}{2})$	$\delta_3(I=\frac{3}{2})$	$\delta_{11}(I=\frac{1}{2})$	$\delta_{31}(I=\frac{3}{2})$
100 MeV (Λdecay)	5.5^o	-2.9^o	-1.2^o	-0.3^o
189 MeV (Σdecay)	8.3^o	-8.3^o	-2.0^o	-1.8^o

Table 1 - Pion-Nucleon Phase Shifts [5]

If we assume the $\Delta I=1/2$ rule only the I=1/2 phase shifts can contribute to the Λ decay. We find

$$\Delta = \delta_1 - \delta_{11} (|\vec{K}| = 100 \text{ MeV}) = 6.7^o$$

while one finds from the decay [6]

$$\Delta = (7.5 \pm 3.9)^o$$

For Σ_-^- only $\Delta I=3/2$ can contribute

$$\Delta = \delta_3 - \delta_{31} (|\vec{K}| = 189 \text{ MeV}) + 180^o =$$
$$= (-6.5 + 180)^o$$

which is to be compared with

$$\Delta = (230 {}^{+31}_{-125})^o$$

from the decay. The phase shifts are small as is to be expected, since the threshold for resonances in the corresponding channels lies considerably higher.

In the following we shall neglect the corrections due to final state interaction. Then one has $\beta = \phi = 0$ and the amplitudes may be chosen to be real. In this case α provides a quadratic equation for s/p

$$\frac{s}{p} = X_{\pm} = \frac{1}{\alpha} \left(\frac{1}{\alpha^2} - 1\right)^{1/2}$$

Note that $|\alpha| \leq 1$. The ambiguity in sign is resolved if one knows the sign of γ.

examples

$\frac{s}{p} = X_+$ for $\alpha > 0, \gamma > 0$ Λ^o_-, Λ^o_o

$\frac{s}{p} = X_-$ for $\alpha < 0, \gamma > 0$ Σ^-_-, Ξ^o_o, Ξ^-_-

$\frac{p}{s} = X_+$ for $\alpha > 0, \gamma < 0$ Σ^+_+

$\frac{p}{s} = X_-$ for $\alpha < 0, \gamma < 0$ Σ^+_o

In this way one obtains the amplitudes listed in table 2 [7].

	$S/10^{-7}$	$P/10^{-7}$	b
Σ^-_-	4.04 ± 0.04	- 0.33 ± 0.84	0.115
Σ^+_+	0.017 ± 0.074	41.4 ± 0.75	0.0975
Σ^+_o	3.38 ± 0.31	25.4 ± 4.08	0.0998
Λ^o_-	3.36 ± 0.05	24.0 ± 1.03	0.0532
Ξ^-_-	4.39 ± 0.06	- 15.3 ± 1.65	0.062

Table 2 - Hyperon Amplitudes [7].

The experimental data fulfill rather well a number of relations for s as well as for p-waves. One set of such relations, eq. (1.15) is easily obtained by assuming that the effective Hamiltonian for the decay transforms like an isospinor. This is the so-called $\Delta I=1/2$ rule.

$$\sqrt{2}\ \Sigma_0^+ + \Sigma_+^+ = \Sigma_-^-$$

$$\Lambda_-^0 = -\sqrt{2}\ \Lambda_0^0$$

$$\Xi_-^- = -\sqrt{2}\ \Xi_0^0 \tag{1.15}$$

Furthermore there is the so-called Lee-Sugawara relation eq. (1.16).

$$2\ \Xi_-^- - \Lambda_-^0 = \sqrt{3}\ \Sigma_0^+ \tag{1.16}$$

For s-waves eq. (1.16) has been derived theoretically under the assumption of octet dominance and CP-invariance [8]. For p-waves the theoretical derivation of eq. (1.16) is less convincing [9].

Our amplitudes are dimensionless. In particular the different sign of $S(\Xi_-^-)$ and $P(\Xi_-^-)$ should be emphasized. Some authors give different signs [20]. Irrespective of sign conventions the present experimental values [6] for the asymmetry parameter α demand, however, that

either $S(\Lambda_-^0)\ P(\Lambda_-^0) > 0$ and $S(\Xi_-^-)(P(\Xi_-^-) < 0$

(our convention)

or $S(\Lambda_-^0)\ P(\Lambda_-^0) < 0$ and $S\ (\Xi_-^-)\ P(\Xi_-^-) > 0$

Radiative corrections to the nonleptonic hyperon decays could cause a small $\Delta I = \frac{3}{2}$ contribution even if the weak Hamiltonian is pure octet. These corrections we expect, however, to be of the order or the fine-structure constant α, that is at most a few percent. Indeed an evaluation for Λ decay estimates radiative corrections to be about 3.5% [10]. Of course the radiative corrections are logarithmically divergent and in the above reference a cut off of the order of 2 GeV has been used.

2. QUARK MODEL APPROACH

After these general remarks let us look into particular models. We shall first consider the quark model, where I shall concentrate on a recent paper of Majerotto and myself [11]. In the quark model one assumes that the hyperon decay results from the decay of one of the quarks which build the hyperon up. In a more refined model the interaction between quarks is also taken into account. If the effective Hamiltonian for the quark decay transforms like $\bar{K}^0$ and K^0, respectively, the $\Delta I=1/2$ rule eq. (1.15) for the hyperon decay is guaranteed. To calculate the Clebsch-Gordan coefficient for any weak quark-quark meson vertex we use the spurion technique. The spurion incorporates the transformation properties of the weak Hamiltonian (i.e. it transforms like K^0 or $\bar{K}^0$). It may be inserted into any leg of the weak meson-quark vertex. We define it by the operator [12]

$$S\lambda = n \quad , \qquad S\bar{n} = \bar{\lambda} \qquad (2.1)$$

To calculate for instance the Clebsch-Gordan coefficient for the quark decay $\lambda \to p+\pi^-$ [12] we evaluate $\lambda \to n \to p + \pi^-$ [13] (Fig. 1a) and $\lambda \to p+K^- \to p+\pi^-$ (Fig. 1b). For the s-wave amplitudes already the simplest model of this kind (Fig. 2), gives good results where only one λ quark decays according to eqs. (2.2) and the other two quarks are spectators [14].

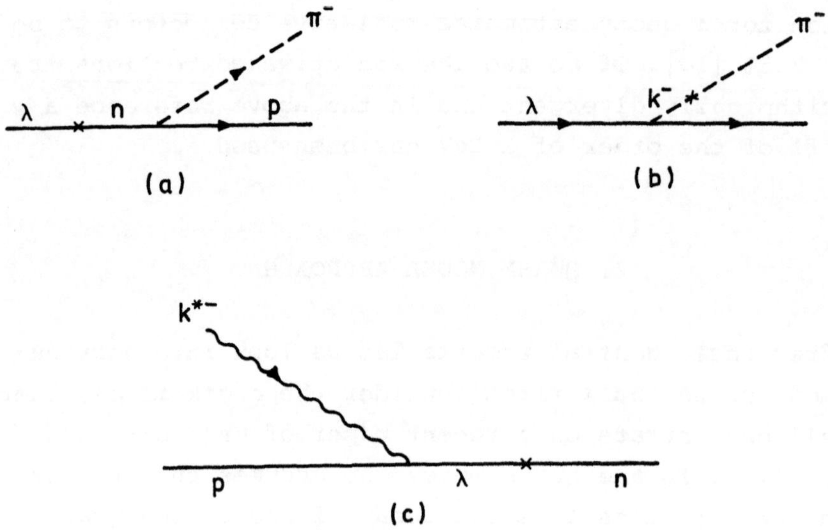

Figure 1 - Diagrams illustrating the spurion insertion. × indicates the spurion.

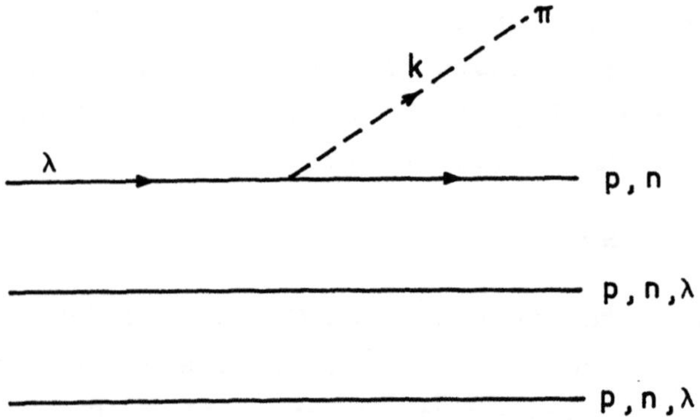

Figure 2 - Single λ-quark decay mechanism for nonleptonic hyperon decays.

$$\lambda \to n + \pi^0 : \lambda_0^0 \tag{2.2a}$$

$$\lambda \to p + \pi^- : \lambda_-^0 \tag{2.2b}$$

$$\lambda_-^0 = -\sqrt{2}\, \lambda_0^0 \tag{2.2c}$$

In particular, from this model it is evident that $S(\Sigma_+^+)=0$ since the λ quark cannot decay into π^+ and another quark. We shall list the corresponding amplitudes in table 3. This model is equivalent to the original SU(6) approach [16]. For p-waves, on the other hand, the decay mechanism, fig. 2, alone cannot work, since it would give $P(\Sigma_+^+)=0$ which is not true. Thus in addition we shall consider vector meson exchange (Fig. 3) [11]. Now two quarks are involved and only the third quark is a spectator. Furthermore the weak vertex need not to be the pion quark vertex, but can also be a vector meson quark vertex (for instance as in fig. 1c). The exchangedvector meson (V) is a member of the nonet ρ, K^*, ω, ϕ. For ω and ϕ we use the ideal mixing

$$\omega = 1/\sqrt{2}(p\bar{p}+n\bar{n})$$

$$\phi = -\lambda\bar{\lambda} \tag{2.3}$$

The Hamiltonian for the strong vectormeson-quark coupling may be written as

$$H = g\bar{\psi}_{q_{\bar{\alpha}}}\Gamma^\mu (V_\mu)^\alpha_\beta \psi_{q_\beta}$$

where $\psi_{q_\alpha}(x)$ $(\alpha=1,2,3)$ are twelve component quark spinors and Γ^μ is defined in eq. (2.6).

Our model combines the single λ quark decay mechanism with the sum of all possible vector meson exchange diagrams.

In particular we achieve $P(\Sigma_-^-)=0$ by a suitable interference of the two mechanisms as well as $P(\Sigma_+^+)\neq 0$ and the

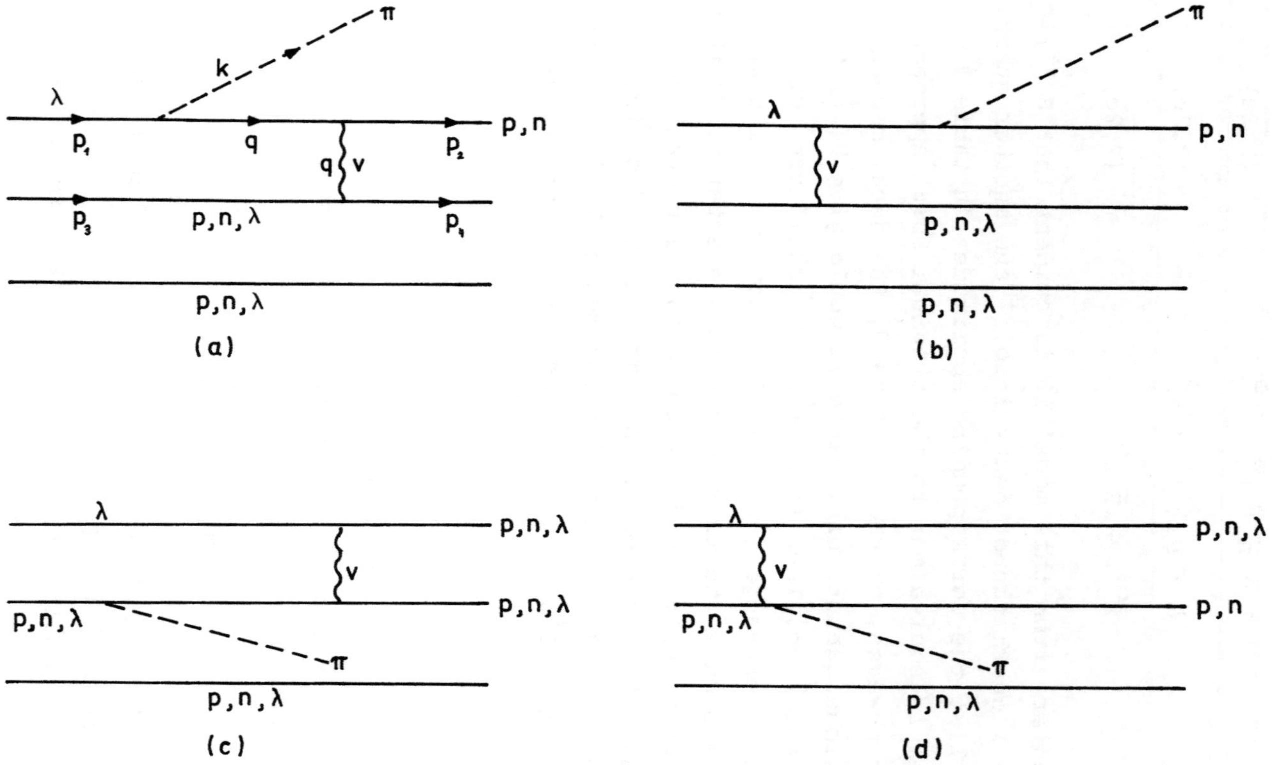

Figure 3 - Diagrams for the hyperon decay amplitudes with meson exchange between two quarks.

	s-wave					p-wave			
	k^o (MeV)	$\langle B_f\|O_A^S\|B_i\rangle$	$\langle B_f\|O_B^S\|B_i\rangle$	$\bar{S}$	experiment	$\langle B_f\|O_A^P\|B_i\rangle$	$\langle B_f\|O_B^P\|B_i\rangle$	$\bar{P}$	experiment
Σ_+^+	232	0	0	0	0.004 ±0.018	0	4	1	1.00 ±0.018
Σ_0^+	232	$\frac{1}{\sqrt{2}}$	0	0.707	0.710 ±0.09	$-\frac{2}{\sqrt{6}}$	$-\frac{2}{\sqrt{3}}$	-0.691	-0.729 ±0.085
Σ_-^-	238	1	0	1	1.00 ±0.009	$-\frac{1}{3}$	$\frac{10}{3}$	0	-0.016 ±0.02
Λ_-^o	172	$\sqrt{3/2}$	0	1.225	0.833 ±0.013	$\sqrt{3/2}$	$-\frac{14}{\sqrt{6}}$	1.05	0.578 ±0.025
Ξ_-^-	197	$\sqrt{3/2}$	$\sqrt{3/2}$	1.373	1.192 ±0.016	$\frac{1}{\sqrt{6}}$	$-\frac{11}{\sqrt{6}}$	-0.327	-0.37 ±0.04

Table3 - Nonleptonic hyperon decay amplitudes. k^o is the energy of the outgoing pion, $\langle B_f|O_A^{S,P}|B_i\rangle$ the contribution of the single λ quark decay mechanism (Fig. 2) for s- and p-waves, respectively, $\langle B_f|O_B^{S,P}|B_i\rangle$ the contribution of the vector-meson exchange graphs (Fig. 3).

$$\bar{S} = \frac{c'\langle B_f|O_A^S+aO_B^S|B_i\rangle}{S(\Sigma_-^-)}$$

are the s-wave amplitudes normalized in such a way that $\bar{S}(\Sigma_-^-) = 1$.

$$\bar{P} = \frac{c\langle B_f|O_A^P+aO_B^P|B_i\rangle}{P(\Sigma_+^+)}$$

are the p-wave amplitudes normalized so that $\bar{P}(\Sigma_+^+) = 1$.

$$\sum_q < B_f | O_A, O_B | B_i >$$

$\sum_q$ denotes the sum over all quarks contained in the initial hyperon B_i and in the final baryon B_f. As an example let me write down the Σ^+ quark wave function [15]

$$\chi_{\Sigma^+} = -1/\sqrt{18}\ (2\overset{\uparrow}{p}\overset{\downarrow}{\lambda}\overset{\uparrow}{p} + 2\overset{\uparrow}{p}\overset{\uparrow}{p}\overset{\downarrow}{\lambda} + 2\overset{\downarrow}{\lambda}\overset{\uparrow}{p}\overset{\uparrow}{p} - \overset{\uparrow}{p}\overset{\downarrow}{p}\overset{\uparrow}{\lambda} - \overset{\uparrow}{p}\overset{\uparrow}{\lambda}\overset{\downarrow}{p} -$$
$$- \overset{\downarrow}{p}\overset{\uparrow}{\lambda}\overset{\uparrow}{p} - \overset{\uparrow}{\lambda}\overset{\downarrow}{p}\overset{\uparrow}{p} - \overset{\uparrow}{\lambda}\overset{\uparrow}{p}\overset{\downarrow}{p} - \overset{\downarrow}{p}\overset{\uparrow}{p}\overset{\uparrow}{\lambda})$$

The minus sign in front corresponds to our De Swart phase convention. The complete wave function is of course the product of a space part $\phi(r_1,r_2,r_3)$ which depends on the quark coordinates times the spin–unitarity spin part χ_B (see for instance χ_{Σ^+} above)

$$\psi_B(x) \leftrightarrow \phi_8(r_1,r_2,r_3)\chi_B$$

$$\chi_B = \sum_{ikl} q_1^i\, q_2^k\, q_3^l\, C_{ikl}^B$$

$$(i,k,l=1,\ldots,6)$$

The index B stands for the baryon considered. C_{ikl}^B gives the weight of the various spin-unitarity spin combinations in the baryon-quark wave function. The above equation means a one to one correspondence between the Dirac spinor $\psi_B(x)$ for the baryon B and its quark wave function $\phi_8\chi_B$.

Now we have for instance

$$\sum_q O_A \chi_{B_i} = K^A_{B_i B_f} \chi_{B_f}$$ + eventually other terms which give zero when we consider $\sum_q \langle B_i | O_A | B_f \rangle$.

$K^A_{B_i B_f}$ is the Clebsch-Gordan coefficient we are interested in. Thus after applying O_A we may go back to the original baryon wave functions.

$$\sum_q \langle B_i | O_A | B_f \rangle \Rightarrow K^A_{B_i B_f} \bar{\psi}_{B_f}(x) \; \psi_{B_i}(x)$$

establishing the relationship with our amplitudes S and P defined in eq. (1.1).

Let us first concentrate on the p-wave amplitudes which are the most delicate ones. To obtain the operators O_A and O_B we have to look at the nonrelativistic limit of the corresponding Feynman diagrams. The operator O_A which corresponds to the single λ-quark decay mechanism can be obtained from the nonrelativistic limit of the Feynman diagram in fig. 2, eq. (2.4). Let me remark that for hyperon decays the approximation of static quarks is very likely to be legitimate, while for meson decays which involve quark pair annihilation and creation it certainly is not.

$$M_A = G(2K^o)^{-1/2} \frac{|\vec{K}|}{2M} \langle q_f | O_A | q_i \rangle = G(2K^o)^{-1/2} \sum_i \chi_q^+ \frac{\vec{\sigma}\vec{K}}{2M} E_i \chi_q \tag{2.4}$$

G is the weak quark decay constant in the process eq. (2.2), K^o the pion energy and M the quark mass. q_i and q_f denote the initial and final quark states, χ_q are six component nonrelativistic spin-unitarity spin quark wave functions, E_i (i=0,-) is an operator defined by

$$E_o \lambda = u \qquad E_- \lambda = p \tag{2.5}$$

which corresponds to the decays (2.2a) and (2.2b), respectively. For vector meson exchange there are four types of graphs as can be seen from fig. 3.

Some of them may appear for every vector meson. We shall write here as an example only the matrix element corresponding to fig. 3a.

$$M_2 = \frac{G}{(2\pi)^4} g^2 M^2 (2K^o p_1^o p_2^o p_3^o p_4^o)^{-1/2} \sum_{ik} \bar{u}(p_2)\Gamma^\mu \frac{\gamma{\cdot}q+M}{(q^2-M^2)} E_i \gamma_5 u(p_1) \times$$

$$\times\ \bar{u}(p_4)\Gamma^\nu F_k u(p_3) \frac{g_{\mu\nu} - q'_\mu q'_\nu/\mu^2}{(q'^2-\mu^2)}\ ,\quad \Gamma^\mu \equiv \gamma^\mu + \mu_q \sigma^{\mu\nu} q'_\nu \qquad (2.6)$$

g is the strong meson-quark coupling constant, $p_1, p_2, p_3, p_4, q, K, q'$ are the four momenta of the quarks and mesons as indicated in fig. 3a. $u(p_i)$ are 12-component quark spinors. The nonrelativistic form of eq. (2.6) defines the operator O_B^a for the vector meson exchange mechanism according to fig. 3a. We find neglecting the quark momenta and terms of the order K^o/M

$$M_B = G(2K^o)^{-1/2} \frac{|\vec{K}|}{2M}\ a\ \langle q,q|O_B^a|q,q\rangle = \qquad (2.7)$$

$$= G(2K^o)^{-1/2}\ a \sum_{i,K} \chi^+_{q_2} \frac{\vec{\sigma}\vec{K}}{2M} E_i \chi_{q_1}\ \chi^+_{q_4} F_K \chi_{q_3}$$

$$a = \frac{g^2}{(2\pi)^4 \mu\, 2K^o V} \qquad (2.8)$$

F_K (K=0,...,8) are operators which give all 9 possible quark transitions corresponding to the coupling of the vector meson nonet ρ, K^*, ω, ϕ to the quarks. μ is the vector meson mass and V the quark wave function normalization volume. We denote the operators corresponding to fig. 3b-d by O_B^b, O_B^c and O_B^d respectively. O_B^b differs from O_A^a in sign. For O_B^c and O_B^d one obtains analogous expressions with $(\vec{\sigma}\vec{K}/2M)E_i$ and F_K interchanged. We call the operator corresponding to the sum of all four graphs O_B

$$O_B = O_B^a + O_B^b + O_B^c + O_B^d \tag{2.9}$$

You may say we have tacitly assumed that the coupling for weak emission of π is equal to the weak coupling of a vector meson in adding all diagrams. But that is not so. If we assume only a kind of factorization (fig. 4) where the weak vertex is proportional to the product of the

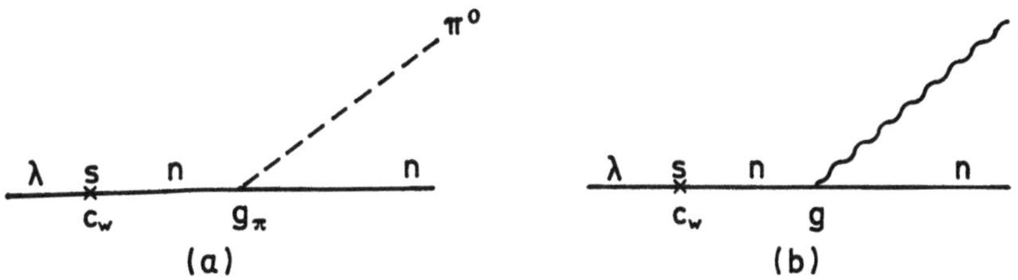

Figure 4 - Factorization of the weak quark decay constant into a spurion-quark vertex c_W and a strong meson-quark vertex g_π,g.

weak spurion transition c_W and the respective strong meson-quark coupling g_π in fig. 4a and g in fig. 4b, every graph of fig. 3 contains the product $c_W g_\pi g^2 \sim G\ g^2$, and we need not to think about the actual relation between the weak pion-quark coupling constant $G \sim c_W g_\pi$ and the weak vector-meson quark coupling constant $G_V \sim c_W g$. Of course, from SU(6) a relation between g_π and g does exist.

To obtain the hyperon decay amplitudes we sandwich these operators between the quark wave functions of initial hyperon B_i and the final baryon B_f [15].

$$P = c \sum_q \langle B_f | O_A + aO_B | B_i \rangle \tag{2.10}$$

The sum runs over all quarks contained in the baryons.

c is the overall normalization factor. By comparison of (1.1) and (2.4) it is found to be

$$c = - G \frac{E_2 + m_2}{2M} \tag{2.11}$$

Numerically one finds from $P(\Sigma_+^+)$

$$c = -1.03 \times 10^{-5} \tag{2.12}$$

The second parameter a is the relative weight of O_B compared with O_A

$$a = a' \frac{\mu}{K^o}$$

with

$$a' = \frac{g^2}{4\pi} \frac{1}{4\pi^3 \mu^3 V} \tag{2.13}$$

Using the universal coupling of the ρ meson to the isospin current we obtain for g

$$g = \frac{f_\rho}{\sqrt{2}} , \quad \frac{f_\rho^2}{4\pi} \approx 2.7$$

We determine a' in such a way as to obtain $P(\Sigma_-^-) = 0$.
This is achieved by

$$a_{\Sigma_-^-} = \frac{1}{10} \tag{2.14}$$

$$a' = K^o{}_{\Sigma_-^-}/10\mu = 0.034 \qquad \text{with } \mu = 0.7 \text{ GeV}$$

It is interesting that this choice corresponds in order of magnitude to a quark wave function normalization volume $V \sim 1/\mu^3$ where μ is the vector meson mass.

Thus both parameters c and a can be understood in principle. The matrix elements $\sum_q \langle B_f | O_A, O_B | B_i \rangle$ in eq. (2.10) are essentially Clebsch-Gordan coefficients. They are listed in table 3. In case one uses the same constant a = 1/10 in eq. (2.10) for all decays one obtains the right order of magnitude and the correct sign, but the numerical values of $P(\Lambda_-^o)$ and $P(\Xi_-^-)$ are off by nearly a factor three. K^o has, however, a wide range of variation for the different decay processes as can be seen from table 3. The factor $1/K^o$ in eq. (2.13) originates in the quark propagator, which, in the nonrelativistic limit, gives

$$\frac{1}{(p_1-K)^2-M^2} \sim \frac{-1}{2MK^o}$$

We shall use the physical values of K^o from table 3 which makes a vary considerably for different decays. This is a dynamical way to break SU(3) symmetry. The results obtained in this way are listed in table 3, where we have normalized the amplitudes in such a way that $\bar{P}(\Sigma_+^+) = 1$. Our amplitudes fulfil the Lee-Sugawara relation eq. (1.16) within 30% . We propose the inclusion of vector-meson exchange diagrams as a general way to calculate corrections to the simple quark model. In particular the dynamical SU(3) symmetry breaking via the quark-propagator should be generally applicable. The first objects to test this assertion are the s-wave amplitudes, for which already SU(6) gives good results [16]. We may again use the Feynman diagrams fig. 1 and fig. 3 but now the weak vertex is parity violating. We can derive s-wave operators O_A^S and O_B^S which are analogous to the p-wave operators but with $\vec{\sigma}\vec{K}/2M$ replaced by -1 . Sandwiching these operators between the relevant baryon states gives the s-wave Clebsch-Gordan coefficients listed in tab. 3. For the relative weight a of O_A^S and O_B^S we use the same values as for p-waves. The overall normalization c' is, however, different. It turns out that $\langle B_f|O_B^S|B_i\rangle$ gives only a slight correction to $S(\Xi_-^-)$ and is zero for the other amplitudes.

Since we propose the vector meson exchange diagrams as a general procedure to calculate corrections to the quark model the question arises, what consequences this would have for leptonic hyperon decays. There the situation is slightly different, since the weak vertex is always reserved for the emission of the lepton pair and hence weak vector meson coupling cannot contribute. Already for the nonleptonic decays the sum of all diagrams with only strong vector-meson coupling gave zero except for Ξ^-_-. In particular I have checked that vector-meson exchange diagrams do not affect the ratio G_A/G_V .

Another question you may ask is: what about pseudoscalar meson exchange? For the nonleptonic decays it has a depressing factor of the order K^o/M where M is the quark mass. It may thus only be important if other parameters are extremely large.

3. REMARKS ON THE CURRENT ALGEBRA APPROACH

A) General Formalism

For nonleptonic hyperon decays Sugawara [17] and Suzuki [18] were the first who applied current algebra and the assumption of the partial conservation of the axial-vector current [19] (PCAC) to calculate the s-wave amplitudes. In the following various refinements of the soft pion approach have been made [20]. In my remarks, to expose the methods of approach, I shall roughly follow a paper by Chan [21] which uses hard pion technique [22]. On the way I shall also mention the earlier approaches. I do not intend, however, to give an extensive review. A very convenient way to treat the nonleptonic hyperon decays is by a fictions scattering process. Let us consider the decay as the inelastic scattering of a spurion S, which comprises all the properties of the weak Hamiltonian (H_W) responsible for the decay, on a hyperon 1 to produce an outgoing baryon 2 and a pion (Fig. 5). Energy momentum conservation can then be written in terms

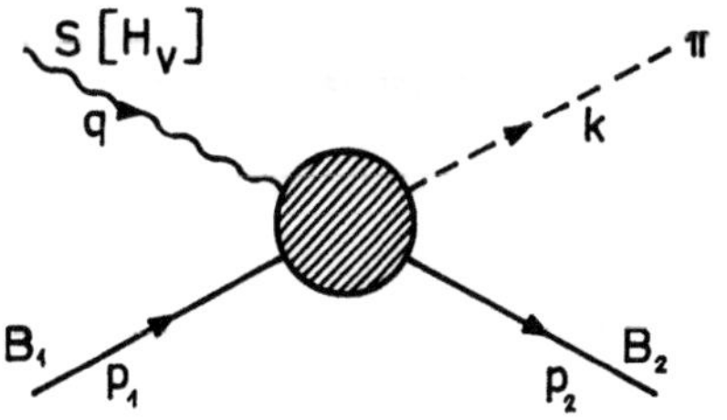

Figure 5 - Spurion-baryon inelastic scattering representing the hyperon decays.

of the baryon momenta p_1, p_2, the pion momentum K and the spurion momentum q as

$$p_1 + q = p_2 + K \tag{3.1}$$

We shall work in the center of mass system. The Mandelstam variables may be defined as

$$s = (p_1 + q)^2 \ , \ u = (p_2 - q)^2 \ , \ t = (p_2 - p_1)^2 \tag{3.2}$$

with the subsidiary condition

$$s+t+u = m_1^2 + m_2^2 + K^2 + q^2 \tag{3.3}$$

Let us start from the off-mass-shell scattering amplitude M for this process in terms of the time ordered product T{}

$$M = (2\pi)^3 \left(\frac{E_1}{m_1}\right)^{\frac{1}{2}} \left(\frac{E_2}{m_2}\right)^{\frac{1}{2}} \frac{(m_\pi^2 - K^2)}{c_\pi} \int d^4x \, e^{ikx} \times \tag{3.4}$$

$$\times \, \langle p_2 | T\{\partial^\mu A_\mu^i(x) \, H_W(o)\} | p_1 \rangle$$

M is readily obtained from $\langle p_2, K | H_W(o) | p_1 \rangle$ using the standard reduction technique and the proportionality of the pion field with isospin index i off the mass-shell to the divergence of the axial vector current A_μ^i (PCAC)

$$\partial^\mu A_\mu^i(x) = c_\pi \phi_\pi^{\prime}(x) \tag{3.5}$$

with

$$c = \frac{2m_N m_\pi^2 g_A}{g_{\pi NN} K(o)} \tag{3.6}$$

where m_N is the nucleon mass, $g_{\pi NN}$ the pion nucleon coupling constant, K(o) the corresponding form factor at zero momentum transfer and $g_A = G_A/G_V$ the renormalization factor of the weak axial vector current in β decay. Numerically one finds from the pion decay rate

$$g_A/K(o) = 1.3 \tag{3.7}$$

Next we use an identity for the time ordered product

$$T\{\partial^\mu A_\mu^i(x),\ H_W(o)\} = \tag{3.8}$$

$$= \frac{\partial}{\partial x_\mu} T\{A_\mu^i(x),\ H_W(o)\} - \delta(x_o)\,[A_o^i,\ H_W(o)]\ .$$

Inserting (3.8) into (3.4) we find

$$M = (2\pi)^3 \left(\frac{E_1}{m_1}\right)\cdot\left(\frac{E_2}{m_2}\right)^{1/2} \frac{m_\pi^2 - K^2}{c_\pi} \{-iK^\mu R_\mu - \tag{3.9}$$

$$- \int d^4x\, e^{iKx}\, \delta(x_o)\, \langle p_2|\,[A_o^i(x),\ H_W(o)]\,|p_1\rangle\}$$

where

$$R_\mu = \int d^4x\, e^{iKx} \langle p_2|T\{A_\mu^i(x),\ H_W(o)\}|p_1\rangle \tag{3.10}$$

As we shall see later the pole terms in the first part of (3.9), $-i\,K^\mu R_\mu$ will be the most important contribution to the p-wave amplitudes, while the equal time commutator essentially gives the s-wave amplitudes.

Now one can of course do all kinds of normal manipulations as are customary for scattering amplitudes. First let us decompose the scattering amplitude M in terms of invariant amplitudes. There are four such amplitudes, twice as many as for ordinary pion nucleon scattering since our spurion S consists of a scalar and a pseudoscalar part.

$$M = \bar{u}(p_2)\{A - \frac{1}{2}\gamma(K+q)B + \gamma_5 C - \frac{1}{2}\gamma(K+q)\gamma_5 D\}\ u(p_1) \quad (3.11)$$

Next we may split off a normalization factor by defining (W is the center of mass energy)

$$F = \frac{(m_1 m_2)^{1/2}}{4\pi W} M \quad (3.12a)$$

and express F in terms of two component Pauli spinors (χ) in the center of mass frame as

$$F = \chi^+\{f_1 + (\vec{\sigma}\vec{K})(\vec{\sigma}\vec{q})f_2 + (\vec{\sigma}\vec{K})f_3 + (\vec{\sigma}\vec{q})f_4\}\chi \quad (3.12b)$$

f_1, f_2, f_3 and f_4 can be expressed in terms of the invariant amplitudes A,B,C,D analogous to the pion nucleon scattering case

$$f_1 = \frac{1}{16\pi W^2}((W+m_2)^2-K^2)^{\frac{1}{2}}((W+m_1)^2-q^2)^{\frac{1}{2}} \times \quad (3.13)$$

$$\times (A - \frac{2W-m_1-m_2}{2} B)$$

$$f_2 = \frac{1}{16\pi W^2}((W-m_2)^2-K^2)^{\frac{1}{2}}((W-m_1)^2-q^2)^{\frac{1}{2}}(-A - \frac{2W+m_1+m_2}{2}D)$$

$$f_3 = \frac{1}{16\pi W^2}((W-m_2)^2-K^2)^{\frac{1}{2}}((W+m_1)^2-q^2)^{\frac{1}{2}}(C + \frac{2W-m_1+m_2}{2}D)$$

$$f_4 = \frac{1}{16\pi W^2}((W+m_2)^2-K^2)^{\frac{1}{2}}((W-m_1)^2-q^2)^{\frac{1}{2}}(-C + \frac{2W+m_1-m_2}{2}D)$$

We may now use a partial wave decomposition for the scalar spurion (parity conserving amplitude (p.c.a.)) and the pseudoscalar spurion (parity violating amplitude (p.v.a.)) separately. For the total partial wave amplitude one has

$$f_{\ell\pm} = 1/2 \int dz((f_1+f_4)P_\ell(z) + (f_2+f_3)P_{\ell\pm1}(z)) \quad (3.14)$$

where $\ell\pm$ stands for final states of orbital angular momentum ℓ and total angular momentum $j=\ell \pm 1/2$. For the p.c.a. we have

$$f_{pc} = \sum_{\ell}(2\ell+1)(\vec{\sigma}\vec{K})(f_{\ell+}\Lambda_{\ell+}+f_{\ell-}\Lambda_{\ell-})P_{\ell}(\cos\delta) =$$

$$= \vec{\sigma}\vec{K}f_3 + \vec{\sigma}\vec{q}f_4 \tag{3.15}$$

while for the p.v.a.

$$f_{pv} = \sum_{\ell}(2\ell+1)(f_{\ell+}\Lambda_{\ell+}+f_{\ell-}\Lambda_{\ell-})P_{\ell}(\cos\delta) = \tag{3.16}$$

$$= f_1 + (\vec{\sigma}\vec{K})(\vec{\sigma}\vec{q})\ f_2$$

$\Lambda_{\pm}$ are projection operators.

$$\Lambda_+ = \frac{(\ell+1)+\vec{\sigma}\vec{L}}{2\ell+1}\ ,\ \Lambda_- = \frac{\ell-\vec{\sigma}\vec{L}}{2\ell+1} \tag{3.17}$$

Inverting (3.15) and (3.16) brings of course back to (3.14). Up to now we have done a lot of formal manipulations. But now comes the essential point. What limiting procedure shall we use to get the decay amplitudes from these off mass-shell scattering amplitudes? The physical values of the variables for the decay process are of course spurion momentum $q^{\mu}=0$, $s=m_1^2$, $u=m_2^2$ and $K^2=m_{\pi}^2$. The decay amplitudes can be obtained by analytic continuation from the scattering amplitudes. In particular if it is easier to evaluate the amplitudes at a point different from the physical point for the decay we may do so if we can hope that the results will not differ much from the physical ones. In the following we shall consider two such approaches, the soft pion and the hard pion technique. The soft pion technique puts the pion momentum $K^{\mu}=0$ and then extrapolates to physical values. The hard pion technique, on the other hand, makes the much weaker assumption that only $K^2=0$ and relies on the PCAC assertion that the invariant amplitudes vary smoothly between $K^2=0$ and $K^2=m_\pi^2$. The latter is for instance checked by the good agreement with experiment of the Goldberger-Treiman relation. To simplify (3.15) and (3.16) we shall insert the physical values $W=m_1$, $K^2=m_\pi^2$, $q^\mu=0$ in all kinematical factors which are in front of the invariant amplitudes A, B, C, D and leave the arguments of the amplitudes themselves for the moment open. Then one easily sees from eqs. (3.13) that $f_2=f_4=0$ and the decay amplitudes S, P, which I defined in the introduction (eq. 1.1), follow from (3.15) and (3.16) adjusting the normalization as

$$S = A(s,K^2,t) - \frac{m_1-m_2}{2}B(s,K^2,t) \qquad (3.18)$$

$$P = C(s,K^2,t) + \frac{m_1+m_2}{2}D(s,K^2,t) \qquad (3.19)$$

Now one assumes that the invariant amplitudes obey unsubtracted dispersion relations in s with t fixed and evaluates R_μ by inserting intermediate states. This assumption is, however, not so crucial since most people anyhow only consider pole terms and resonances as intermediate states (Fig. 6) suppressing all other possible states. If t is not fixed also meson pole diagrams as in fig. 7 must be considered. For the weak Hamiltonian it is customary to assume a current-current form

$$H_W = \frac{G}{\sqrt{2}}\{j_\mu, j^{\mu^+}\} \qquad (3.20)$$

with j^μ of the Cabibbo form

$$j_\mu(x) = \ell_\mu + \cos\delta\ J_\mu + \sin\delta\ S_\mu \qquad (3.21)$$

where ℓ_μ is the lepton current, J_μ the strangeness preserving and S_μ the strangeness changing charged weak hadron current, which belongs to an SU(3) octet and con-

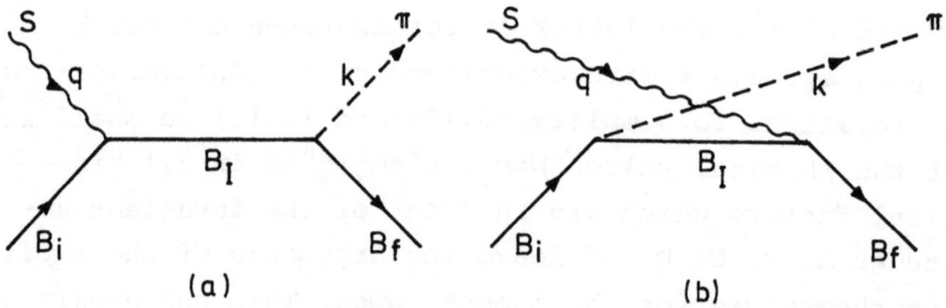

Figure 6 - Born-terms for nonleptonic hyperon decays.

tains a vector (V) and an axial vector part (A). δ is the Cabibbo angle. The strangeness changing part relevant for nonleptonic decay is something like

$$H_W^{NL}(\Delta S=1) = \frac{G}{\sqrt{2}} \sin\delta \cos\delta (\{J_\mu, S_\mu^+\} + \\ + \{S_\mu, J_\mu^+\} = H_W^{(8)} + H_W^{(27)} \qquad (3.22)$$

For $\Delta S=1$ only 8 and 27 out of the decomposition $8 \otimes 8 = 1 \oplus 8_A \oplus 8_S \oplus 10 \oplus \overline{10} \oplus 27$ can contribute. 10 and $\overline{10}$ are out since they are not self-adjoint, while 1 cannot change the strangeness S. It should be remarked that $H_W^{NL}(\Delta S=1)$ transforms like a U spin doublet.

B) Soft Pion Technique

If we would have assumed the soft pion technique, that is $K_\mu = 0$, from the very beginning, life would have been much easier. In fact we could have started right from eq. (3.9) and could have thrown everything away in $-iK^\mu R_\mu$ except pole terms which are singular at $K^\mu=0$ and thus survive the limit $K^\mu \to 0$. Charge conjugation invariance and SU(3) symmetry imply that only the scalar part of H_W couples to the baryon octet so that the Born approximation contributes only to the p-wave amplitudes

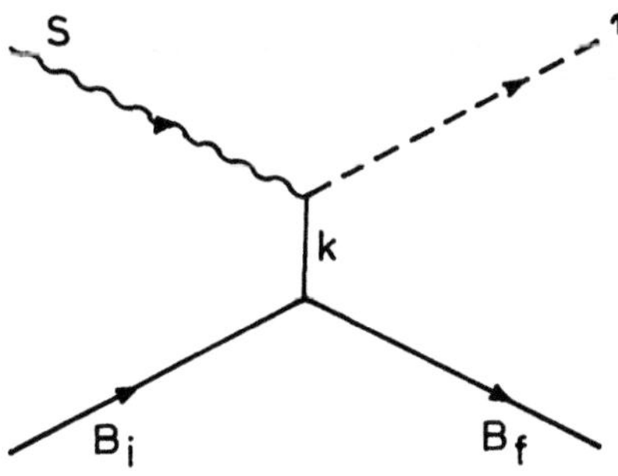

Figure 7 - Meson pole term for nonleptonic hyperon decays.

[23], [31]. Also the equal time commutator would be simplified a lot. Because of

$$\int d^3x \, A_o^i(x) = Q_A^i(t) \tag{3.22}$$

only the equal time commutator of H_W^{NL} with the axial isospin charge Q_A^i would have remained for which we know the result from the SU(2) × SU(2) algebra, namely,

$$[Q_A^i(o), H_W^S(o)] = [I^i, H_W^P(o)] \tag{3.23}$$

$$[Q_A^i(o), H_W^P(o)] = [I^i, H_W^S(o)] \tag{3.24}$$

where I^i are the isospin generators and the subscripts S and P at H_W mean scalar and pseudoscalar part, respectively.

If we use again SU(3) symmetry and charge conjugation invariance we find that the expectation value of (3.23) between baryon states vanishes [18], [31] and thus the equal time commutator contributes only to the s-wave amplitudes.

$$M^{(S)} = -(2\pi)^3 \frac{E_1E_2}{m_1m_2} \frac{m_\pi^2}{C_\pi} \langle p_2| [I^i, H_W^S] |p_1\rangle \tag{3.25}$$

The isospin operator I^i applied to a baryon state B gives again a baryon state and thus we have

$$M^{(S)} \sim \langle B|H_W^S|B'\rangle$$

The s-wave amplitude is proportional to the expectation value of the weak Hamiltonian between two baryon states. (3.25) is the s-wave formula of Sugawara and Suzuki. To obtain the $\Delta I=1/2$ rule we have to drop H_W^{27} or at least its $I=3/2$ part. If we call the reduced matrix elements for antisymmetric and symmetric coupling of H_W^{8S} f and d, respectively, we have

$$S(\Xi^-_-) = -\sqrt{2}\, S(\Xi^0_0) = \frac{1}{\sqrt{6}}\,(3f-d)$$

$$S(\Lambda^0_-) = -\sqrt{2}\, S(\Lambda^0_0) = -\frac{1}{\sqrt{6}}(3f+d)$$

$$S(\Sigma^+_+) = 0$$

$$S(\Sigma^-_-) = -\sqrt{2}\, S(\Sigma^+_0) = -\ f+d \qquad (3.26)$$

The d/f ratio entering here is for the weak spurion and need not be the same as the one for strong interactions (D/F=1.7). In fact, for instance, in reference [20] d/f = -0.75 is used. Moreover even the absolute value of the s-wave amplitude can be determined by evaluating explicitly the matrix elements $\langle B|H_W^S|B'\rangle$.

This can be done using the current form eq. (3.20) and inserting a complete set of intermediate states between the currents [28], [29]

$$\langle B|J_\mu S_\mu^+|B'\rangle = \sum_n \langle B|J_\mu|n\rangle\langle n|S_\mu^+|B'\rangle$$

It turns out that taking only the baryon octet and decuplet as intermediate states is quite a good approximation. Furthermore $\langle B|H_W^{S27}|B'\rangle$ is found to be small due to a cancellation of octet and decuplet contributions. Thus allowing for a number of approximations it is possible to calculate the absolute value of the decay amplitude from first principles. As I have already mentioned in the introduction this is a major success of the theory. For the p-wave amplitudes the simplest approximation is to take only the Born terms (Fig. 6 with B_I a member of the baryon octet). Now the weak d/f ratio from the baryon-spurion vertex as well as the strong D/F ratio from the pion-baryon vertex appear. Since the matrix element of the parity violating part of H_W between single baryon states is zero , [31]

$$<B_2|H_W^P|B_1> = 0$$

If one assumes either octet dominance for H_W and CP invariance [23] or alternatively SU(3) and charge conjugation invariance only [18] the matrix element of the parity conserving part

$$<B_2|H_W^S|B_1> \neq 0$$

contributes.

This means that the same spurion vertex determines s-waves and p-waves. In other words with the above approximation <u>current algebra establishes a relation between the s-wave and the p-wave amplitudes</u>. As an illustration I give $P(\Lambda_-^o)$ [20], [32] ($\alpha = \frac{D}{D+F}$)

$$P(\Lambda_-^o) = (1-\alpha)\left(\frac{2}{3}\right)^{\frac{1}{2}}(d-f)\,\frac{2m_N}{m_\Sigma - m_N} + \frac{1}{\sqrt{6}}(3f+d)\,\frac{2m_N}{m_\Lambda - m_N}$$

The results for the p-wave amplitudes obtained in this way are however by about a factor 2 too small. Many proposals to remedy this situation have been made [30]. One is, for instance, to include meson pole diagrams as in fig. 7 [24].

We shall take the view that the assumption of SU(3) invariance is suspect since the p-wave amplitudes obviously depend very crucially on SU(3) breaking mass differences.

c) Hard Pion Technique

Because of the ambiguities concerning the extrapolation from $K^\mu = 0$ to $K^2 = m_\pi^2$ [20] let us now use hard instead of soft pion technique. This is done in the paper of Chan [21]. Because of lack of time I shall only sketch his results. He evaluates the invariant amplitudes for $K^\mu \neq 0$, but $K^2 = 0$, $s = m_1^2$ and $t=0$ and considers two models. In the first model he allows for SU(3) breaking by taking

$$<B_2|H_W^P|B_1> \neq 0 \tag{3.27}$$

and he also keeps H_W^{27}. It turns out that the terms which result from (3.27) are proportional to $(m_1-m_2)/2m_1$. How much they really contribute depends on the d/f ratio for the weak spurion. He takes only baryon pole terms and can express the fourteen decay amplitudes in terms of six reduced matrix elements. These correspond to the symmetric 8_S as well as the antisymmetric coupling 8_A of the octet parts H_W^{8P} and H_W^{8S} and the coupling of the 27 parts H_W^{27P} and H_W^{27S} of the weak Hamiltonian. With these six parameters and a strong D/F ratio equal to 2 he can fit the experimental data within 10-20%. Incidentally $<B_2|H_W^{27S}|B_1> = 0$. It is no surprise that with so many parameters he can fit the data well. His choice corresponds to a different d/f ratio for the parity violating and the parity conserving spurion as well as different overall normalization for S and P in addition to the inclusion of H_W^{27}. The merit of the hard pion technique is, however, to remove the ambiguity in the extrapolation of the pole terms. Moreover if one is more ambitious one may calculate all the reduced matrix elements [25]. In the second model Chan assumes SU(3) to be good but he includes in addition to the baryon pole terms as intermediate states the $3/2^+$ decuplet and the $1/2^-$ $Y_o^+(1405)$ unitary singlet. For these additional states the SU(3) and charge conjugation invariance arguments mentioned previously do of course not apply and he has

$$<10|H_W^P|8> \neq 0$$

Again he can fit all amplitudes nicely with the strong D/F = 1.75 and six parameters assuming octet dominance for H_W. In comparison I would like to remind you that the quark model in the previous section needed only three parameters.

If one is more ambitious and wants to calculate the parameters explicitly one should be careful about applying SU(3) invariance. Also in the quark model I presented in section 2 SU(3) breaking was an important ingredient. The dependence on the mass difference should be studied carefully.

At the same time the effect of resonances and possibly even two body intermediate states must be considered in addition to the Born terms.

All what I still would like to mention is a paper by Good and Nieto [26]. Similar ideas have also been published by Nishijima [27]. These authors say that the experimental data show such striking regularities as the $\Delta I=1/2$ rule and the Lee-Sugawara relation while it is very hard to explain them theoretically. So may be there is some previously unexpected symmetry principle which gives $P(\Sigma_-^-) = 0$ and $P(\Sigma_o^+) = 1/\sqrt{2}\ P(\Sigma_+^+)$. The rule they propose is: assume that

$$<\psi_f^n|H_W^S|B_i^m> = 0 \quad \text{if} \quad m \neq n \tag{3.28}$$

where ψ_f^n is the final state of the decay, it is final baryon + pion, and m,n identify the SU(3) representation. Thus for the p-wave hyperon decays they demand that the final state has to be an octet, which means only I=1/2 and I=1 are allowed. For the Λ decays rule (3.28) gives $\Sigma \rightarrow (N_\pi)$ I=1/2 which gives immediately $P(\Sigma_-^-)=0$, since $N+\pi^-$ has I=3/2, and $P(\Sigma_o^+) = \frac{1}{\sqrt{2}}P(\Sigma_+^+)$. For λ decays (3.28) requires I_f=1/2 which implies the ΔI=1/2 rule. For Ξ decay $\Xi \rightarrow \Lambda\pi$ one has to consider the "crossed" decay $\Lambda \rightarrow \Xi\pi$ and use crossing to return to the original decays. In this way one can establish the ΔI=1/2 rule for all p-wave decays. Via current algebra one can then infer ΔI=1/2 also for s-wave decays.

For an explicit calculation of the p-wave amplitudes all pole terms where either the direct or the crossed diagram fig. 6a and 6b, respectively, contain an unallowed final state are left out. Proceeding this way and using the relation between s- and p-wave amplitudes obtained from current algebra and the pole approximation one obtains quite satisfactory values for the asymmetry parameters.

FOOTNOTES AND REFERENCES

1. See for instance J. W. CRONIN and O. E. OVERSETH, Phys. Rev. 129, 1795 (1963).
2. G. KÄLLEN, Elementarteilchenphysik (Hochschultaschenbücher-Verlag, 1965) p. 503 .
3. Compare L. B. OKUN, Weak Interaction of Elementary Particles (Pergamon Press Ltd., 1965), p.165 .
4. J. J. De SWART, Rev. Mod. Phys. 35, 916 (1963).
5. L. D. ROPER, R. M. WRIGHT and B. T. FELD, Phys. Rev. 138, B190 (1965).
6. N. BARASH-SCHMIDT et al. UCRL-8030 Pt.1 Jan.1969,Rev.

7. J. P. BERGE, Proc. 13th Int. Conf. on High Energy Physics, Berkeley (1966), p. 47. Since 1966 the value of $\alpha(\Xi^-_-)$ changed from -0.39 ± 0.03 to -0.41±0.04 [6] which is taken into account in table 2. In contract to the above reference our amplitudes are dimensionless.
8. M. GELL-MANN, Phys. Rev. Lett. 12, 155 (1964).
9. H. SUGAWARA, Prog. Theor. Phys. 31, 213 (1964).
10. F. ABBUD and C. N. YANG, Phys. Rev. Lett. 18, 980 (1967); A. A. BELAVIN and I. M. NARODETZKY, Phys. Lett. 26B, 668, (1968). The author would like to thank Dr. O. NACHTMANN for bringing this reference to his attention.
11. D. FLAMM and W. MAJEROTTO, "p-wave nonleptonic hyperon decays in the quark model", to be published in the Proceedings of the Topical Conference on Weak Interactions, CERN, Geneva, Jan. 1969.
12. We denote the quark doublet by p,n and the singlet by λ.
13. One may also view this process as the inelastic scattering of an energy-momentum zero spurion on a quark to produce a meson.
14. I. OHBA, Prog. Theor. Phys. 37, 1245 (1967).
F. CHAN, Nuovo Cimento 45, 236 (1967).
S. BADIER, Phys. Lett. 24B, 157 (1967).
M. NAKAGAWA and N. N. TROFIMENKOFF, Nuovo Cim. 52A, 961 (1967).
15. W. THIRRING, Acta Phys. Austriaca, Suppl. II, 205 (1966). R. van ROYEN and V. F. WEISSKOPF, Nuovo Cim. 50A, 617 (1967). To be conform with our phase convention the signs of the quark wave functions for $\pi^+, K^o, K^+, \Sigma^+$ and Ξ^o are opposite to the ones in the above references.
16. An extensive compilation of references on nonleptonic hyperon decays up to 1967 can be found in K. FUJII et al., Prog. Theor. Phys. 38 (1967).

17. H. SUGAWARA, Phys. Rev. Lett. 15, 870, 997 (1965).
18. M. SUZUKI, Phys. Rev. Lett. 15, 986 (1965).
19. M. GELL-MANN and M. LÉVY, Nuovo Cim. 16, 705 (1960).
20. See for instance L. S. BROWN and C. M. SOMMERFIELD, Phys. Rev. Lett. 16, 751 (1966).
21. F. C. P. CHAN, Phys. Rev. 171, 1543 (1968).
22. H. J. SCHNITZER, Phys. Rev. 158, 1471 (1967).
23. M. GELL-MANN, Phys. Rev. Lett. 12, 155 (1964).
B. W. LEE and A. R. SWIFT, Phys. Rev. 136, B228 (1964).
24. Compare S. A. BLUDMAN, "Nonleptonic Weak Interactions", 1966 Cargese Lectures.
25. Compare B. W. LEE, "Theory of Nonleptonic Hyperon Decays - A Survey" to be published in the Proceedings of the Topical Conference on Weak Interactions at CERN in January 1969.
26. M. L. GOOD and M. M. NIETO, Phys. Rev. Lett. 20, 624 (1968).
27. K. NISHIJIMA, Prog. Theor. Phys. 40, 912 (1968).
28. Y. HARA, Prog. Theor. Phys. 37, 710 (1967).
29. S. NUSSINOV and G. PREPARATA, Preprint SLAC-PUB-455.
30. See for instance C. ITZYKSON and M. JACOB, Nuovo Cim. 43A, 655 (1967).
31. The argument persists if H_W contains a part which transforms like a 27-plet [18], [24].
32. A list of amplitudes including the I=1/2 27 contribution is, for instance, in ref. [26].

FINITE ENERGY SUM RULES - Use and Interpretation*†

By

D. HORN

Physics Department, Tel Aviv University
Tel Aviv, Israel

The finite energy sum rules (FESR) are a tool for investigating the structure of scattering amplitudes. Together with this technique there goes a certain philosophy. We try to explain both in this lecture. We do not present a complete survey of the subject and, therefore, many important contributions are omitted. We prefer to discuss several cases that can serve as a good example of the technique and a foundation for the philosophy.

We first derive the FESR and explain the wrong and right signature cases. Then we discuss the applications to $A'^{(-)}$ at $t = 0$ and to $B^{(-)}$ amplitude of πN scattering. Attention is paid to the problem of various fixed poles. After that we turn to the concept of duality and its uses in bootstrap calculations. The last sections are devoted to Veneziano's formula and recent developments. Also here we do not give a summary of recent works but mention just those points that shed light on the main concepts discussed in this paper.

---

* Work sponsored by the Air Force Office of Scientific Research, Office of Aerospace Research, United States Air Force, under AFOSR grant number EOOAR-68-0010, through the European Office of Aerospace Research.

† Lecture given at the VIII. Internationale Universitätswochen für Kernphysik, Schladming, February 24 - March 8, 1969.

1. FESR FOR REAL ANALYTIC FUNCTIONS WITH A RIGHT HAND CUT

For many years we have been using the dispersion relations in particle physics. A typical unsubtracted dispersion relation is

$$F(\nu) = \frac{1}{\pi}\int_{\nu_0}^{\infty} \frac{\mathrm{Im}\, F(\nu')d\nu'}{\nu'-\nu} \tag{1}$$

which tells us the following:

$F(\nu)$ is a real analytic function defined in the complex ν plane cut from ν_0 to ∞ along the real axis. $F(\nu)$ is purely real for real ν below ν_0 and its discontinuity through the cut is $2i\, \mathrm{Im}\, F(\nu)$. Also $F(\nu) \to 0$ as $\nu \to \infty$.

This last property enables us to write an unsubtracted dispersion relation. If $F(\nu)$ does not tend to zero we have to introduce subtractions in the dispersion relation. We use these relations in the analysis of various physical amplitudes. The variable ν might represent physical variables like the energy or the momentum transfer in some scattering process. The information that we get from experimental data is necessarily confined to a certain finite region of ν. It is therefore important to formulate our theoretical tools in such a way that this obvious limitation is taken into account. This is done by the finite energy sum rules (FESR).

The FESR are consistency conditions that have to be obeyed by the function $F(\nu)$ as a result of its analytic structure and asymptotic expansion [1]. Suppose $F(\nu)$ is a real analytic function in the cut ν-plane that can be expanded in an asymptotic power series. We note that each term in the series can be written in such a form that it is also a real analytic function in the cut ν-plane: The function $-(e^{-i\pi}\nu)^{\alpha}/(\sin\pi\alpha)$ has thei maginary part ν^{α} at the cut from 0 to ∞. We can therefore write asymptotically

$$F(\nu) \approx \sum_i - \frac{c_i (e^{-i\pi}\nu)^{\alpha_i}}{\sin\pi\alpha_i} \tag{2}$$

Let us now form the integral $\oint F(\nu)d\nu$ along the contour of fig. 1. The integration from $N-i\varepsilon$ to 0 and then to $N+i\varepsilon$ will give

$$2i \int_{\nu_0}^{N} \operatorname{Im} F \, d\nu \quad .$$

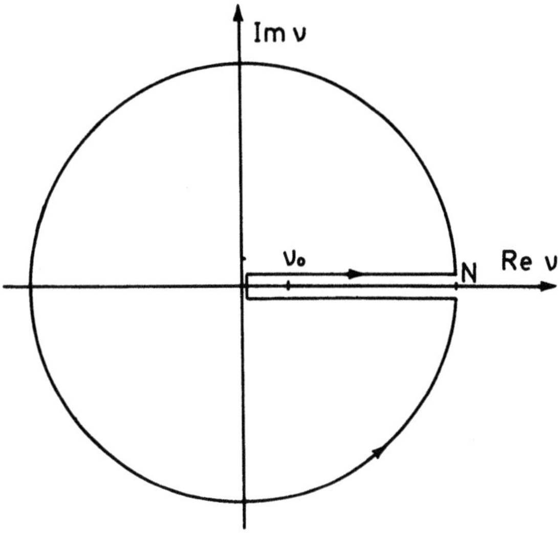

Figure 1 - The contour of integration used to derive the FESR

For the integration on the circle we use the asymptotic expansion (2). All of that has to be equated to zero according to Cauchy's theorem. Performing this calculation we find

$$\int_{\nu_0}^{N} \operatorname{Im} F(\nu)d\nu \approx \sum_i \frac{c_i N^{\alpha_i+1}}{\alpha_i + 1} \tag{3}$$

Obviously we can repeat the same procedure for any $\nu^n F(\nu)$ where n is an integer. This new function has the same analytic structure as $F(\nu)$ and its discontinuity is just ν^n disc(F(ν)). We find therefore

$$S_n \equiv \frac{1}{N^{n+1}} \int_{\nu_o}^{N} \nu^n \; \text{Im}\; F(\nu) d\nu \approx \sum_i \frac{c_i \; N^{\alpha_i}}{\alpha_i + n+1} \tag{4}$$

The S_n are the various FESR and have to hold for any function F(ν) with the above mentioned analytic behaviour and asymptotic expansion. Note that one gets the same type of FESR independently of the value of the highest α_i. Consequently any function that obeys a dispersion relation with an arbitrary number of subtractions will obey also the FESR. If the leading power in the asymptotic series is low enough we can find from eq. (4) all possible superconvergence relations by letting $N \to \infty$. After all, the superconvergence relations correspond to antisubtractions and they can therefore be determined by α_{max} just as the subtractions are determined by it.

From the derivation of the FESR presented here it is obvious that they can be generalized just by multiplying F(ν) by an arbitrary known function. That function must of course be simple enough so that the integration along the contour at $|\nu| = N$ can be carried out. A particular example found in the literature is the function ν^n with arbitrary real continuous n rather than the integer values that we have used here [2]. Now the discontinuity along the cut will be more complicated because the ν^n factor has a cut too. We find therefore a cut from 0 to ∞ involving both ReF(ν) and Im F(ν). The question of whether more information can be obtained by using a more elaborate technique has to be investigated in each case separately. Mostly the reliable information is given by total cross sections that contribute to ImF. In that case no generalization of the FESR of eq. (4) will add any new information. However in cases when we have a separate good experimental information on ReF we might expect generalizations of (4) to be useful in practical applications. In the present paper we will concentrate on FESR of the kind of eq. (4).

Eq. (4) is true for integer or zero values of n. A similar form can be derived for negative powers too. However in that case a pole is added at the origin. It leads to the appearance of the corresponding derivative of F at the origin:

$$S_{-m-1} \equiv N^m \int_{\nu_o}^{N} \frac{\text{Im } F(\nu)\, d\nu}{\nu^{m+1}} \approx \sum_i \frac{c_i N^{\alpha_i}}{\alpha_i - m} + N^m \frac{\pi f^{(m)}(o)}{m!} \qquad (5)$$

In the applications that we are going to consider we will usually find a cut plus a simple pole (Born term). This pole has to be included in the left side of the equations (4) and (5) as a δ-function contribution to Im F.

To check the consistency of our equations let us ask ourselves what happens if to our function $F(\nu)$ we add some polynomial - say simply the term ν^k with integer k. Although that new function does not satisfy the dispersion relation it still has to fit into the framework of the FESR. We have now to add to the asymptotic series a term with $\alpha_i = k$ and $c_i = -\tan\pi k$. Since this c_i vanishes it does not alter any of the FESR in (4). As far as (5) is concerned we will find a contribution to the asymptotic expansion of S_{-k-1}; however it cancels against the addition to $F^{(k)}(o)$. Hence we find that all equations are consistent with the change. We also learn another lesson: The integer power in the expansion is given smoothly by the limit $\alpha_i \to k$ and can be dealt with like any other term in the sums.

Before ending this section let us give a pictorial representation of the imaginary part of a function of the kind we are talking about. Fig. 2 shows $\text{Im}F(\nu)$ in the complex plane that is cut from $\nu = 0$ to $+\infty$. The Schwarz reflection principle that applies to this real

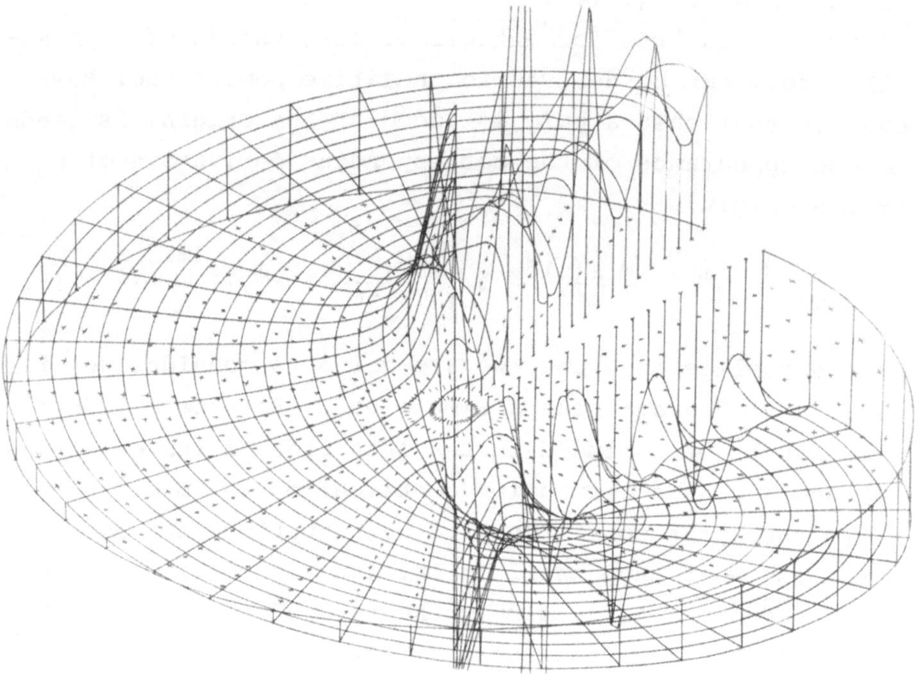

Figure 2 - Inverted plot of Im $\psi(z)$ on a cut z plane.
Thanks are due to Mr. B. Wolf for producing this computer drawing.

analytic function forces Im F to have opposite values above and below the cut and to vanish along the real axis where no cut is present. Another interesting feature of this function that can be seen here is a resonant structure near the origin that dissolves gradually into a smooth behaviour further away. We will return to this point in section 6 .

2. FESR FOR SCATTERING AMPLITUDES

A physical scattering amplitude has usually two cuts - the right hand cut (RHC) and the left hand cut (LHC). Let us decompose such an amplitude $G(\nu)$ into a symmetric and an antisymmetric part:

$$G(\nu) = \frac{1}{2}\left[G^{(+)}(\nu) + G^{(-)}(\nu)\right]$$
$$G^{(\pm)}(\nu) = G(\nu) \pm G(-\nu) \tag{6}$$

We can now define a function $g^{(+)}(\nu)$ that has the RHC of $G^{(+)}(\nu)$ and a function $g^{(-)}(\nu)$ that has the RHC of $G^{(-)}(\nu)$. Moreover :

$$G^{(+)}(\nu) = g^{(+)}(\nu) + g^{(+)}(-\nu)$$
$$G^{(-)}(\nu) = g^{(-)}(\nu) - g^{(-)}(-\nu) \tag{7}$$

The functions $g^{(+)}(\nu)$ and $g^{(-)}(\nu)$ are of the type described in the first section. In Regge theory we expect them by analogy to potential scattering, to have an asymptotic power series

$$g(\nu) \xrightarrow[\nu\to\infty]{} \sum_i - \frac{\beta_i}{\Gamma(\alpha_i+1)} \frac{e^{-i\pi\alpha_i}\nu^{\alpha_i}}{\sin\pi\alpha_i}$$

for high energy (ν) and low momentum transfer (t). The parameters α_i and β_i can depend on t. $\alpha_i(t)$ represents the location of the moving Regge pole in the j-plane.

The poles that appear in $g^{(+)}$ and $g^{(-)}$ are called poles of positive and negative signature respectively. From (7) follows

$$G^{(+)}(\nu) \to \sum_i \frac{\beta_i}{\Gamma(\alpha_i+1)} \frac{-1-e^{-i\pi\alpha_i}}{\sin\pi\alpha_i} \nu^{\alpha_i}$$
$$G^{(-)}(\nu) \to \sum_i \frac{\beta_i}{\Gamma(\alpha_i+1)} \frac{1-e^{-i\pi\alpha_i}}{\sin\pi\alpha_i} \nu^{\alpha_i} \tag{8}$$

A positive signature Regge pole will contribute a real pole to the amplitude when α_i is even and a negative signature one when α_i is odd. In the opposite case, namely an odd positive signature α or an even negative signature one, we find a pure imaginary contribution. This is called a wrong signature point.

We will generally assume that both β and α are essentially real. Both can in principle develop imaginary parts for t larger than the threshold in the t-channel, however these seem to be small. The reality of α can be deduced from the fact that the curvature of the trajectories is small as well as from the narrow widths of the resonances that lie on it. Neglecting the imaginary part of the Regge parameters implies neglecting the effects of the double spectral function and therefore enlarges the domain of convergence of the partial wave expansion of Im $G(\nu)$ (the big Lehmann ellipse).

We note that while the amplitudes $G^{(\pm)}(\nu)$ can be constructed directly from the measurable amplitude $G(\nu)$, that is not true for $g^{(\pm)}(\nu)$. These latter functions might have parts that cancel out when constructing $G^{(\pm)}(\nu)$. Thus if $g^{(+)}(\nu)(g^{(-)}(\nu))$ has odd (even) terms in ν they will cancel out in $G^{(+)}(\nu)$ $(G^{(-)}(\nu))$. If such an odd (even) term is a pole in g it might lead to a regular contribution in G.

Such odd or even function of ν would be recognized in Regge language as fixed poles since each such function can be expanded in a Laurent series in ν. Since they occur at the wrong signature points they will be referred to as wrong signature fixed poles. They can therefore affect the amplitudes $G^{(\pm)}(\nu)$ at most as regular terms of the kind ν^n or ν^{-n} with imaginary coefficients. We have to allow also for the possibility that their $\frac{\beta}{\Gamma(\alpha+1)}$ vanishes, in which case they will contribute zero to $G^{(\pm)}(\nu)$ and a finite term to $g^{(\pm)}(\nu)$. Therefore using $G^{(\pm)}(\nu)$ only we will not be able to recognize such terms. Thus starting with $G^{(-)}(\nu)$ ($G^{(+)}(\nu)$) we can derive only the even (odd) FESR [3]. The other kind, namely the odd (even) FESR might include such wrong signature fixed poles that will not affect at all the physical amplitude. We will call the first type right-signature FESR and the second one wrong-signature FESR. Most of the recently published papers on this subject deal only with the right-signature FESR.

What can we say about the wrong-signature FESR? Let us first discuss the case of positive integer moments n. In that case the FESR can be evaluated and the Regge parameters derived. We know that we might encounter wrong-signature fixed poles that are not observed in the scattering amplitudes. Therefore these FESR cannot be used in the opposite direction, namely, putting in information about Regge poles from high energy data to try to construct the amplitude satisfying the FESR. The case of negative moments S_{-n-1} is more complicated. For right-signature FESR the value of $g^{(n)}(o)$ is just $\frac{1}{2}G^{(n)}(o)$, therefore, by knowing the amplitude $G(\nu)$ near $\nu=0$ we can construct the sum rules,. However, for wrong-signature FESR, the $g^{(n)}(o)$ is some undetermined number whereas $G^{(n)}(o)=0$. The freedom in choosing these terms corresponds to adding to the function g an arbitrary polynomial that will have the opposite parity to that of G so that it will cancel out on the way from g to G.

3. THE S_0 FESR OF THE A' AMPLITUDE OF πN CEX

As the first example of the application of FESR we will use the one that also appeared first in the literature [4]. Here we want to use a FESR as a tool to check the consistency of the experimental data with the Regge asymptotic behaviour.

The reason for picking πN charge exchange (CEX) as our example is that this process should be dominated by the ρ Regge amplitude. The only Regge trajectory that has the right quantum numbers to be exchanged is that of the ρ. Our FESR can therefore tell us how good the ρ fit is and whether other trajectories of the same type have to be introduced. There is of course another good reason for choosing the πN case - here there exists the most accurate and reliable experimental data. Therefore if we want to learn a lesson about physics we should better start here.

The πN scattering is conventially [5] described by two amplitudes A' and B. A' corresponds to no helicity flip in the t channel and B to helicity flip in the t channel. The differential cross section can be described in terms of these amplitudes as

$$\frac{d\sigma}{dt} = \left(\frac{M}{4\pi W}\right)^2 \left[\left(1-\frac{t}{4M^2}\right)|A'|^2 + \frac{t}{4M^2}\left(s-\frac{(M+\nu_L)^2}{1-\frac{t}{4M^2}}\right)|B|^2\right] \tag{9}$$

We see that in the forward direction (t=0) B cannot contribute (no helicity flip because of angular momentum conservation). Using the optical theorem one can relate A' at t=o to the total cross-section. If we turn to CEX this amplitude is denoted by $A'^{(-)}$, and the optical theorem leads to

$$2 \text{ Im } A'^{(-)} \ (t=o) = k[\sigma_T(\pi^-p)-\sigma_T(\pi^+p)] = k\Delta\sigma \tag{10}$$

k denotes the momentum of the pion in the laboratory frame and $\nu = \nu_L$ its energy (In general

$$\nu = \nu_L + \frac{t}{4M} = \frac{s-u}{4M}.)$$

It can be further shown that $A'^{(-)}$ is an antisymmetric amplitude in ν, therefore S_o is a right-signature FESR and we might expect the dominance of the negative -signature ρ trajectory. Fig. 3 shows a comparison between

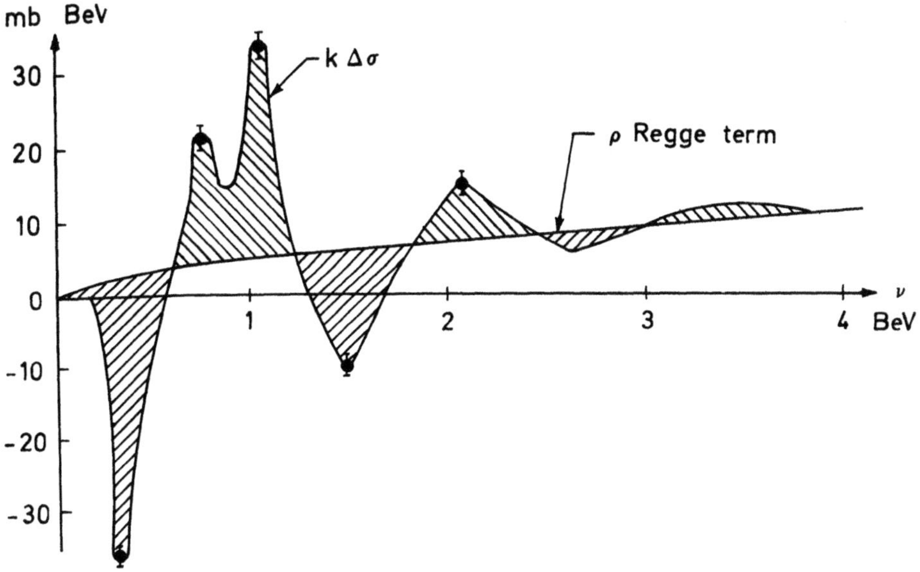

Figure 3 - A comparison of $k[\sigma_T(\pi^-p)-\sigma_T(\pi^+p)]$ with the Regge term. These two terms appear on the l.h.s. and the r.h.s. of $S_o(A'^{(-)})$ respectively. Compare with eq. (11).

$2\ \mathrm{Im}A'^{(-)}(t=0)$ as derived from (10) and the continuation of the ρ fit to the total cross section at high energies (6-20 BeV). The S_o FESR states that

$$\int_{RHC}^{N} \mathrm{Im}\, A'^{(-)} d\nu = -\frac{\pi g^2 \mu^2}{4M^2} + \frac{1}{2}\int_{\mu}^{N} k[\sigma_T(\pi^- p) - \sigma_T(\pi^+ p)]d\nu \approx \frac{\beta_\rho}{\Gamma(\alpha_\rho+1)} \frac{N^{\alpha_\rho+1}}{\alpha_\rho+1} \quad (11)$$

The integration over the RHC includes in addition to the area of the experimental curve a contribution of the nucleon Born term that amounts to $-\pi g^2\mu^2/4M^2 \approx -1.4\mathrm{mb\,BeV}^2$.

Since the right hand side of (11) is equivalent to what one would get from integrating from 0 to N over $\beta_\rho/\Gamma(\alpha_\rho+1)\nu^{\alpha_\rho}$ which is the asymptotic ρ-Regge fit to $\mathrm{Im}A'^{(-)}$ and, since the Born term is small, eq. (11) states essentially that the areas under the two curves of fig. 3 should be equal. It is easy to see from the curve that for $N \approx 3{,}5$ BeV that is true. A more accurate calculation verifies this fact.

One might ask how do we know where to choose N. The answer to that is the key to understanding the application of the formalism of section 1. What the FESR state is that if a certain function F with the specified analytic structure has an asymptotic expansion, then also the various moments of its discontinuity have a similar asymptotic expansion. Hence eq. (11) has to hold asymptotically. This means that when one plots the LHS of (11) as a function of N one finds some oscillating behaviour that eventually smoothes out. (As a matter of fact it has quite similar features to fig. 3). This final smooth curve is the asymptotic function. The deviation of the true curve of the LHS from the smoothed out behaviour at each N are a measure of the error at that point N. In comparing the two sides of eq. (11) we have to take this error into account. It is clear that choosing N=1 BeV will introduce such a big error in the LHS that comparing it with the RHS will be meaningless. However the higher N gets the smaller this error becomes. Going far into the asymptotic region will of course lead us to no new information . The optimal choice of N is such that the error is already small but the low energy contribution to the integral is still important.

The big success of eq. (11) means that one is able to check with low energy data a results that previously was obtainable from high energy data only. Analyticity brought about this miracle. It gave us the connection specified by the S_o FESR. That and something else - the fact that the ρ-Regge pole is really dominating the $A'^{(-)}$ amplitude in the region of the chosen N. It might very well be that a lower ρ' can be missed by the fit to the high energy data and can be discovered by the S_o that is based on the low energy data. That might still be the case if we will have better data (smaller systematic errors) on the integral between 1 to 3.5 BeV. What can be stated now is that the single ρ Regge fit to $A'^{(-)}(t=0)$ is consistent with S_o. It is also consistent with the higher FESR.

4. FESR APPLIED TO $B^{(-)}$

In section 3 we derived our experimental information from total cross sections. Using other detailed experimental results about the differential crcss sections, polarizations and various correlations we might hope to be able to construct the functions A' and B in the whole physical regions. This is attempted by the people who construct the phase shifts of πN. Nowadays we have phase shifts available up to $\nu \simeq 2$ BeV. We can use these phase shifts to construct A' and B from the various FESR, and test them [6]. One might of course wonder whether the low N forced upon us by the experimental information (2 BeV) can lead to any useful results. We have already seen that in the case of $A'^{(-)}$ at t=0 it introduces such a big error that the check of the S_0 FESR is meaningless. However that does not say that such a low N will necessarily always be bad. We have to check each amplitude in question and ask ourselves how low an N can we choose for it. It so happens that the $B^{(-)}$ leads to a satisfactory accuracy (low errors) even with N near 1 BeV.

Let us see what makes $B^{(-)}$ a better candidate than $A'^{(-)}$ at t=0. A look at fig. 3 will convince us that the various resonances at the low energy region contribute significantly with changing signs. Comparison of eq. (10) with a simple Clebsch-Gordan table tells us that a resonance with I=3/2 will contribute a negative quantity to Im $A'^{(-)}$ (t=0) and a resonance with I=1/2 a positive quantity. Thus the first dip in fig. 3 corresponds to the 1238 I=3/2, then come peaks due to the I=1/2 resonances between 1400-1700 followed by the dip of the Δ(1950) etc. The big resonances cause the big oscillations of the function and delay the smoothing out process. If we can find an amplitude in which all resonances appear with the same sign then we can use FESR with lower N. The reason is of course that in this case the resonances combine to give a high average function with small wiggles around it. That is the case for Im$B^{(-)}$(t=0). Here by accident (or perhaps on purpose?) all the dominant resonances have such quantum numbers that they contribute positively. Since they also overlap considerably it turns out that even N=1BeV is not an unreasonable choice.

Whereas in section 3 we checked the right form of the asymptotic behaviour, namely $\alpha_\rho(o)$, we can now try to look for the t-behaviour of the asymptotic amplitude. In Regge theory one expects the A' amplitude to be given by $P_\alpha(\nu)$ which goes asymptotically to ν^α, whereas the spin flip B amplitude is given by $P'_\alpha(\nu)$. Therefore one expects B to go like $\alpha\nu^{\alpha-1}$. That means that the contribution of the Regge term should vanish when α passes through zero. High energy data seems to verify this behaviour since a dip in $\frac{d\sigma}{dt}$ is observed for t in the region of -0.5 to -0.6 BeV^2, which is where a more or less linear ρ trajectory would cross zero. The interesting question is whether the low energy data would show a similar behaviour via the FESR.

Since the $B^{(-)}$ amplitude is symmetric in ν we consider S_1 which is the first right-signature FESR. It turns out that using as input phase shift data with as low a cutoff as $N \approx 1$ BeV one finds a linearly decreasing function for decreasing t that passes through zero around $t \approx -0.4\ BeV^2$ (the accuracy is of order 0.1 BeV^2). This is an amazing support of the high energy behaviour by the low energy region! Moreover, the same behaviour is encountered in S_3 as well as the higher odd FESR. This is expected from the Regge point of view but very amazing as far as the resonance region is concerned.

Let us ask ourselves how the zero in S_1 comes about. Detailed calculations show that there are two significant contributions - those of the N Born term and the $\Delta(1238)$ - that cancel each other at that t-value. The rest of the integral is small compared to each of these contributions. On the other hand the vanishing of S_3 comes from a completely different mechanism. Due to the additional ν^2 in the integrand the contributions of N and $\Delta(1238)$ are significantly reduced. It turns out that the contribution of each of the higher resonances in the integrand vanishes at a t-value between -0.3 to -0.6 . The combined effect is that the integral vanishes around -0.4. Fig. 4 shows the relevant part of the Mandelstam plot. We see the first zeros of the contributions of the various resonances to $A'^{(-)}$ and $B^{(-)}$. All zeros occur together at about the same value of t. Thus each resonance separately seems to follow the trend specified by the Regge form. These miraculous results opened the door for wild speculations. However, before turning loose our imagination, let us discuss some of the open questions that a detailed analysis of $B^{(-)}$ leads to.

5. FIXED POLES IN $B^{(-)}$

An amazing result comes from evaluating S_o, the first wrong-signature FESR, for $B^{(-)}$ [6]. The integral on the LHS turns out to be between 5 to 10 times bigger than the expected ρ Regge pole contribution over the whole range of t between 0 to -1 BeV^2. In terms of the s-channel input we note that this drastic effect is due to the nucleon Born term. This term occurs at $\nu_L = -0.01$ BeV. Because of the smallness of its $\nu = \nu_L + (t/4M)$ it is overpowering in S_o of $B^{(-)}$, moderately important in $A'^{(-)}$ and $B^{(-)}$, and negligible in higher moment FESR. Overpowering nucleon dominance in $B^{(-)}$ means that its high energy tail that goes like $1/\nu$ brings about a fixed pole at $\alpha = 0$ (here the Regge terms behave like $\nu^{\alpha-1}$!) This is the wrong-signature point in which a fixed pole will affect S_o only. Since this is also a non-sense point (β vanishes at α=0) such a fixed pole will not appear in the symmetric $B^{(-)}$ amplitude. In other words this term that possesses such a pompous name - the non-sense wrong-signature fixed pole at α=0 - ist just a constant term

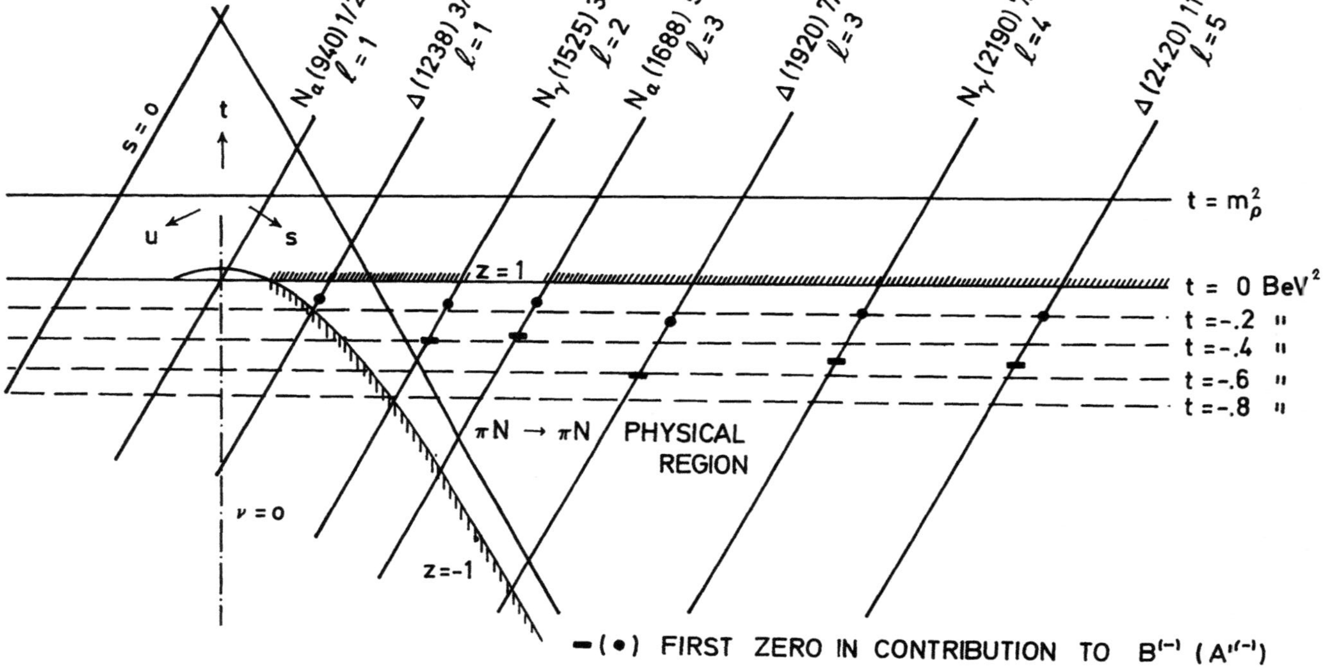

Figure 4 - Mandelstam plot for πN scattering

that appears in S_o only. It is no pole in any scattering amplitude, still it would be nice to understand it on general grounds.

A word of caution about how fixed the fixed pole is. It should be clear by now that there is a finite accuracy to within which we can determine the location of any pole by the use of our techniques. Since all results behave smoothly in the variables α we cannot say whether the effect observed is really due to a fixed pole, or to a slowly moving pole, or to some moving cut - provided it is concentrated around $\alpha = 0$. The way of determining α is by fitting several FESR simultaneously with the same Regge parameters. Obviously every such fit that has to explain S_o will have to include an enhancement around $\alpha \simeq 0$. If we call it a fixed pole it is just because this seems to be the simplest creature we can invent to explain this phenomenon in Regge language.

One might wonder how good is a single ρ-Regge pole fit to the right-signature FESR of $B^{(-)}$. The answer is that the parameters as deduced from high energy data do not provide a satisfactory fit and one has to introduce a ρ' moving about half a unit of α below the ρ trajectory [6]. Using the recent phase shifts (like the CERN [7] ones extending to $\nu_L \approx 2$ BeV) we can reach a wide range in t and see whether more surprises wait for us there [8]. Before finding out what they are we have to have a criterion for how far in t we can get. Using the phase shifts we limit ourselves to some finite polynomial in $\cos\theta_s$ (and, therefore, in t) determined by the highest partial wave taken into account in the calculation. As we go down in t more and more of our integrand will emerge in the non-physical region ($\cos\theta_s < -1$) and will grow according to its angular momentum. A look at fig. 4 shows us that at $t = -0.8\ \mathrm{BeV}^2$ all the resonances below the N (1525) are already outside the physical region. We can however expect that as long as the point N is still well in the physical region our sum rules make sense. In practice we just look at the t-behaviour of the integral. It usually behaves smoothly (oscillates mildly) until some point at which it starts to blow up in either the positive or negative direction. For the right-signature FESR of $B^{(-)}$ calculated with the CERN phase shifts, this point is $t \approx -2.4$ BeV^2. We will therefore consider the results along 2/3 of this region as significant.

Fig. 5 shows the behaviour of S_1, S_3 and S_5. (The S_n plotted here are by a factor N bigger than the definition (4)). In order to be sure that we don't see any spurious effect we shall ask ourselves the following question:

What features seen in fig. 5 hold also if we change the upper limit of integration N? With the phase shifts at our disposal we can change N between 1 and 2 BeV. That will change the various curves in fig. 5. However the fact that all of them pass through zero around $t = -0.4$ BeV^2, the fact that S_3 and S_5 show the oscillation seen on the curves and the very important fact that $|S_1|$ is so much bigger than $3|S_3|$ and $5|S_5|$ over the whole range between -1 to -2 BeV^2, all these facts remain unchanged. We have to explain why $|S_1|$ is so much bigger than the other $|S_n|$. The explanation is that there exists a very important effect in the angular momentum plane around $\alpha=-1$. Maybe it means that there exists a fixed pole at $\alpha = -1$ or a very slowly moving pole around $\alpha = -1$. It might even be the ρ that tends to $\alpha = -1$, however then we must have additional moving poles in order to explain the oscillatory behaviour of S_3 and S_5. We did not expect to find this pole at $\alpha = -1$. We note that here it is a right-signature point which means that

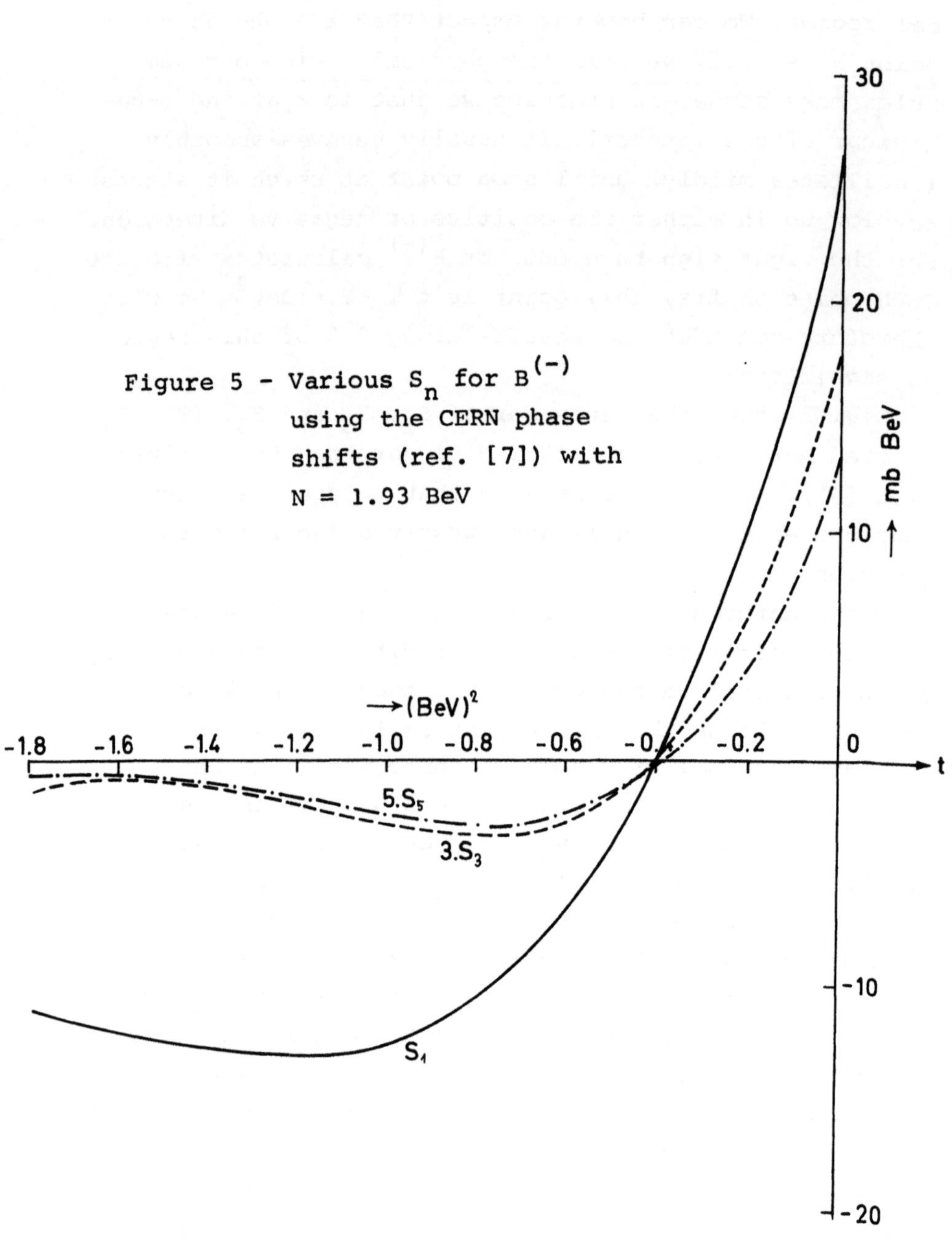

Figure 5 - Various S_n for $B^{(-)}$ using the CERN phase shifts (ref. [7]) with N = 1.93 BeV

it will affect also the amplitude $B^{(-)}$. As a matter of fact it just adds a real positive ν^{-2} behaviour to the B amplitude. Now $|B^2|$ will no longer vanish at $\alpha=0$, nevertheless it will still be minimal at that point. That means that the old explanation of the dip in $d\sigma/dt$ as showing the effect of α_ρ passing through zero still holds.

The fixed poles that we discussed in this section might belong to a new family of creatures that we have to include in the Regge analysis. Maybe even the Pomeranchukon belongs to this family. We know that the Pomeranchuk trajectory seems to be different in slope from the other moving Regge poles. Leaving the question of the Pomeranchukon aside for the moment it seems that these fixed poles are not the most important features of the high energy scattering. Those important features, as observed in the gross behaviour of $d\sigma/dt$, come from the already known moving poles like the ρ trajectory, the A_2 trajectory, the P' trajectory etc. We will therefore keep all the annoying facts that we learned about the $B^{(-)}$ amplitude somewhere back in our memory and turn our attention from now on to the usual moving Regge poles.

6. DUALITY AND ALL THAT

We saw that the FESR give us connections between the behaviour in the low energy region and the behaviour in the high energy region of a scattering amplitude. The high energy region of the amplitude is usually expanded in terms of Regge poles. These are poles in the t channel and we observe them here in the region of t around zero. Therefore one could say that the FESR connect the low s channel behaviour with the t channel effects. This connection between the s channelaand the t channel has become known as Duality [9].

One can give a much stronger and more speculative definition of this duality if one notes that the low energy region in the s channel might be dominated by resonances. In fig. 3 we clearly see the effects of resonances in building up the amplitude Im $A'^{(-)}$ at low energy. Explicit calculations show that using resonances only or using phase shifts for πN CEX is equivalent at least in the region up to 1 BeV. Therefore one can say that the calculations for πN CEX given us a connection between the resonance structure and the Regge structure; so, instead of speaking about duality between the s and t channels, one can talk about duality between resonances and Regge.

The duality between the s and the t channel is a general consequence of the FESR and therefore this mild definition is generally true. However the duality between resonances and Regge, checked here in the case of πN CEX, is of a much more speculative nature, and we have to keep this point in mind. The most enthusiastic boost to this line of thought is given by the cooperative effect of all the resonances in building up the zero in the $B^{(-)}$ amplitude. We have already noted, while discussing fig. 4, that each resonance separately vanishes almost at the right point. Therefore one could even expect something stronger, namely, that each resonance separately incorporates some of the Regge features. This was particularly stressed by Schmid [10] who showed that taking the ρ Regge amplitude by itself and projecting out of it the various partial waves, one finds circles in the Argand diagram looking similar to those expected from resonances. If we define a resonance as a pole in s then we cannot expect that a partial wave integration of the Regge amplitude will give the complete behaviour of resonaces simply because this partial wave integration does not have a pole in s. Nevertheless we might expect that some main features still remain. It is of course an open question whether the particular behaviour seen for the amplitude that we looked at should hold for any sattering amplitude we wish. That is a weak point and we will return to it later. Let us just say that like in other places in theoretical particle physics, from some point on, we have to make some speculation and hope for the best.

Another word of caution is due here. We considered as an example πN scattering. This process has a definite symmetry on exchanging $s \leftrightarrow u$ because we get back to $\pi N \to \pi N$. This is because π is an antiparticle of itself. If, on the other hand, we consider $\bar{K}N \to \bar{K}N$, the $s \to u$ crossing leads to $KN \to KN$ which has a completely different resonance structure. Here the physical scattering amplitude G has a cut in s determined by the $\bar{K}N$ system, and a cut in u determined by the KN system. In writing down the FESR we construct first $G^{(+)}$ and $G^{(-)}$ which now have a RHC that is the sum or the difference of the two absorptive parts respectively. Now the concept of duality has in principle to apply to a correspondence not between low s on one hand and low t on the other, but rather between low s and low u structure on one hand and the low t structure on the other.

Nevertheless, since many leading Regge poles show exchange degeneracy, namely, equal values of $\alpha(t)$ and $\beta(t)$ for two Regge poles of opposite signature (like ρ and A_2) one can go back to an s-t duality. Let us see that explicitly on the following example. Consider the scattering amplitude G describing $K^- p \to \bar{K}^o n$ in the s-channel. Its LHC will be given by the imaginary part measured in the crossed process $K^o p \to K^+ n$. The sums and the differences of these two imaginary parts will correspond to the RHC of $G^{(+)}$ and $G^{(-)}$ respectively. + and - correspond here to positive and negative charge conjugation as well as signature (the A_2 and the ρ Regge poles respectively). However, since the imaginary part of $K^op \to K^+n$ is so much smaller than that of $K^-p \to \bar{K}^on$ (at t=O compare $\sigma_T(K^+p) - \sigma_T(K^+n)$ with $\sigma_T(K^-p) - \sigma_T(K^-n)$), the RHC of of $G^{(+)}$ and $G^{(-)}$ are essentially equal. That means that $g^{(+)}$ and $g^{(-)}$ are equal and so are their Regge expansions. This is exchange degeneracy. Here we can then say that since $Im(K^op \to K^+n) \approx O$ a simple s-t duality holds.

In terms of the resonance approximation one can simply blame the smallness of $Im(K^op \to K^+n)$ on the absence of resonances of exotic quantum numbers (Y = 2). So one could say that the assumption of "resonances only" plus s-t duality works well in this case. Nevertheless there is still a case where trouble occurs : if we were to consider $K^-p \to K^-p$ in the s-channel with $K^+p \to K^+p$ in the u channel we realize that although resonances are absent from K^+p it still has the regular Pomeranchukon contribution. Now if we were to apply the arguments of the former paragraph and rely on resonance saturation we would be led to the conclusion that $K^+p \to K^+p$ is purely real. It is therefore clear that saturation by resonances leads us into trouble. Harari suggested [11], as a way out of this problem, to assume that the nonresonating background that we have to add is connected via the FESR to the Pomeranchukon contribution only. This connection was not rigorously proved, nevertheless it is consistent with the various calculations.

Fig. 2 can serve as a mathematical example of the duality. It is an inverted plot of Im $\psi(z)$ over the z complex plane.

$$\psi(z) = \frac{\Gamma'(z)}{\Gamma\ (z)} = -\ \gamma - \sum_{n=o}^{\infty} \left(\frac{1}{z+n} - \frac{1}{n+1}\right) \qquad (12)$$

$\psi(z)$ is a real analytic function and its singularities are given by simple poles at every non positive integer.

For large z, ψ can be expanded as

$$\psi(z) \rightarrow \log z - \frac{1}{2z} - \frac{1}{12z^2} \cdots \qquad (13)$$

Going along any ray in the complex plane we can see how the wiggling structure of Im $\psi(z)$ dissolves into the asymptotic behaviour. Fig. 2 was constructed by cutting out of the function an angular section close to the real axis. This gives the illusion of a cut Riemann sheet. In other words one could say that in order to reproduce something similar to a scattering amplitude one has simply to perform a conformal mapping on that function such that a sector enclosing the poles is mapped onto the real axis. The resulting function is a real analytic function in a cut plane dominated by poles (on the second sheet) and shows a violent structure near the origin which dissolves gradually into a smooth behaviour.

Duality as described above becomes especially useful in bootstrap calculations. In the previous sections we have used the finite energy sum rules as a tool for checking Regge parameters with low energy scattering data. If one believes that the low energy data is dominated by a certain pattern of resonances then one can use Regge as input and determine, say, the couplings of these resonances. Alternatively one could use a certain pattern of resonances and couplings as an input for some arbitrary scattering amplitude and determine the resulting Regge behaviour. You can easily see the ideal bootstrap program we are aiming at : One can ask for an over-all consistent solution of even an infinite set of FESR in which the same type of Regge trajectories are taken as an input into the imaginary part via their low energy resonances as the Regge trajectories that are used for describing the high energy behaviour.

Bootstrap programs with FESR are much simpler than

older bootstrap programs with N/D equations. The formulation of the problem with FESR leads to algebraic equations in the variables that we are looking for, namely, algebraic equations for the trajectories and for their coupling constants. That has to be compared with the integral and differential equations that appear in bootstrap formulations in different theories. This is the reason that now, only a year or so after it became fashionable to play with FESR, we already have a solution for such a bootstrap problem. The solution is an explicit function in which resonances and Regge live happily and consistently together.

It may be claimed that the assumption that an arbitrary scattering amplitude is dominated by resonances only (even with Harari's ansatz) is too severe. That might certainly be the case. The important point to understand here is that we are looking for the systematics that lie behind that assumption. After we find a solution and we understand its systematics we can use it as some first order, or perhaps zero order approximation for a physical scattering amplitude. It has to be realized that there are really two different problems here. One is the consistency between the s and the t behaviour and the other is the consistency between the internal degrees of freedom namely, the SU(2) and SU(3) indices of the external as well as the exchanged particles. It will therefore seem advantageous to consider first a problem in which we get rid of the question of the internal degrees of freedom. Such a problem is given for instance by the scattering amplitude $\pi\pi \to \pi\omega$. This reaction, as well as its generalizations, was considered in a series of papers by Ademollo et al. [12]. It is described by one single scattering amplitude and includes only the ρ trajectory in either the s, t or u channel. This scattering process, that is so far from experimental feasibility, may serve as a good theoretical laboratory because of its unique internal quantum numbers. After working some time on the problem these authors were led to an asymptotic form

$$A \to \frac{\bar{\beta}}{\pi}\,\Gamma(1-\alpha(t))\,(-\alpha(s))^{\alpha(t)-1} + (s \to u)$$

which gives the usual $\nu^{\alpha(t)-1}$ for linear trajectories. That led Veneziano [13] to a successful guess of a closed form for the scattering amplitude of that process which is valid for all s and t values and is completely symmetric in s, t and u. Veneziano's formula is

$$A = \frac{\bar{\beta}}{\pi}\left[\frac{\Gamma(1-\alpha(s))\,\Gamma(1-\alpha(t))}{\Gamma(2-\alpha(s)-\alpha(t))} + (s \to u) + (t \to u)\right] \qquad (14)$$

$$\alpha(s) = a + bs$$

It is successful because it shows the expected behaviour in all regions of s and t.

7. THE VENEZIANO FORMULA

Let us look at the term

$$f(s,t) = \frac{\Gamma(1-\alpha(s))\,\Gamma(1-\alpha(t))}{\Gamma(2-\alpha(s)-\alpha(t))}$$

Two important properties of the function $\Gamma(z)$ are that it has poles for every nonpositive integer and no zeros at all. That means that $1/\Gamma(z)$ is analytic. Therefore the term that we are looking at has poles wherever $\alpha(s)$ or $\alpha(t)$ pass through positive integers, as expected in Regge theory. Since $\alpha(s)$ is real we find narrow-width resonances (real poles) in the physical region. One way to avoid it is to make a conformal mapping of f(s,t) in the complex s plane similar to the treatment of $\psi(z)$ in fig. 2. Then we can look also at the asymptotic behaviour when $s \to \infty$

and t remains fixed and find the expected form $\Gamma(1-\alpha(t))\times$ $\times(-\alpha(s))^{\alpha(t)-1}\cdot f(u,t)$ will give the symmetric u behaviour so that one finds the right ν behaviour with the right signature, and f(s,u) vanishes asymptotically. Let us remark here that an alternative view that can be found in the literature is to add to $\alpha(s)$ an imaginary part growing with s. That will lead to similar results as far as the single f(s,t) is concerned, however now α is no longer a real function of its variable and one finds that various poles of the total amplitude (14) develop rising imaginary parts in their residues. Therefore we prefer the interpretation provided by the conformal mapping, applied to the total amplitude.

A pole at $\alpha(t) = n$ does not correspond to a particle of spin n only. It turns out that it also implies the existence of all the particles with spins lower than n (the daughters). This can be simply seen by making a partial wave decomposition of the explicit $\cos\theta_s$ or t function at the pole in $\alpha(s) = n$. It turns out that the condition $\alpha(s) + \alpha(t) + \alpha(u) = 2$ kills all the odd daughters in eq. (14). An example of such an "odd daughter" is a 1^{--} ρ-like particle at $\alpha_\rho(s) = 2$, namely near the A_2 mass. Such a particle is not seen, and that is a good motivation for requiring this equality. On the other hand $\alpha(s) + \alpha(t) + \alpha(u) = 3a+b(3m_\pi^2+m_\omega^2)$. Therefore if the equality holds then the formula of Veneziano, eq. (14), cannot hold for different external masses, when the external π or ω is replaced by its Regge recurrence. This is part of the basic problem of including all external particles in the theory, which has not yet been solved.

The introduction of internal degrees of freedom will allow us to generalize the Veneziano formula to an arbitrary four point function. That can be easily achieved within the quark model classification of particles. One has to attach a term of the kind

$\Gamma(n_1-\alpha_1(t))\Gamma(n_2-\alpha_2(s))/\Gamma(m-\alpha_1(t)-\alpha_2(s))$ to an internal symmetry combination that has an appropriate structure. (The various factors n_1, n_2 and m regulate the appearance of the first poles and the asymptotic behaviour of the amplitude).

The appropriate symmetry structure is provided by the "duality diagram" 14 of fig. 6. The various lines

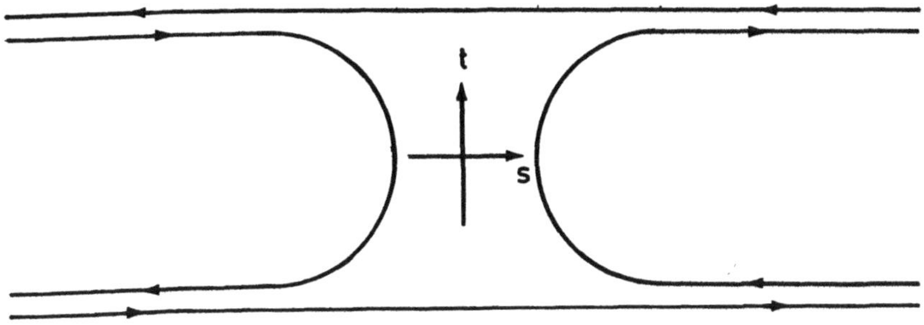

Figure 6 - Duality diagram for meson-meson scattering in the quark model.

represent different quarks and thus build the four external particles (mesons in this picture) in question. We see that the topological structure of this diagram is suitable for that type of function considered here. The identity of the quark-combination found in the t-direction will tell us the quantum numbers of the particles lying on $\alpha_1(t)$, and similarly we will find in the s-direction the interpretation of $\alpha_2(s)$. This works also for meson-baryon scattering, however troubles are encountered for baryon-baryon scattering (appearance of exotic quantum numbers).

If we apply that technique to, say, $K^+p \to K^+p$ we find that only a u-t diagram is possible. That means that for high s we have a purely real function going like $s^{\alpha(t)}$. Thus we find an exchange-degenerate ρ-A_2 as well as ω-P' combination and the Pomeranchukon is missing. We see that the Pomeranchuk trajectory is not automatically included in the solution that was found. This is in accord with our discussion of Harari's ansatz in the previous section. Thus, in the present state of the game, the Pomeranchukon has to be "added by hand" to the amplitude, blaming the approximation of "resonances only" for its disappearance.

Before closing this section let us still devote some time to the Veneziano formula for $\pi\pi$ elastic scattering. This being an elastic scattering amplitude provides us with an additional strong condition, namely, that the residues of all the poles in the various partial waves be positive. One describes the scattering amplitude by one function

$$a(s,t) = -c\,\frac{\Gamma(1-\alpha(s))\,\Gamma(1-\alpha(t))}{\Gamma(1-\alpha(s)-\alpha(t))}$$

such that the three isostates are

$$A^{o} = \frac{3}{2}[a(s,t)+a(s,u)] - \frac{1}{2}a(t,u)$$

$$A^{1} = a(s,t) - a(s,u)$$

$$A^{2} = a(t,u)$$

α represents here an exchange degenerate trajectory of ρ and f^{o}. This representation of the $\pi\pi$ amplitude was used by Lovelace [15] to show that it obeys Adler's consistency condition of $\alpha(o) \approx 1/2$. It turns out that if the external pion masses vanish, then miraculously enough, the residues of the poles become positive [16].

This might be the first hint that the simple Veneziano formula might include many elements of truth, much more than expected at first sight. The inclusion of unitary into the game and, with it, the scattering amplitude for high resonances, is still an open field.

8. CONCLUDING REMARKS

The purpose of this talk was to derive the FESR, study them for certain physical amplitudes and see their relevance and implications to recent bootstrap programs. We did not have time to summarize all work done in this rapidly growing field, and let us mention some of the topics we skipped.

We saw that FESR can be used to check Regge parameters by the use of the best available data for low energy scattering amplitudes. We discussed πN CEX, however much work was done also on the other πN amplitudes, on π photoproduction, on KN reactions etc. In most cases the question at hand was the behaviour of Regge trajectories at crucial points (like a non-sense point or a conspiracy point).

We mentioned only briefly the Schmid circles in the Argand diagram. That was a subject for endless discussions last year, mainly showing the inability to define in an unequivocal testable way what a resonance is. Applications of these ideas led to predictions of exchange degeneracies between trajectories which follow also from adopting the Veneziano formula to various processes. In the last section we presented only those results of that formula that were mostly relevant to other parts of this talk.

Let us conclude with some general remarks. The bootstrap program does not have a unique answer (at least in the present stage when unitarity is not included). Various modifications and sums of the Veneziano formula were already suggested. It is not clear yet in what sense such a formula is an approximation to the scattering amplitude. The predictions are of the kind that various trajectories have equal slopes and have definite regularities in their intercepts. However the Pomeranchuk trajectory as well as all odd effects like the fixed poles of $B^{(-)}$ that we discussed, are not in yet. This is still mainly a speculative game and is therefore open for new ideas. On the other hand, the use of FESR for Regge analysis is nowadays an accepted and widely used tool, and will presumably stay like that for years to come. It helps us to find out how good the Regge representation is, and will hopefully lead us to better descriptions of the scattering amplitudes disclosing their true structure.

FOOTNOTES AND REFERENCES

1. D. HORN and C. SCHMID, CALT-68-127 (1967), unpublished. A general analysis of theory and applications which we follow quite closely here is given by R. DOLEN, D. HORN and C. SCHMID, Phys. Rev. 166, 1768 (1968).
2. Y. C. LIU and S. OKUBO, Phys. Rev. Lett. 19, 190 (1967); V. BARGER and R. J. N. PHILLIPS, Phys. Lett. 26B, 730 (1968).
3. Note that an odd (even) amplitude has an odd (even) real part and an even (odd) imaginary part. Therefore the right-signature FESR, that can be derived by integrations from -N to N are even (odd) respectively. The wrong signature FESR lead to Schwarz superconvergence relations; J. H. SCHWARZ, Phys. Rev. 159, 1269 (1967).

4. A. LOGUNOV, L. D. SOLOVIEV and A. N. TAVKHELIDZE, Phys. Lett. 24B, 181 (1967); K. IGI and S. MATSUDA, Phys. Rev. Lett. 18, 625 (1967); D. HORN and C.SCHMID, ref. 1. These papers were based on a method introduced and applied by K. IGI, Phys. Rev. Lett. 9,76 (1962) to an investigation of the Pomeranchukon.
5. We use the notation of V. SINGH, Phys. Rev. 129, 1889 (1963) and G. F. CHEW, M. L. GOLDBERGER, F. E. LOW and Y. NAMBU, Phys. Rev. 106, 1337 (1957).
6. R. DOLEN, D. HORN and C. SCHMID, Phys. Rev. Lett. 19, 402 (1967); Phys. Rev. 166,1768 (1968).
7. A. DONNACHIE, R. G. KIRSOPP and C. LOVELACE, Phys. Lett. 26B, 161 (1968).
8. R. AVIV and D. HORN, Phys. Rev. Lett. 21, 704 (1968).
9. G. F. CHEW and A. PIGNOTTI, Phys. Rev. Lett. 20, 1078 (1968); G. F. CHEW, Comments on Nuclear and Particle Physics, 2, 74 (1968).
10. C. SCHMID, Phys. Rev. Lett.20, 689 (1968).
11. H. HARARI, Phys. Rev. Lett. 20,1395 (1968).
12. M.ADEMOLLO, H. R. RUBINSTEIN, G. VENEZIANO and M. A. VIRASORO, Phys. Rev. Lett. 19, 1402 (1962); Phys. Lett. 27B, 99 (1968); Phys. Rev.176, 1904 (1968).
13. G. VENEZIANO, Nuovo Cim. 57A, 190 (1968).
14. J. L. ROSNER, Phys. Rev. Lett. 22, 689 (1969), H. HARARI, Phys. Rev. Lett. 22, 562 (1969).
15. G. LOVELACE, Phys. Lett. 28B, 264 (1968).
16. J. SHAPIRO and J. YELLIN, preprint 1968; F. DRAGO and S. MATSUDA, preprint 1969.

MODIFIED QUARK MODELS*†

By

D. HORN

Physics Department, Tel-Aviv University,
Tel-Aviv, Israel

We discuss the implications of the splitting of the A_2 resonance. It leads us to theories where either a 1^{-+} or another 2^{++} resonance are present. The 1^{-+} is included in the Gell-Mann-Zweig (4) scheme that we describe in some detail. An additional 2^{++} can be provided by many models, one of which - the giant quark model - is shortly discussed.

1. INTRODUCTION

The usual quark model [1] is a simple scheme that predicts in a straight forward way mass formulas, decay and even scattering amplitudes of the hadrons. Moreover, a simple quark field theory leads to the well-known current algebra. The model might be viewed as a simple parametrization of the data in terms of some underlying pattern given by the quarks. These quarks are not real particles - they are just a handy abstract tool. The mesons are de-

† Seminar given at the VIII. Internationale Universitätswochen für Kernphysik, Schladming, Feb. 24 - March 8, 1969.

* Work sponsored by the Air Force Office of Scientific Research, Office of Aerospace Research, United States Air Force, under AFOSR grant number EOOAR-68-0010, through the European Office of Aerospace Research.

scribed by a $q\bar{q}$ pair with an arbitrary angular momentum, and the baryons by qqq. Although there may be doubts with respect to the classifications of various resonances, it was not until recently that this model got into trouble. We refer of course to the splitting of the A_2 resonance [2]. This splitting means that there are either two close 2^{++} states (notation: J^{PC}) or, perhaps one 2^{++} with a near by 1^{-+}. If the first case is the correct one then our problem is to find two such close 2^{++} states, and that does not come out naturally in the usual quark model. The alternative is even worse, since there is no 1^{-+} state in the quark model! That means that we either have to modify the quark model or, perhaps to give up the quark game.

People sometimes argue that it is after all only one failure of the quark model and why worry so much and invent new complicated models to account for one event. The answer is simple: It is enough to have one counter-example to destroy a theory. Since we see a strong and definite effect it cannot be waved away as a small correction. And now, if we don't want to give up a simple substructure we should better find a systematic modification of the usual quark model. In the present lecture we will discuss the question of meson classification only.

2. THE A_2 DOUBLET

Since the splitting of the A_2 is the experimental issue that brought about this discussion let us start by describing it. The Rosenfeld tables tell us that there exists an A_{2L} at 1270 MeV with a width of 26 ± 7 and A_{2H} at 1315 MeV with a width of 29 ± 7 . The splitting was first observed in missing mass experiments. Bubble chamber experiments observe two peaks in the $\rho\pi$ mode and a similar situation is indicated in $\eta\pi$. There are conflicting reports with respect to the $K\bar{K}$ mode. The quantum numbers of the possible resonances are determined by the observed decays. A list of the relevant modes is given in table 1.

	$\pi\rho$	$\pi\eta$	$K\bar{K}$
2^{++}	+	+	+
$0^{-+}, 1^{++}, 2^{-+}$	+	-	-
1^{-+}	+	+	-
0^{++}	-	+	+

Table 1 - Various allowed (+) and forbidden (-) decay modes of even C candidates (I=1).

If both $\pi\rho$ and $\pi\eta$ modes are seen for the two peaks it is an indication of either two 2^{++} or a combination of 2^{++} and 1^{-+}. However, a 1^{-+} would not decay into $K\bar{K}$. It is interesting to note that different experimental groups got different results with respect to the $K\bar{K}$ mode. This might be the manifestation of a 0^{++} that would not be produced peripherally ($\pi N \rightarrow A_2 N$ via ρ exchange), but would appear in an annihilation process ($p\bar{p}$). Recently, Lipkin and Meshkov [3] pointed out that in the SU(3) limit the 1^{-+} decay into $\pi\eta$ is forbidden. That seems like a theoretical argument in favour of the two 2^{++} assign - ment. There exists also an important experimental reason for favouring this choice - the CERN missing mass experiment [2] shows such a sharp dip between the two A_2 peaks that it seems to come from an interference of two Breit-Wigner resonances. This interference can take place only if the two have the same quantum numbers. Let us point out here that such an interference can rise in either one of two ways: A . Two slightly separated resonances with destructive interference between them. B. Two coinciding resonances, a wide and a narrow one, interfering destructively. If the first possibility is the right one then we expect to find a similar structure for the whole nonet of 2^{++} resonances from SU(3) considerations. However in the second case it might happen that a shift between the two interfering resonances occurs for other members of the nonet. Then we should find a big bump with a small one nearby and that latter one might be unnoticeable in certain cases.

3. THE 1^{-+} QUESTION

Let us first concentrate on the possibility that a 1^{-+} state was observed here. We note that such a state does not occur in the quark model. If one builds all mesons out of a $q\bar{q}$ structure with arbitrary L, then the parity of the state is given by $P=(-1)^{L+1}$ and the charge conjugation $C=(-1)^{L+S}$. Thus for S=0 we find $P=-C=(-1)^{L+1}$ and, for S=1, $P=C=(-1)^{L+1}$. The states that are not included in this structure are 0^{--} and the series 0^{+-}, 1^{-+}, 2^{+-}, 3^{-+} etc. This is connected with the fact that we have no 1^{-+} generator in $[U(6)\times U(6)]_{\beta} \times O(3)_{L}$. The generators can be written in terms of the quark fields as $\int\psi^{+}\Gamma\psi\, d^3x$ (we omit the SU(3) λ matrices). We find that $\Gamma=1,\beta,\vec{\sigma},\beta\vec{\sigma}$ have the quantum numbers 0^{+-}, 0^{++}, 1^{++} and 1^{+-} respectively. Both $\vec{S}$, $\vec{L}$ and $\vec{J}$ are 1^{++} operators. The 1^{+-} one is an operator connecting different C parts of one representation (e.g. the pseudo-scalars and the vector mesons in $(\underset{\sim}{6},\bar{\underset{\sim}{6}})$). On the other hand we know a very important 1^{-+} operator in physics - the boost of the Lorentz group. Can there be any connection between the boost and internal symmetry? A connection can be given by the Toller symmetry: For elastic scattering of pairwise equal mass particles one finds an SL(2,C) symmetry at the point t=0. Since there exists here a 1^{-+} step operator connecting adjacent Regge poles in a single Toller pole, one might wonder whether one should include also an internal 1^{-+} symmetry operator in the structure of the hadrons. After all, the particles are manifestations of these Regge poles. (It might of course happen that the "odd daughters" do not produce particles and then this argument does not follow). The Gell-Mann-Zweig model [4] does exactly that - it adds a 1^{-+} operator to the internal symmetry group.

4. THE GELL-MANN-ZWEIG O(4) MODEL

To the 1^{++} angular momentum generator we add another 1^{-+} operator such that they form together an O(4) algebra. That can be done separately for the $\vec{S}$ and $\vec{L}$. Let us denote the 1^{-+} operators by $\vec{S}'$ and $\vec{L}'$ respectively. Then we can define

$$\vec{s}_1 = \frac{1}{2}(\vec{S} + \vec{S}') \qquad \vec{s}_2 = \frac{1}{2}(\vec{S} - \vec{S}')$$

$$\vec{\ell}_1 = \frac{1}{2}(\vec{L} + \vec{L}') \qquad \vec{\ell}_2 = \frac{1}{2}(\vec{L} - \vec{L}')$$

where

$$[\vec{L},\vec{S}] = [\vec{L}',\vec{S}'] = 0 \ .$$

By definition we know that under parity $P\vec{S}P^{-1} = \vec{S}$, $P\vec{S}'P^{-1} = -\vec{S}'$ etc. Therefore $P\vec{s}_1P^{-1} = \vec{s}_2$, $P\vec{s}_2P^{-1} = \vec{s}_1$ and $C\vec{s}_1C^{-1} = \vec{s}_1$, $C\vec{s}_2C^{-1} = \vec{s}_2$. Similar results hold for the $\vec{\ell}$ operators.

The classification of states will be given by (s_1,s_2) (ℓ_1,ℓ_2) S and L. Let us look at a structure of an (s_1,s_2) family. It includes spin values from $S = s_1+s_2$ to $S = |s_1-s_2|$. In a usual quark model we have two S values in a $q\bar{q}$ system, namely, S=0 and S=1. The first one will be given by $s_1=s_2=0$ and the second one will be the highest member of $s_1=s_2=\frac{1}{2}$, which includes another S=O state with opposite parity. The story of the L is different. Here we allow arbitrary high states. However, we realize that to any representation (ℓ_1,ℓ_2) there corresponds an (ℓ_2,ℓ_1) reached by the parity operation. These two form a parity doublet. In order to avoid these states we require $\vec{L}.\vec{L}'=0$, which means $\ell_1=\ell_2$.

Let us now find what the P and C quantum numbers of the various states are. We look first at the L ground state $\ell_1=\ell_2=L=0$. Here we find the usual 0^{-+} in $s_1=s_2=0$ and a 1^{--} and 0^{+-} in $s_1=s_2=\frac{1}{2}$. This 0^{+-} is a new additional state. We can therefore specify $P=(-1)^{1+s_1+s_2-S}$ and $C=(-1)^{s_1+s_2}$. The fact that P changes by one unit within the O(4) multiplet and C remains the same comes of course from the 1^{-+} nature of the S lowering operator. Now we can include the ℓ-states. In the first excitation we find a 1^{--} contribution for the usual P and C of $(-1)^L$. In addition there is a 0^{+-} contribution. Similarly for every $\ell_1=\ell_2=\frac{L}{2}$ we find a $P=C=(-1)^L$ contribution changing by steps of P=-1,C=1 in the multiplet. Hence we can finally write $P=(-1)^{1+L+s_1+s_2-S}$, $C=(-1)^{\ell_1+\ell_2+s_1+s_2}$.

All these states can be embedded in one representation of a spectrum generating algebra. Thus the simple L excitations in the usual quark model can be embedded in an O(1,3) that is built out of a 1^{++} $\vec{L}$-operator and a 1^{--} non-compact raising and lowering operator. To these we add now the compact 1^{-+} $\vec{L}'$ operator and build an O(1,4). The generators of O(1,4) include an additional non-compact 0^{+-} that comes from the commutation relations of the 1^{--} with $\vec{L}'$. The relevant representation of O(1,4) includes the (ℓ_1,ℓ_2) states (0,0), $(\frac{1}{2},\frac{1}{2})$, (1,1) etc., with the L^{PC} decomposition shown in fig. 1 .

On the tower of states in fig. 1 we have now to superimpose the 0^{-+} of $s_1=s_2=0$ as well as the 1^{--} and 0^{+-} of $s_1=s_2=\frac{1}{2}$. This will then give us the rich structure provided by this model. In particular looking at the $\ell_1=\ell_2=\frac{1}{2}$, $s_1=s_2=\frac{1}{2}$ we find the old members of the $q\bar{q}$ structure - 2^{++}, 1^{++}, 0^{++} - together with the new two 1^{-+} states and one 0^{++}. If we would not have generalized the spin part into O(4) (and we are free to play the game also

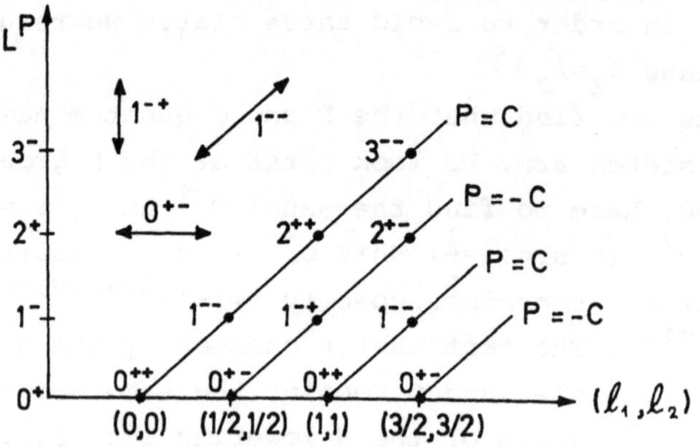

Figure 1 - The O(1,4) representation including the expected L^{PC} excitations.

this way) then we should consider only the 1^{--} contribution from the spin. That would lead to just one new 1^{-+} state.

5. 2^{++} DOUBLETS

After we saw the O(4), that accommodates 1^{-+} states let us look at the alternative of having a 2^{++} doublet. An additional 2^{++} can come in the quark model from a $q\bar{q}$ L=3 or a $qq\bar{q}\bar{q}$ configuration. We would however expect the L=3 mode to lie higher and the $qq\bar{q}\bar{q}$ explanation has the disadvantage of involving mesons with "exotic" quantum numbers. It might be that this will after all be correct if we attribute to these states weaker couplings in accord with explanation B in the end of the second section.

Let us however discuss shortly another possible speculation that leaves the $q\bar{q}$ SU(3) structure untouched. That is given by the giant quark model [5].

The fact that guides our intuition here is that a solution to current algebra [6] can be given by a simple quark with all possible angular momentum states. Let us therefore introduce such a basic quark with internal angular momentum $\vec{K}$, as the basic building block. It will have all possible J^P states $\frac{1}{2}^+$, $\frac{3}{2}^-$, $\frac{1}{2}^-$, $\frac{5}{2}^+$, $\frac{3}{2}^+$ etc. We might expect that the lowest meson states are given by the usual $q_{\frac{1}{2}+}\ \bar{q}_{\frac{1}{2}-}$ combination. The higher states $q_{\frac{1}{2}+}\ \bar{q}_{\frac{1}{2}-}$, L=1 can be now accompanied by $q_{\frac{1}{2}+}\ \bar{q}_{\frac{3}{2}+} \pm q_{\frac{3}{2}-}\ \bar{q}_{\frac{1}{2}-}$ combinations. That will introduce new 2^{++}, 2^{+-}, 1^{++} and 1^{+-} states. The 2^{++} could serve as one of the partners in the doublet, however then also the other new states have to be found. Here we implicitly assumed that the mass of a configuration q_{J_1}, $\bar{q}_{J_2}$ L is roughly proportional to L and K with a splitting such that $J=K+\frac{1}{2}$ would lie lower than $J=K-\frac{1}{2}$. Note that whenever $J_1 \neq J_2$ we have doublets in charge conjugation and that introduced the unconventional additional 2^{+-} state. Obviously we expect also the $q_{\frac{1}{2}+}\ \bar{q}_{\frac{1}{2}+} \pm q_{\frac{1}{2}-}\ \bar{q}_{\frac{1}{2}-}$ to lie close by. These are additional 1^+ and 0^- states with the two possible C values. If one accepts alternative B of section 2 then it might be that all the new states are characterized by smaller widths than the conventional quark model ones.

On the basis of the various S,K and L operators one can build also a decay symmetry in addition to the spectrum symmetry described until now. That can be simply achieved by using the usual W-spin operators defined as $W_\pm = S_\pm^q - S_\pm^{\bar{q}}$ and $W_z = S_z^q + S_z^{\bar{q}}$. In addition one has to require the conservation of L_z as well as $K_z = K_z^q + K_z^{\bar{q}}$ for decays in the z-direction . The predictions obtained from a decay symmetry can in principle be used as a guide for the classification of the particles.

In conclusion we might say that the splitting of the A_2 points towards a new type of structure in hadron spectroscopy. We might expect other such doublets to occur, in particular for the corresponding strange mesons, and look for a much richer spectrum than expected from L excitation of the simple quark model. What the details of that structure are - that is the interesting question that we discussed here. Many other speculations are possible and future will tell which, if any, is correct.

The author would like to thank Prof. Y. Ne'eman for many fruitful discussions about the algebraic structure of the Gell-Mann-Zweig model. The discussion in section 4 represents the author's interpretation of this unpublished model.

REFERENCES

1. M. Gell-Mann, Physics Letters 8, 214 (1964); G.ZWEIG, CERN preprint TH 402 (1964) unpublished.
2. See e.g. the review of B. FRENCH in 14th Intern. Conference on High Energy Physics, Vienna 1968, p.91; H. BENZ et al., Phys. Letters 28B, 233 (1968).
3. J. H. LIPKIN and S. MESHKOV, Phys. Rev. Lett. 22, 212 (1969); C. A. LEVINSON, H. J. LIPKIN and S. MESHKOV, Nuovo Cim. 32, 1376 (1964).
4. M. Gell-MANN and G. ZWEIG, unpublished; A review is contained in the Rapporteur talk of H. HARARI in 14th Int. Conf. on High Energy Physics, Vienna 1968,p.195.
5. D. Horn, Nuovo Cimento, in press.
6. R. DASHEN and M. GELL-MANN, Phys. Rev. Lett. 17, 371 (1966). The solution for a degenerate mass giant quark

can be found there. The general solution is discussed in M. GELL-MANN, D. HORN and J. WEYERS, Proceedings of the Heidelberg International Conference on Elementary Particles 1967, p.479.

HAMILTONIAN APPROACH TO QUANTUM FIELD THEORY†

By

John R. KLAUDER

Bell Telephone Laboratories, Incorporated
Murray Hill, New Jersey

1. INTRODUCTION

In these notes I should like to discuss some general features of a Hamiltonian quantum field theory, as well as to present some recent results regarding existence questions for certain model field theories. Especially exciting among these recent results are those pretaining to a rigorous Hamiltonian formulation of a relativistic scalar field with quartic self-interaction in a two-dimensional space time [1]. These promising results clearly make the study of model theories attractive, and their study as Hamiltonian theories especially so.

We begin with a brief reminder of free field properties, and a heuristic discussion of problems that arise in extending these concepts to theories with interesting interactions. To gain suitable perspective we next deal with a number of general features that are expected to accompany any Hamiltonian theory. The great progress that has been made over previous analyses has been made possible by a heavy use of modern methods of functional analysis. While we cannot hope to reproduce the

† Lecture given at the VIII. Internationale Universitätswochen für Kernphysik, Schladming, 24 Feb.- 8 March, 1969.

intricate steps involved, we shall present a review of some aspects of operator theory so as to make understandable the content of the theorems that have been proven. Finally we discuss some of the known features of the models themselves.

A. Free Scalar Field Theory

As is well known the quantum theory of the free scalar field may be reduced to a collection of independent harmonic oscillator problems by the introduction of fields defined on momentum space. To begin with we may introduce the field of annihilation and creation operators, $a(\vec{k})$ and $a^{\dagger}(\vec{k})$, respectively. They fulfill the commutation relation

$$[a(\vec{k}),\ a^{\dagger}(\vec{k}')] = \delta(\vec{k}-\vec{k}') \ . \tag{1}$$

The number density operator $a^{\dagger}(\vec{k})\ a(\vec{k})$ is positive and the Hamiltonian is a weighted linear superposition which reads

$$\mathcal{H} = \int \omega(k)\ a^{\dagger}(\vec{k})\ a(\vec{k})\ d\vec{k} \ . \tag{2}$$

In the relativistic case

$$\omega(k) = (k^2+m^2)^{1/2} \quad ,$$

although we may imagine more general functions such that $\omega(k) \geq m > 0$. The existence of $\mathcal{H}$, demanded on physical grounds ensures the existence of the number operator $N = \int a^{\dagger}(\vec{k})a(\vec{k})\ d\vec{k}$ since clearly

$$0 \leq \langle\Psi|N|\Psi\rangle \leq m^{-1}\langle\Psi|\mathcal{H}|\Psi\rangle \quad .$$

The existence of a ground state $|0\rangle$ of zero energy, $\mathcal{H}|0\rangle = 0$, again plausible on physical grounds ensures that $N|0\rangle = 0$ and moreover that

$$a(\vec{k})\,|0\rangle = 0 \tag{3}$$

for all $\vec{k}$. If we further assert the "completeness" of the operators $a(\vec{k})$ and $a^{\dagger}(\vec{k})$, corresponding to the physically plausible remark that these are all the degrees of freedom of the system, we have essentially formulated the quantum theory of the free field. The completeness assumption is formalized by the statement that every operator is a function of the a and $a^{\dagger}$ operators, or equivalently only multiples of unity commute with all a and $a^{\dagger}$. In turn these properties are equivalent to the assumption that apart from a numerical factor $|0\rangle$ is the only vector which satisfies (3), which can be viewed as a consequence of the physical requirement of a unique ground state. We have intentionally drawn our basic assumptions from physically motivated notions.

Fock Space Formulation

It is common to realize the free field Hamiltonian, the creation and annihilation operators, etc., as transformations on Fock space $\mathfrak{F}$. It is important to remark that there is no loss in generality involved in this realization for it is known under the foregoing assumptions that all such representations are unitarily equivalent to that on Fock space. In forming Fock space we deal with symmetric, square integrable, momentum space functions of n variables, $\phi_n(\vec{k}_1,\ldots,\vec{k}_n)$. For $n = 0$, $\phi_o \in C$, the space of complex numbers. The vectors $\Phi \in \mathfrak{F}$ are represented as sequences,

$$\Phi = \bigoplus_{n=o}^{\infty} \phi_n = \{\phi_o, \phi_1(\vec{k}_1), \phi_2(\vec{k}_1,\vec{k}_2), \ldots\} ,$$

where

$$(\Phi,\Phi) = ||\Phi||^2 = \sum_{n=o}^{\infty} ||\phi_n||^2 = |\phi_o|^2 +$$

$$+ \sum_{n=1}^{\infty} \int |\phi(\vec{k}_1,\ldots,\vec{k}_n)|^2 d\vec{k}_1 \ldots d\vec{k}_n \quad .$$

The creation and annihilation operators are realized through the relations

$$(a(\vec{k})\Phi)_{n-1}(\vec{k}_1,\ldots,\vec{k}_{n-1}) = \sqrt{n}\ \phi_n(\vec{k},\vec{k}_1,\ldots,\vec{k}_{n-1}) \tag{4}$$

$$(a^\dagger(\vec{k})\Phi)_n(\vec{k}_1,\ldots,\vec{k}_n) = \frac{1}{\sqrt{n}}\sum_{j=1}^{n}\delta(\vec{k}-\vec{k}_j)\phi_{n-1}(\vec{k}_1,\ldots,\hat{\vec{k}}_j,\ldots,\vec{k}_n),$$

where the variable $\vec{k}_j$ is omitted. The number operator N and the Hamiltonian $\mathcal{H}$ have the realizations

$$(N\Phi)_n(\vec{k}_1,\ldots,\vec{k}_n) = n\ \phi_n(\vec{k}_1,\ldots,\vec{k}_n) \quad ,$$

$$(\mathcal{H}\Phi)_n(\vec{k}_1,\ldots,\vec{k}_n) = \left[\sum_{j=1}^{n}\omega(k_j)\right]\phi_n(\vec{k}_1,\ldots,\vec{k}_n) \quad .$$

The ground state $|0\rangle$ has the representation

$$\Omega_0 = \{1,0,0,\ldots\}$$

and is clearly nondegenerate. It is evidently the no-particle state since $N\ \Omega_0 = 0$.

Like the Hamiltonian, other important operators for a free field act by simple superposition of their one-particle properties. For instance, the total momentum operators, or space translation generators, are represented by

$$(\vec{\mathcal{P}}\ \Phi)_n(\vec{k}_1,\ldots,\vec{k}_n) = (\sum_{j=1}^{n}\vec{k}_j)\phi_n(\vec{k}_1,\ldots,\vec{k}_n) \quad ,$$

and abstractly given as

$$\vec{\mathcal{P}} = \int a^\dagger(\vec{k})\vec{k}\ a(\vec{k})\ d\vec{k} \quad . \tag{5}$$

When applicable the angular momentum generators, and the relativity transformation ("boost") generators, in the case of a relativistic theory, are determined by similar formulas.

Local Field Description

In the standard fashion we may introduce the canonical field and momentum operators, $\phi(\vec{x})$ and $\pi(\vec{x})$, according to the prescription

$$\phi(\vec{x}) = (2\pi)^{-s/2} \int e^{i\vec{k}.\vec{x}}[a(\vec{k}) + a^{\dagger}(-\vec{k})][2\omega(k)]^{-1/2}d\vec{k} ,$$

$$\pi(\vec{x}) = (2\pi)^{-s/2} \int e^{i\vec{k}.\vec{x}}[-ia(\vec{k})+ ia^{\dagger}(-\vec{k})][\omega(k)/2]^{1/2}d\vec{k} ,$$

where s denotes the number of space dimensions. These operators are Hermitian, and in view of (1) satisfy the usual canonical commutation relations

$$[\phi(\vec{x}), \pi(\vec{y})] = i\delta(\vec{x}-\vec{y}) .$$

We may express the Hamiltonian $\mathcal{H}$ in terms of the canonical operators in the form

$$\mathcal{H} = \frac{1}{2} \int :\pi^2(\vec{x}) + [\vec{\nabla}\phi(\vec{x})]^2 + m^2\phi^2(\vec{x}) : d\vec{x}$$

in the important case that $\omega^2 = k^2 + m^2$. Here the double set of colons denotes normal ordering, which for each monomial in creation and annihilation operators reorders the expression (ignoring noncommutativity) so that all creation operators stand to the left of all annihilation operators. It is evident that the ground state expectation value of a normally ordered operator vanishes, the only exception being the constant operator involving neither creation nor annihilation operators. For the scalar field, normal ordering removes the (infinite) ground state energy. In the present case, the space translation generators (5) are given in terms of the field operators as

$$\vec{\mathcal{P}} = \int : \pi(\vec{x})\vec{\nabla}\phi(\vec{x}) : d\vec{x} . \qquad (6)$$

Apart from normal ordering, these are just the classical Hamiltonian and space translation operators with q-number field and momentum operators.

The $\vec{\mathcal{P}}$ operators evidently have the general properties that $\vec{\mathcal{P}}|0\rangle = 0$ and

$$[\vec{\mathcal{P}}, \phi(\vec{x})] = -i\vec{\nabla}\phi(\vec{x}) \quad ,$$

$$[\vec{\mathcal{P}}, \pi(\vec{x})] = -i\vec{\nabla}\pi(\vec{x}) \quad ,$$

while it follows that $[\vec{\mathcal{P}}, \mathcal{H}] = 0$. We note also that $\mathcal{H}|0\rangle = 0$ and

$$[\mathcal{H}, \phi(\vec{x})] = -i\pi(\vec{x}) \quad , \qquad (7)$$

which is a characteristic feature of a canonical theory. In fact, it would seem that the field operator $\phi(\vec{x})$ and the Hamilton $\mathcal{H}$ characterize the theory for by (7) they permit the determination of $\pi(\vec{x})$, and thereby of $\vec{\mathcal{P}}$, etc. There is much to be said for such a program and we shall see traces of it in our subsequent discussion.

B. Free Spinor Field Theory

The foregoing discussion for the scalar field can be applied to the case of spinor fermion fields without too much change. To be concrete, let us note that a free fermion field may be decomposed into independent fermion oscillators characterized by annihilation and creation operators, $b(\vec{k})$ and $b^{\dagger}(\vec{k})$, respectively, in which we have suppressed any additional indices. These operators satisfy the canonical anticommutation relations

$$\{b(\vec{k}), b(\vec{k}')\} = \{b^{\dagger}(\vec{k}), b^{\dagger}(\vec{k}')\} = 0 \quad ,$$

$$\{b(\vec{k}), b^{\dagger}(\vec{k}')\} = \delta(\vec{k}-\vec{k}') \quad .$$

The number density operator $b^\dagger(\vec{k})b(\vec{k})$ is positive, and the Hamiltonian for a free fermion field is given by

$$\mathcal{H} = \int \omega(k) b^\dagger(\vec{k}) b(\vec{k}) d\vec{k} \ . \tag{8}$$

As in the boson case

$$0 \leq \langle\Psi|N|\Psi\rangle \leq m^{-1}\langle\Psi|\mathcal{H}|\Psi\rangle$$

where

$$N = \int b^\dagger(\vec{k}) \ b(\vec{k}) \ d\vec{k} \quad ,$$

and thus the existence of $\mathcal{H}$, plausible on physical grounds, ensures the existence of N. The existence of a ground state $|0\rangle$ of zero energy ensures that $N|0\rangle = 0$ as well as

$$b(\vec{k})|0\rangle = 0 \tag{9}$$

for all $\vec{k}$. The specification of the theory is completed by the physically motivated assumption that $b(\vec{k})$ and $b^\dagger(\vec{k})$ are a complete set of operators, or equivalently that apart from a numerical factor $|0\rangle$ is the only vector fulfilling (9). As in the scalar field the physically motivated requirements uniquely specify the theory up to unitarily equivalent realizations.

Fock Space Formulation

The Fock space formulation of the free fermion field proceeds from the introduction of completely antisymmetric, square integrable, momentum space functions of n variables, $\chi_n(\vec{k}_1,\ldots,\vec{k}_n)$. For n=0, $\chi_0 \in C$. The vectors $X \in \mathcal{G}$, the fermion Fock space (a Grassman algebra) are given by

$$X = \bigoplus_{n=0}^{\infty} \chi_n = \{\chi_0, \chi_1(\vec{k}_1), \chi_2(\vec{k}_1,\vec{k}_2), \ldots\} \quad ,$$

where

$$(X,X) = ||X||^2 = \sum_{n=0}^{\infty} ||\chi_n||^2 =$$

$$= |\chi_0|^2 + \sum_{n=1}^{\infty} \int\cdots\int|\chi_n(\vec{k}_1,\ldots,\vec{k}_n)|^2 d\vec{k}_1 \ldots d\vec{k}_n \quad .$$

The annihilation and creation operators have the respective realizations

$$(b(\vec{k})X)_{n-1}(\vec{k}_1,\ldots,\vec{k}_{n-1}) = \sqrt{n}\,\chi_n(\vec{k},\vec{k}_1,\ldots,\vec{k}_{n-1}) \; ,$$

$$(b^\dagger(\vec{k})X)_n(\vec{k}_1,\ldots,\vec{k}_n) =$$

$$= \frac{1}{\sqrt{n}} \sum_{j=1}^{n} (-1)^{j-1}\delta(\vec{k}-\vec{k}_j)\chi_{n-1}(\vec{k}_1,\ldots,\hat{\vec{k}}_j,\ldots,\vec{k}_n)$$

where the variable $\vec{k}_j$ is omitted. The number operator N and the Hamiltonian $\mathcal{H}$ have the realizations

$$(NX)_n(\vec{k}_1,\ldots,\vec{k}_n) = n\chi_n(\vec{k}_1,\ldots,\vec{k}_n)$$

$$(\mathcal{H}X)_n(\vec{k}_1,\ldots,\vec{k}_n) = \left[\sum_{j=1}^{n}\omega(k_j)\right]\chi_n(\vec{k}_1,\ldots,\vec{k}_n) \; .$$

The ground state $|0\rangle$ has the representation

$$\Lambda_0 = \{1,0,0,\ldots\} \quad ,$$

and is clearly nondegenerate; Λ_0 is also the fermion no-particle state. As in the scalar field case, other important operators have similar realizations. In particular, the total momentum operators, or space translation generators

$$\vec{\mathcal{P}} = \int b^\dagger(\vec{k})\vec{k}b(\vec{k})d\vec{k}$$

are represented by

$$(\vec{\mathcal{P}}X)_n(\vec{k}_1,\ldots,\vec{k}_n) = \left[\sum_{j=1}^{n}\vec{k}_j\right]\chi_n(\vec{k}_1,\ldots,\vec{k}_n) \quad .$$

The angular momentum generators and boost generators, when applicable, have similar representations.

Local Field Description

The fermion field is usually taken as non-Hermitian which necessitates the introduction of an independent set of fermion oscillator variables. We may assume they are already contained within the set of b variables since we have suppressed any distinguishing indices. We single out half of these variables and hereafter distinguish them by carets as $\hat{b}(k)$ and $\hat{b}^\dagger(k)$. In terms of these various operators we set

$$\psi(\vec{x}) = (2\pi)^{-s/2} \int e^{i\vec{k}.\vec{x}}\{b(\vec{k})+\hat{b}^\dagger(-\vec{k})\}[2\omega(k)]^{-1/2}d\vec{k}$$

for s spacial dimensions. It follows that

$$\{\psi(\vec{x}),\psi(\vec{y})\} = \{\psi^\dagger(\vec{x}),\psi^\dagger(\vec{y})\} = 0 \; ,$$

$$\{\psi(\vec{x}),\psi^\dagger(\vec{y})\} = \delta(\vec{x}-\vec{y}) \quad ,$$

which is the usual configuration space form of the fermion anticommutation rules. Observe that $\psi(\vec{x})$ is not the Pauli adjoint for a spinor field. The Pauli adjoint is defined by

$$\bar{\psi}(\vec{x}) \equiv \psi^\dagger(\vec{x})\beta \; ,$$

where β is a nonsingular Hermitian matrix satisfying $\beta^2 = 1$; the Pauli adjoint will be needed in forming relativistic models for fermion fields. Normal ordering, again denoted by colons, is extended to fermion fields and for each monomial in the fields ψ and $\bar{\psi}$ reorders the expression so that all creation operators are to the left of all annihilation operators and introduces an overall sign $(-1)^P$ where P is the number of operators permuted to reach the ordered form. Apart from the constant operator the vacuum state expectation vanishes for normally ordered operators.

C. Interacting Field Theories

The central problem in quantum field theory is the formulation and solution of interacting field theories. Some of the standard interaction forms of interest are the quartic self interactions of a fermion or boson field and the trilinear Yukawa coupling of both fields. The total Hamiltonian $\mathcal{H} = \mathcal{H}_0 + V$, where $\mathcal{H}_0$ is either (2) or (8), or both, and V is the interaction term which formally in the three cases may be taken for illustration as

$$V = \int V(\vec{x})\, d\vec{x} \quad , \tag{10}$$

where

$$V(\vec{x}) = \lambda : [\bar{\psi}(\vec{x})\psi(\vec{x})]^2 : \; ,$$

$$V(\vec{x}) = \lambda : \phi^4(\vec{x}) : \; ,$$

$$V(\vec{x}) = \lambda : \bar{\psi}(\vec{x})\psi(\vec{x})\phi(\vec{x}) : \; .$$

Here λ denotes the coupling constant. On the one hand, the local polynomial of the fields is dictated by relativistic covariance, and yet, on the other hand, none of the quantities V defines an operator. While normal ordering leads to $\langle 0|V|0\rangle = 0$ in the three cases, all three have the property that $||V|0\rangle||^2 = \langle 0|V^2|0\rangle = \infty$. This is evident from the translation invariance of $|0\rangle$ which leads to

$$||V|0\rangle||^2 = \int d\vec{x}d\vec{y}\langle 0|V(\vec{x})V(\vec{y})|0\rangle =$$

$$= \int d\vec{x}d\vec{y}\; F(\vec{x}-\vec{y}) = \infty \quad .$$

Although V is normally ordered, the term with all creation operators in $V(\vec{y})$ and the term with all annihilation operators in $V(\vec{x})$ gives rise to an non-vanishing $F(\vec{x})$ and $\int d\vec{x} F(\vec{x})$. Vacuum polarization and infinite space volume lead to the failure of V to be an operator. The introduction of a spacial cutoff by means of a real non-negative C^∞ function $g(\vec{x})$ which vanishes for sufficiently large $|\vec{x}|$ changes V into

$$V(g) = \int g(\vec{x})\, V(\vec{x})\, d\vec{x} \quad .$$

However, in essentially all cases V(g) is still not an operator because $V(\vec{x})$ is so singular that even the exceptionally nice smoothing over space by $g(\vec{x})$ is inadequate to fully regularize it. The only exceptional case is the self-interacting scalar field in a one-dimensional space (two-dimensional space time). This is an important exception and we shall subsequently exploit this fact to assist in the construction of a relativistic quantum theory for such a model.

For all other interactions a further regularization is required which amounts to introducing a momentum space cutoff. Let us imagine replacing the fields ϕ, ψ and $\bar{\psi}$ in V by the smeared out fields

$$\phi_\kappa(\vec{x}) = \int \mathcal{K}(\vec{x}-\vec{y})\,\phi(\vec{y})\,d\vec{y} \quad ,$$

and similarly for ψ_κ and $\bar{\psi}_\kappa$, where

$$\mathcal{K}(\vec{x}) = (2\pi)^{-s} \int e^{i\vec{k}.\vec{x}} \chi(\vec{k})\, d\vec{k} \quad .$$

Here $\mathcal{K}(\vec{x})$ is a real C^∞ function which decreases for large $|\vec{x}|$ faster than any inverse power, i.e., $\mathcal{K}$, and therefore the momentum space cutoff χ, are elements of Schwartz's test function space S . With the fields so regularized, the interaction is replaced by

$$V_\kappa(g) = \int g(\vec{x})\, V_\kappa(\vec{x})\, d\vec{x}$$

where $V_\kappa(\vec{x})$ is one of the following

$$:[\bar{\psi}_\kappa(\vec{x})\psi_\kappa(\vec{x})]^2: \ , \ :\phi_\kappa^4(\vec{x}):, \ :\bar{\psi}_\kappa(\vec{x})\psi_\kappa(\vec{x})\phi_\kappa(\vec{x}): .$$

The full Hamiltonian is given by

$$\mathcal{H}_\kappa(g) = \mathcal{H}_0 + V_\kappa(g) \quad ;$$

but, perhaps surprisingly, even though we have severely butchered the interaction so as to be able to define $V_\kappa(g)$, there is in fact no a priori guarantee that $\mathcal{H}_\kappa(g)$ has the necessary properties to qualify as a Hamiltonian. Any number of pathologies might arise some of which we shall illustrate later. Recent work of A. M. Jaffe, O. E. Lanford III and A. S. Wightman [2] has shown that such pathologies do not arise for these, as well as for higher order, doubly cutoff interactions. Armed with a properly defined, butchered theory, an attempt can be made to remove the cutoffs, i.e., $\chi(\vec{k}) \to 1$ and $g(\vec{x}) \to \lambda$, making due allowance in $\mathcal{H}_\kappa(g)$ for necessary counter terms. Yet as the cutoffs are removed, what is it that should converge and in what sense? We can not expect that $\mathcal{H}_\kappa(g) \to \mathcal{H}$ as defined in (10) since, as we have noted, such Hamiltonians make no sense. Moreover, Haag's theorem [3] states, so to say, that convergence of $\mathcal{H}_\kappa(g)$ takes place, if at all, on a renormalized Hilbert space which carries the physical representation of the canonical operators. It is on this space that $\mathcal{H}$ will be realized. Under physically reasonable conditions it will follow, for example, that

$$[\mathcal{H}, \phi(\vec{x})] = -i\pi(\vec{x}) \quad ,$$

$$[\phi(\vec{x}), \pi(\vec{y})] = i\,\delta(\vec{x}-\vec{y}) \quad .$$

In the physical representation, moreover, the operators ϕ and π are inequivalent to the free field operators defined on Fock space [4]. Hence, the convergence must actually change the representation of the field operators to an inequivalent one, a neat trick indeed!

To secure the desired weak form of convergence we can appeal to a complicated chain involving the C^* algebraic formof quantum theory [5] and the Gelfand-Naimark-Segal representation theory for states on these algebras [6]. Such a tortuous route has been successfully carried out for the scalar field with quartic interaction in two dimensions [1]. Before delving into these techniques let us discuss several general properties required of the physical theory involving the physical representation of the canonical operators.

2. EXPECTED PROPERTIES OF THE PHYSICAL THEORY

A. Canonical Operators Defined

According to remarks made previously, the interacting theory is expected to involve representations of the canonical operators inequivalent to the Fock representation. The study of such representations as well as various possible candidates for Hamiltonians is evidently worthwhile, for it may bypass or eliminate a complicated analysis of approximate theories.

Boson Fields

Since the smeared field operators

$$\phi(f) = \int \phi(\vec{x})\, f(\vec{x})\, d\vec{x} \ ,$$

$$\pi(g) = \int \pi(\vec{x})\, g(\vec{x})\, d\vec{x}$$

are unbounded operators it is convenient to study the Weyl form of the commutation relations. Let $V[f]$, $W[g]$

denote two unitary families of operators for which

$$V[f]V[f'] = V[f+f'],\ V[0] = I\ ;$$

$$W[g]W[g'] = W[g+g'],\ W[0] = I$$

which are defined for suitable real test functions f and g, say $f,g \in \mathcal{S}$. We impose the requirement that

$$V[f]W[g] = e^{i(f,g)}\ W[g]V[f]$$

for all such f and g, where

$$(f,g) = \int f(\vec{x})\ g(\vec{x})\ d\vec{x}\ .$$

If we require that

$$\langle\Lambda|W\ \tau g|\psi\rangle\ ,\ \langle\Lambda|V[\tau f]|\psi\rangle \tag{11}$$

are continuous in τ for each fixed f, $g \in \mathcal{S}$ and Λ, $\Psi \in \mathcal{H}$, then Stone's theorem ensures us that

$$V[f] = e^{i\phi(f)},\ W[g] = e^{-i\pi(g)}$$

where $\phi(f)$ and $\pi(g)$ are self-adjoint generators. On a suitable domain these unbounded operators satisfy the relation

$$[\phi(f)\ ,\ \pi(g)] = i(f,g) \tag{12}$$

We set

$$U[f,g] \equiv \exp\{-\tfrac{i}{2}(f,g)\}V[f]W[g] \equiv \exp\{i[\phi(f)-\pi(g)]\}\ .$$

Fermion Fields

In some ways the fermion operators are easier to study. If we set

$$\psi(h) = \int h^{*}(\vec{x})\,\psi(\vec{x})\,d\vec{x}\ ,$$

$$\psi(h)^{\dagger} = \int h(\vec{x})\psi^{\dagger}(\vec{x})d\vec{x},$$

then it follows that

$$\{\psi(h),\psi(h')\} = \{\psi(h)^{\dagger},\psi(h')^{\dagger}\} = 0 \qquad (13a)$$

$$\{\psi(h),\psi(h')^{\dagger}\} = (h,h') = \int h^{*}(\vec{x})h'(\vec{x})d\vec{x} \ . \qquad (13b)$$

In particular

$$\psi(h)\psi(h)^{\dagger} + \psi(h)^{\dagger}\psi(h) = (h,h) \equiv |h|_2^2$$

from which it follows that $\psi(h)$ is bounded. Recall that the operator norm is given by

$$\|B\| = \sup_{\substack{\phi \\ \|\phi\|=1}} \|B\phi\| = \sup_{\substack{\phi,\lambda \\ \|\phi\|=\|\lambda\|=1}} |(\phi,B\lambda)|$$

which leads to

$$\|B^{\dagger}\| = \|B\|$$

One readily finds that

$$\|\psi(h)\| = \|\psi(h)^{\dagger}\| = |h|_2$$

so that no serious domain questions arise in the study of (13) since the smeared operators are all bounded operators.

B. Aspects of Representations

For finitely many degrees of freedom, either boson or fermion or both, there is but one irreducible representation of the canonical operators apart from unitary equivalence. For example, for the boson degrees of freedom the usual Schrödinger representation is equivalent to any other irreducible one [7]. In case of a field, however, where one deals with an infinite number of degrees of freedom the situation is completely different. For both boson and fermion fields there are uncountably many inequivalent, irreducible representations [8]. The choice of the correct representation for the problem becomes an essential ingredient in its formulation and solution; the wrong representation may lead to the wrong theory or even worse to no theory at all! For example, to formulate the problem of a spin system in a magnetic field one must know what value of spin to assume, i.e. what representation of SU(2) to choose. In a number of problems (e.g., the hydrogen atom) the rotation group enters reducibly so as to exhibit a number of inequivalent representations, any one of which is inadequate to carry the correct physics. In canonical field theories, where a number of inequivalent representations also occur, it is conceivable that the right representation for a given problem is a reducible one. This situation actually arises in certain soluble model field theories with nontrivial interaction [9].

Relativistic Application

In this regard two recent theorems are of considerable interest. The first deals with fermion fields and is due to Powers [10]. For simplicity we present a paraphrased version.

Theorem : Let $\psi(\vec{x},t)$ denote a fermion field which (i) transforms covariantly under a unitary representation of the inhomogeneous Lorentz group in s+1 space-time dimensions, (ii) fulfills, along with its adjoint $\psi^{\dagger}(\vec{y},t)$, the canonical equal-time anticommutation relations (13), (iii) corresponds to an irreducible representation of the anticommutation relations at any one time, and (iv) fulfills a few technical assumptions such as the applicability of $\psi(h)$ and $\psi^{\dagger}(h)$ to the vacuum and one-particle states. Granted these conditions, then the theorem states that the field $\psi(\vec{x},t)$ obeys a linear field equation schematically given by

$$i\,\dot{\psi} = L\psi + M\psi^{\dagger}$$

where L and M are linear transformations, so long as the spacial dimension $s \geq 2$. The argument is largely geometric, depending on the fact that in a relativistic theory fermion (boson) operators separated by space-like intervals anticommute (commute).

A related theorem for scalar fields has recently been proven by Sinha and Emch [11]. Their result is not so strong in view of the unbounded nature of ϕ and π.

Theorem : Let $\phi(\vec{x},t)$ denote a scalar field, with $\pi(\vec{x},t) = \dot{\phi}(\vec{x},t)$, which (i) transforms covariantly under a unitary representation of the inhomogeneous Lorentz group in s+1 space-time dimensions, (ii) fulfills the canonical equal-time commutation relations (12), (iii) corresponds to an irreducible representation of the commutation relations at any one time, and (iv) fulfills several technical assumptions. Granted these conditions, then the theorem states that the field $\phi(\vec{x},t)$ obeys a linear field equation schematically given by

$$\ddot{\phi} = N\phi + c\ ,$$

where N is a linear transformation and c a multiple of the unit operator, so long as the spacial dimension $s \geq 3$. Again the argument is largely geometric and in many ways similar to that for the fermions. It is highly plausible that the two results can be combined to predict trivial interactions for certain irreducible and canonical relativistic theories involving both fermions and bosons.

Cyclic Representations

In any case, relativistic or not, in which canonical field operators exist they form either an irreducible or a reducible family of operators. In the latter case we can distinguish cyclic and noncyclic representations. In a cyclic representation for the scalar field, for example, there exists a vector $|0\rangle$ for which the set of vectors of the form

$$|f,g\rangle \equiv U[f,g]\,|0\rangle$$

span the Hilbert space $\mathcal{H}$, and we say that the representation of ϕ and π is cyclic. It is a further assumption that the cyclic vector $|0\rangle$ is the ground state for the problem at hand. It may happen that the restricted set of vectors

$$|f\rangle = |f,0\rangle = U[f,0]\,|0\rangle = V[f]\,|0\rangle$$

span $\mathcal{H}$, in which case we say that the representation of ϕ is cyclic. Cyclic representations, of either type, are characterized by expectation functionals such as

$$E(f,g) = \langle 0|f,g\rangle = \langle 0|U[f,g]\,|0\rangle,$$

$$E(f) = \langle 0|f\rangle = \langle 0|V[f]\,|0\rangle \quad .$$

Let us consider $E(f,g)$ for the moment; corresponding remarks apply to $E(f)$ when ϕ is cyclic.

C. Choice of Representation

Translation, Cluster and Inequivalence

In order for the vector $|0\rangle$ to be invariant under space translation it is necessary and sufficient that

$$E(f_a,g_a) = E(f,g) \tag{14}$$

for all $\vec{a}$ and $f,g \in \mathcal{S}$ where

$$f_a(\vec{x}) \equiv f(\vec{x}-\vec{a}), \quad g_a(\vec{x}) \equiv g(\vec{x}-\vec{a}) \quad .$$

Since Euclidean invariance is "next to Godliness" we assume that E satisfies (14). Uniqueness of the space-translation-invariant state can be deduced from the physical requirement that measurements made in remote regions of space should be statistically independent. In turn a unique translationally invariant state exists if and only if E fulfills the cluster decomposition property, namely that

$$\lim_{|\vec{a}|\to\infty} E(f'+f_a, g'+g_a) = E(f',g')E(f,g)$$

for arbitrary test functions in $\mathcal{S}$. Given such a representation it follows that

$$\lim_{|\vec{a}|\to\infty} \langle f_1,g_1|U[f_a,g_a]|f_2,g_2\rangle =$$

$$= \langle f_1,g_1|f_2,g_2\rangle\langle 0|U[f,g]|0\rangle$$

since for arbitrarily large $\vec{a}$ the probability amplitude factorizes.Recall that a set of bounded operators B_n is said to weakly converge to B, in symbols

$$\text{w-}\lim_{n\to\infty} B_n = B \quad ,$$

provided that

$$\lim_{n\to\infty} \langle\Lambda|B_n|\Psi\rangle = \langle\Lambda|B|\Psi\rangle \qquad (15)$$

for all $\Lambda, \Psi \in \mathcal{H}$. For uniformly bounded operators $\|B_n\| \leq b$ it suffices to prove that the forms

$$\langle\Lambda| B_n|\Psi\rangle = B_n\{\Lambda,\Psi\} \to B\{\Lambda,\Psi\}$$

for all $\Lambda , \Psi \in \mathcal{J}$, a total set, i.e., any set whose span is $\mathcal{H}$. From this it follows that B exists and that (15) holds. For a representation with ϕ and π cyclic the vectors $|f,g\rangle$ constitute such a total set, and thus we see that

$$\underset{|\vec{a}|\to\infty}{\text{w-lim}}\, U[f_a,g_a] = E(f,g)\ I\ ,$$

a multiple of the unit operator which we call a "tag". Suppose we have two such representations which are unitarily related according to

$$U'[f,\ g] \ =\ V^{\dagger}\ U[f,\ g]\ V$$

for all f, $g \in \mathcal{S}$, and thus for all f_a, g_a. It follows that

$$\underset{|\vec{a}|\to\infty}{\text{w-lim}}\, U'[f_a,g_a] = \underset{|\vec{a}|\to\infty}{\text{w-lim}}\, V^{\dagger}U[f_a,g_a]\ V$$

$$= E(f,g)V^{\dagger}IV$$

$$= E(f,g)I' \equiv E'(f,g)I'$$

Evidently, equivalent representations have identical tags. Unequal tags - for any pair of functions f,g - correspond to inequivalent representations. Hence distinct functionals E(f,g) label inequivalent representations of the canonical commutation relations. A similar conclusion holds for E(f) provided that $|0\rangle$ is cyclic for the field ϕ.

Space-Time Translation Generators in a Reducible Representation

If the representation of ϕ and π is reducible then we have the common decomposition

$$\phi = \phi_1 \oplus \phi_2\ ,\quad \pi = \pi_1 \oplus \pi_2\ .$$

If the space-translation generator $\vec{\mathcal{P}} = \vec{\mathcal{P}}\ (\pi,\phi)$, as for example in (6), the necessarily

$$\vec{\mathcal{P}} = \vec{\mathcal{P}}_1 \oplus \vec{\mathcal{P}}_2 \quad .$$

But this decomposition is inconsistent with the requirement of a unique translationally invariant state $|0\rangle = |0\rangle_1 \oplus |0\rangle_2$ which is cyclic for ϕ and π. Consequently in a reducible representation $\vec{\mathcal{P}}$ is not given as a function solely of ϕ and π. In fact, any operator for which $|0\rangle$ is a nondegenerate eigenstate, such as the Hamiltonian $\mathcal{H}$, cannot be expressed solely in terms of ϕ and π.

Matrix Elements of Interest

Although $\vec{\mathcal{P}}$ may not be representable in terms of ϕ and π it is a known operator since its action on $|f,g\rangle$ states is simply given by

$$\langle f_1,g_1|e^{i\vec{a}.\vec{\mathcal{P}}}|f,g\rangle = \langle f_1,g_1|f_a,g_a\rangle$$

which involves just the functional E and an appropriate phase factor. Unfortunately the action of the Hamiltonian is not so simply expressed.

Only partial information of a general nature is known. Time reversal invariance of $|0\rangle$ is assured if

$$E(f,-g) = E(f,g)$$

which to first order in g leads to

$$\langle 0|V[f]\pi(\vec{x})|0\rangle = -\tfrac{1}{2}f(\vec{x})\langle 0|V[f]|0\rangle \quad .$$

Since $V[f]\pi(\vec{x}) = [\pi(\vec{x})-f(\vec{x})]V[f]$ we learn that

$$\langle f'|\pi(\vec{x})|f\rangle = \tfrac{1}{2}[f'(\vec{x})+f(x)]\langle f'|f\rangle$$

$$= \tfrac{1}{2}[f'(\vec{x})+f(\vec{x})]E(f-f') \quad .$$

Because $i[\mathcal{H},\phi(\vec{x})] = \pi(\vec{x})$ and $\mathcal{H}|0\rangle = 0$ we further find

$$\langle f'|\mathcal{H}|f\rangle = \frac{1}{2}(f',f)E(f-f') \qquad (16)$$

as first shown by Araki [12] . If $|f,g\rangle$ spans $\mathcal{H}$, but not $|f\rangle$ this relation is a submatrix condition. However, if $|0\rangle$ is cyclic for ϕ, then this relation gives the matrix elements of $\mathcal{H}$ for a dense set of vectors (finite linear sums of $|f\rangle$). Does this relation then determine $\mathcal{H}$? Not always, as we shall see in the next section. Nevertheless it shows the intimate connection between the Hamiltonian and the representation of the field operators; recall that under our assumptions distinct expectation functionals correspond to inequivalent representations.

The relation (16) has an analog for a finite number of degrees of freedom, say one. In that case let us adopt the Schrödinger representation. Then, if we denote the ground state of the Hamiltonian by $\phi_o(x)$, which is a real function in virtue of time reversal invariance, we have $(f',f \to p',p)$

$$\int\phi_o(x)e^{-ip'x}\left\{-\frac{1}{2}\frac{\partial^2}{\partial x^2}+V(x)\right\}e^{ipx}\,\phi_o(x)dx =$$

$$= \frac{1}{2}p'p\int\phi_o(x)e^{i(p-p')x}\,\phi_o(x)\,dx$$

which is almost immediately evident.

It is hard to read out of the relation (16) just what theory or model problem it corresponds to. An alternate set of matrix elements sheds light on this question. Let us set

$$\vec{P}(f,g) = \langle f,g|\vec{\mathcal{P}}|f,g\rangle$$

Then $\vec{P}(0,0) = 0$ since $\vec{\mathcal{P}}\,|0\rangle = 0$, and we have

$$\frac{\delta}{\delta f(\vec{x})}\vec{P}(f,g) = \langle f,g|i[\vec{\mathcal{P}},\phi(\vec{x})]|f,g\rangle = \langle f,g|\vec{\nabla}\phi(x)|f,g\rangle =$$

$$= \langle 0|\vec{\nabla}[\phi(\vec{x})+g(\vec{x})]|0\rangle = \vec{\nabla}g(\vec{x}) \quad .$$

A similar calculation shows that

$$\frac{\delta}{\delta g(\vec{x})} \vec{P}(f,g) = - \vec{\nabla} f(\vec{x}) \ ,$$

all of which lead to the unique solution

$$\vec{P}(f,g) = \int f(\vec{x}) \vec{\nabla} g(\vec{x}) d\vec{x} \ .$$

This equation has the functional form of the classical generator of space translations where $f(\vec{x}) \equiv \pi_{cl}(\vec{x})$ and $g(\vec{x}) \equiv \phi_{cl}(\vec{x})$. This remark holds true whether or not the representation of ϕ and π is irreducible.

For the Hamiltonian we set

$$H(f, g) \equiv \langle f,g | \mathcal{H} | f,g \rangle \ ,$$

where $H(0,0) = 0$ since $\mathcal{H}|0\rangle = 0$. From the relation

$$\frac{\delta}{\delta f(\vec{x})} H(f,g) = \langle f,g | i[\mathcal{H}, \phi(\vec{x})] | f,g \rangle = \langle f,g | \pi(\vec{x}) | f,g \rangle =$$

$$= \langle 0 | \pi(\vec{x}) + f(\vec{x}) | 0 \rangle = f(\vec{x}) \ ,$$

we learn that

$$H(f,g) = \tfrac{1}{2} \int f^2(x) dx + W(g) \ ,$$

where $W(0) = 0$. In the case of a relativistic theory the form of $W(g)$ can be further spelled out. In particular, it can be shown [13] that

$$H(f,g) = \tfrac{1}{2} \int \{ f^2(\vec{x}) + [\vec{\nabla} g(\vec{x})]^2 + m_0^2 g^2(\vec{x}) \} d\vec{x} + \int V[g(\vec{x})] d\vec{x} \ , \tag{17}$$

where m_0^2 is the bare mass defined as usual in terms of the spectral weight function of the two-point function,

$$m_0^2 = \int m^2 \rho(m^2) \, dm^2 \ ,$$

and V remains undetermined. If the theory involved a quartic self-interaction of the field $\phi(\vec{x})$, then it follows that $V[g(\vec{x})] = \lambda g^4(\vec{x})$. In short, the diagonal matrix elements of the Hamiltonian (17) correspond to the classical Hamiltonian and would determine quite directly which particular model a given $\mathcal{H}$ and canonical field representation describe.

The general properties we have described here should belong to any explicit model theory which fulfills the necessary prerequisites. They can be used both as touchstones in any given calculations, or as partial or complete determinants of the model itself. Some progress has been made in constructing model theories directly in the renormalized Hilbert space and with the physical representation of the canonical operators based on the preceding arguments [14], and further work along these lines should be carried out.

3. OPERATORS ON HILBERT SPACE

One of the most successful approaches to constructing model theories begins with the regularized form of the theory defined on Fock space. The analysis of a particular model as the limit of the sequence of approximate theories can become technically quite involved, and it is appropriate that we examine some properties of operators that are useful in such analyses.

A. Definitions

We begin with some definitions and standard properties for (linear) operators. An operator A is a linear transformation defined for a set of vectors $\underset{\sim}{D}(A)$ which make up its domain. These vectors are imaged by A into a set of vectors $\underset{\sim}{R}(A)$ which make up its range. A bounded operator B with domain $\underset{\sim}{D}(B)$ and bound $||B||$ can always be extended to a bounded operators with domain $\mathcal{H}$ while preserving the bound $||B||$. If $\underset{\sim}{D}(B)$ is dense in $\mathcal{H}$ the extension is uniquely determined by linearity and continuity. Thus if $\psi_n \in \underset{\sim}{D}(B)$ is a fundamental sequence converging to $\psi \in \mathcal{H}$, $\psi_n \to \psi$, we may define $B\psi = \lim B\psi_n$. Here we may adopt strong convergence, in symbols $B\psi_n \to B\psi$, which holds if $||B\psi_n - B\psi|| \to 0$. An operator A is said to be closed if for any fundamental sequence $\psi_n \in \underset{\sim}{D}(A), \psi_n \to \psi \in \mathcal{H}$, for which $A\psi_n \to \phi \in \mathcal{H}$, then $\psi \in \underset{\sim}{D}(A)$ and $A\psi = \phi$. An operator is closable if it can be extended (i.e., $\underset{\sim}{D}(A)$ extended and A defined thereon) so that A becomes closed. Every bounded operator B is closable since B can be extended by linearity and continuity to the closed subspace spanned by $\underset{\sim}{D}(B)$; if $\underset{\sim}{D}(B)$ is dense this extension gives a closed operator with domain $\mathcal{H}$. The closure of an operator A with domain $\underset{\sim}{D}(A)$ is denoted by $A^{\sim}$. If A is closed, $A=A^{\sim}$, then we say that a core for A is any domain D so that $A=(A|D)^{\sim}$, where $A|D$ denotes the restriction of A to D. An unbounded operator A is one for which $||A\psi||/||\psi||$ is not uniformly bounded for all $\psi \in \underset{\sim}{D}(A)$. Unbounded operators need not be closable. As an example consider the scalar field annihilation operator $a(\underset{\sim}{k})$ defined on Fock space. Consider vectors of the form

$$\Psi = \bigoplus_{n=0}^{\infty} (n!)^{-1/2} \psi(\vec{k}_1)\psi(\vec{k}_2) \ldots \psi(\vec{k}_n) ,$$

where $\psi(\vec{k})$ is a square integrable, continuous function, which constitute a special subset of vectors in Fock space. It follows directly from (4) that

$$a(\vec{k})\Psi = \psi(\vec{k})\Psi$$

so that the Ψ is an eigenvector for $a(\vec{k})$ with eigenvalue $\psi(\vec{k})$ - apart from normalization these are just the well-known coherent states [15]. Since such Ψ are total in $\mathcal{H}$, it follows that $a(\vec{k})$ is a densely defined operator. But it cannot be closed, for let $\psi_n(\vec{k})$ be a sequence of continuous square integrable functions converging in L^2 norm to the continuous function $\psi(\vec{k})$ but converging pointwise to a function $\hat{\psi}(\vec{k})$ having a jump at a single point $\vec{k}$. Thus with $\Psi_n \varepsilon \underset{\sim}{D}(a(\vec{k}))$ we have $\Psi \varepsilon \underset{\sim}{D}(a(\vec{k}))$, and yet $a(\vec{k})\Psi_n \to \hat{\psi}(\vec{k})\Psi \neq a(\vec{k})\Psi = \psi(\vec{k})\Psi$.

Let A be a densely defined operator and ask for what pair of vectors λ and ϕ the relation

$$(\lambda,\psi) = (\phi,A\psi) \tag{18}$$

holds for all $\psi \varepsilon \underset{\sim}{D}(A)$. For each pair satisfying (18) we set $\lambda = A^+\phi$ and the set of ϕ for which (18) holds is $\underset{\sim}{D}(A^+)$, the domain of A^+, and the set of λ is $\underset{\sim}{R}(A^+)$, the range of A^+. In all cases A^+ is a closed operator for $(A^+\phi_n,\psi) = (\phi_n,A\psi) \to (\phi,A\psi) = (\lambda,\psi) \equiv (A^+\phi,\psi)$. The adjoint operator A^+ may be defined only on the zero vector; this is precisely the case for $a^+(\vec{k})$, the creation operator at a point in Fock space. However, if A is closed or closable, A^+ is densely defined. When A^+ is densely defined we can define $A^{++} = (A^+)^+$, which is necessarily a closed operator that is densely defined. Under these conditions A^{++} is a closed extension of A, and is the minimal such extension. A symmetric operator T is a densely defined operator for which $(\phi,T\psi)=(T\phi,\psi)$ for all $\phi,\psi \varepsilon \underset{\sim}{D}(T)$. As a consequence, $\underset{\sim}{D}(T)\subseteq\underset{\sim}{D}(T^+)$, and on $\underset{\sim}{D}(T)$ we have $T=T^+$. This may be written as $T\subseteq T^+$. Thus a symmetric operator always has a closed extension T^+, but this extension may not be symmetric. On the other hand T^{++} always provides a closed symmetric extension of T, since $T\subseteq T^+$ leads to $T^+ \supseteq T^{++}$ and to $T^{++}\subseteq(T^{++})^+$. Let $T=T^{++}$ be a closed symmetric operator, $T\subseteq T^+$. If T has no proper closed symmetric extension then T is called maximal symmetric. If $T=T^+$, i.e., equal operators on identical domains, then T is called self-adjoint. These are the operators of primary importance in physical applications. It is these that possess complete spectral resolutions of unity, generate unitary groups of transformations, etc., all important aspects in any quantum theory.

B. Symmetric and Self-Adjoint Operators

Criteria for determining whether or not an operator fulfills the delicate conditions to be self-adjoint are of basic significance. Turning the unitary group problem around we have the fundamental theorem of Stone which states that every one-parameter unitary group U(t) such that $U(t)U(t')=U(t+t')$, $U(0)=I$, $U(t)^{+}=U(-t)$, which is weakly continuous in t, is necessarily given by $U(t)=$ $=\exp(iTt)$ where T is a self-adjoint operator. Another criterion asks for solutions of

$$(T\pm i)\psi_{\pm} = 0 , \qquad (19)$$

where T is a symmetric operator. The dimensionality of the two subspaces $d_{\pm}$ of solutions $\psi_{\pm}$ is written as (d_{-},d_{+}) and called the deficiency indices of T. If (a) $d_{+} > 0$, $d_{-} > 0$ there are proper symmetric extensions of T which decrease each deficiency index by the same integral value; if (b) $d_{+} = 0$, $d_{-} > 0$ or $d_{+} > 0$, $d_{-}=0$, the operator is maximally symmetric and possesses no self-adjoint extensions; if (c) $d_{+}=d_{-}=0$, the operator is self-adjoint. The resolvent operator $R(z)=(T-z)^{-1}$ provides another handle which will be discussed below.

There are standard examples of the foregoing situations [16]. Let $T=-i\partial/\partial x$ on $L^2(a,b)$. Then the functions $\psi_{\pm} = \exp(\pm x)$ satisfy (19) and thus T has deficiency indices (1,1). Such operators admit self-adjoint extensions but there are a one-parameter family of distinct extensions. If $T=-i\partial/\partial x$ on $L^2(0,\infty)$ then only $\psi_-=\exp(-x)$ is square integrable and T has deficiency indices (1,0). This T is maximally symmetric and has no self-adjoint extension. Lastly if $T=-i\partial/\partial x$ on $L^2(-\infty,\infty)$ then there are no square integrable solutions to (19) and the deficiency indices are (0,0).

It proves useful to introduce the concept of an essentially self-adjoint operator T with domain $\mathcal{D}(T)$. A symmetric operator T is essentially self-adjoint if its adjoint T^+ is symmetric or self-adjoint or if its closure T^{++} is selfadjoint. For example, the operator $T=-i\partial/\partial x$ on $L^2(-\infty,\infty)$ is essentially self-adjoint on test functions in $\mathcal{S}$. However, this operator is <u>not</u> essentially self-adjoint on test functions in $\mathcal{S}$ which also vanish at x=0 although this still provides a dense set of elements for L^2 .For in this case the adjoint operator $T^+ =$ $= -(i/x)[(\partial/\partial x)x-1]$ and $T^+ \supset T^{++}$. In particular T^+ has normalizable eigenvectors (e.g., $\phi = e^{-ax}$, $x > 0$, $\phi = 0$, $x < 0$, where Re $a > 0$) with complex eigenvalues and is not symmetric.An important criterion for essential self-adjointness is provided by Nelson's theorem [17] . Let T be a symmetric operator with an invariant domain $\mathcal{D}$ (i.e., $T\mathcal{D} \subseteq \mathcal{D}$) containing a dense set of analytic vectors, namely vectors ψ for which

$$\Sigma(n!)^{-1}||T^n\psi||z^n < \infty$$

converges for all $|z| < C$, for some $C > 0$. Then T is essentially self-adjoint on $\mathcal{D}$.

Weaker still are the criterion on symmetric operators that ensure that a self-adjoint extension exists. There is the important theorem of Friedrichs which states that every semi-bounded symmetric operator, $T \geq b$, possesses at least one self-adjoint extension [16]. Another theorem [16] pertains to a "real" transformation such as a real symmetric transformation on L^2 (e.g., Schrödinger's Hamiltonian $-\frac{1}{2}\partial^2/\partial x^2+V(x)$) . Reality ensures that $d_+=d_-$ so that self-adjoint extensions exist.

When is A+B self-adjoint?

Suppose A and B are self-adjoint operators. What can be said about C=A+B defined on the domain $\mathcal{D}(A) \cap \mathcal{D}(B)$? In general C is not even a closed operator (set B=-A, then C=0 on $\mathcal{D}(A)$, a domain which is a proper subset of $\mathcal{H}$ if A is unbounded). In general, C is not even essentially self-adjoint on this domain (set $A = -\partial^2/\partial x^2$, $B=-x^4$ on $L^2(-\infty,\infty)$ this quantum mechanics problem has several selfadjoint extensions - the relative signs of A and B are important in this regard). There is a standard theorem due to Rellich [18] that applies in special cases.

<u>Theorem</u>: Let A be self-adjoint, and let B be symmetric and satisfy

$$||B\psi|| \leq a\,||\psi||+ b||A\psi||$$

for all $\psi \in \mathcal{D}(A) \subseteq \mathcal{D}(B)$, where $b < 1$. Then C=A + B is self-adjoint and $\mathcal{D}(C) = \mathcal{D}(A)$. In particular A+B is self-adjoint if B is bounded (b=0).

We prove this theorem below.

Actually this theorem applies to cases where B may have no self-adjoint extension, an aspect which is evidently overwhelmed by A. Of more direct interest to us is another result obtained recently by Glimm and Jaffe [19].

<u>Theorem</u>: Let A and B be self-adjoint operators. If there exists a constant c such that

$$||A\psi||^2 + ||B\psi||^2 \leq c\,\{||(A+B)\psi||^2 + ||\psi||^2\}$$

for all $\psi \in \mathcal{D}(A) \cap \mathcal{D}(B)$, then C=A+B is self-adjoint on

$\mathcal{D}(C) = \mathcal{D}(A+B) = \mathcal{D}(A) \cap \mathcal{D}(B)$, provided $\mathcal{D}(C)$ is dense.

Remark. We may rewrite the conditions of this theorem as

$$A^2 + B^2 \leq c(A+B)^2 + c \ . \tag{20}$$

In applications to field theory models Glimm and Jaffe have further assumed that $A \geq 0$ and that

$$0 \leq aA + B + b \tag{21a}$$

$$0 \leq \varepsilon A^2 + \left[A^{1/2}, [A^{1/2}, B]\right] + b \tag{21b}$$

where a and ε are positive constants such that $2a+\varepsilon < 1$. These two conditions lead to

$$A^2+B^2 = (A+B)^2 - 2A^{1/2}BA^{1/2} - \left[A^{1/2}, [A^{1/2}, B]\right] \leq$$

$$\leq (A+B)^2 + (2a+\varepsilon)A^2 + 2Ab + b \ ,$$

which implies (20).

C. Resolvents

These theorems are consequences of fundamental properties of resolvent operators. If C is a closed operator, then

$$R(z) = (C-z)^{-1}$$

denotes its resolvent. The values of z where $R(z)$ is bounded are called the resolvent set $P(C)$, and the complementary set is called the spectrum $\Sigma(C)$. For example, if C is self-adjoint, $P(C)$ contains all points with $\operatorname{Im} z \neq 0$; for C maximally symmetric with deficiency indices $(0,n)$, $n > 0$, $P(C)$ is the open upper half plane $\operatorname{Im} z > 0$, and conversely; for C symmetric with non-zero indices (n,m), $P(C)$ is empty. Any resolvent $R(z)$ satisfies the resolvent equation

$$R(z_1) - R(z_2) = (z_1 - z_2)\, R(z_1)\, R(z_2)\ . \tag{22}$$

Conversely, any R(z) fulfilling the resolvent equation is a resolvent $R(z) = (C-z)^{-1}$ if and only if $R(z)\psi = 0 \rightarrow \psi = 0$ for a single z. Note that it follows directly from (22) that the null space $\underset{\sim}{N} = \{\psi : R(z)\psi = 0\}$ is independent of z. When $\underset{\sim}{N} = 0$, R(z) has a densely defined inverse and we set $C = R(z)^{-1} + z$, which may be shown to be independent of z. Let C_n be a sequence of self-adjoint operators and let $R_n(z) = (C_n - z)^{-1}$ be uniformly bounded for some z (e.g., Im $z \neq 0$). Suppose further that

$$\|\{R_n(z) - R(z)\}\psi\| \rightarrow 0 \tag{23}$$

for all $\psi \in \mathcal{H}$ (or for a total set). The strong convergence implies R(z) satisfies the resolvent equation and so if R(z) has a densely defined inverse, then $R(z) = (C-z)^{-1}$. In this case we say C_n converges to C in the sense of resolvents. The condition of densely defined inverse may be replaced by the condition that R(z) be invertible for an unbounded set of z's with $\|zR(z)\|$ bounded uniformly in z. If $R(z)^+ = R(z^*)$, then C is self-adjoint. If C_n is uniformly semibounded, $C_n \geq b$, then C is self-adjoint.

Proof of Rellich's Theorem.

We employ these properties to establish Rellich's theorem. Let B_n be a sequence of bounded operators for which $\|(B_n - B)\phi\| \rightarrow 0$ on a dense domain, where

$$\|B_n\psi\| \leq a\,\|\psi\| + b\,\|A\psi\| \tag{24}$$

for all $\psi \in \underset{\sim}{D}(A)$, and where $b < 1$. Clearly B_n converges to a symmetric hence closable operator with domain

$\mathcal{D}(B)$, and for any $\psi \in \mathcal{H}$ the vector $\phi = (A-z)^{-1}\psi$, $\text{Im}z \neq 0$, lies in this domain thanks to (24). Let us consider $R_n(z) = (C_n - z)^{-1}$, $C_n = A + B_n$, which satisfies

$$R_m(z) - R_n(z) = R_n(z)(C_n - C_m)R_m(z) .$$

We rewrite

$$(C_n - C_m)R_m(z) = (B_n - B_m)(A + B_m - z)^{-1} =$$

$$= [(B_n - B_m)(A-z)^{-1}][(A-z)(A+B_m-z)^{-1}] .$$

We note first that

$$||(A+B_m-z)(A-z)^{-1}\psi|| \geq \Big| ||\psi|| - ||B_m(A-z)^{-1}\psi|| \Big| .$$

From (24) and the relation $||R_m(z)|| < |\text{Im}z|^{-1}$ we learn (for $|\text{Im } z| > |\text{Re } z|$) that

$$||B_n(A-z)^{-1}\psi|| \leq (a|\text{Im } z|^{-1} + b)||\psi|| = c\, ||\psi||$$

which for large enough z can be chosen so that $c < 1$. Thus

$$(A-z)(A+B_n-z)^{-1}$$

is a uniformly bounded operator since

$$||(A+B_n-z)(A-z)^{-1}\psi|| \geq \Big| ||\psi|| - ||B_n(A-z)^{-1}\psi|| \Big| \geq (1-c)||\psi||$$

is bounded away from zero. Moreover $(B_n - B_m)(A-z)^{-1}$ converges in norm which shows that (23) converges to zero (in fact uniformly). Since $(B-B_o)\, R_o(z)$ is bounded, it can, for sufficiently large z, be bounded by a constant less than unity. When this is the case the relation

$$R(z) = R_o(z) + R_o(z)(B-B_o)R(z)$$

has a convergent series solution

$$R(z) = R_o(z) \sum_{n=o}^{\infty} [(B-B_o)R_o(z)]^n . \qquad (26)$$

From the foregoing we conclude that R(z) has a densely defined inverse and that $R(z)^+ = R(z^*)$. Thus $R(z)=(C-z)^{-1}$ and C is self-adjoint. Such examples describe analytic perturbations, so-called because of the convergence of (26).

Proof of Glimm-Jaffe Theorem

To establish the Glimm-Jaffe theorem we study

$$\{R_n(z) - R_m(z)\}\psi = R_n(z)(B_n-B_m)M^{-1}M\, R_m(z)\, \psi$$

for a dense set of ψ. We note first that $||R_n(z)|| \leq |\text{Im } z|^{-1}$. We choose M so that

$$||(B_n-B_m)M^{-1}|| \to 0 ,$$

and we choose ψ in a dense domain (or total set) for which $||MR_m(z)\psi||$ is uniformly bounded in m and z. Under the stated assumptions $R_n(z) \to R(z)$ strongly. Since for $|\text{Re } z| < |\text{Im } z|$,

$$||zR_n(z)\psi|| \leq |z||\text{Im } z|^{-1}||\psi|| \leq 2||\psi|| ,$$

it follows that $||zR_n(z)||$ and $||zR(z)||$ are uniformly bounded by 2. To show that $R(z)\psi = 0 \to \psi = 0$ we study

$$||zR(z)\psi + \psi|| \leq ||zR(z)\psi - zR_n(z)\psi|| + ||zR_n(z)\psi + \psi||$$

for arbitrary $\psi \in \mathcal{H}$. If n is chosen sufficiently large the first term on the right side can be made arbitrarily small uniformly in z (always assuming $|\text{Re } z|<|\text{Im } z|$). If $|z|$ is chosen sufficiently large the second term can be made arbitrarily small as well. Hence for arbitrary ψ

$$\lim_{|z|\to\infty} ||zR(z)\psi + \psi|| = 0 ,$$

which means that $R(z)\psi = 0 \to \psi = 0$ since the null space $\underset{\sim}{N}$ is independent of z.

Now we apply the assumed condition of the Glimm-Jaffe theorem to pick M. Let B_n be a sequence of bounded self-adjoint operators for which $||(B_m-B_n)\psi|| \to 0$ on a dense domain. We suppose that

$$||A\psi||^2+||B_n\psi||^2 \le c\{||\psi||^2+||C_n\psi||^2\}$$

where $C_n = A+B_n$.

Thus for $\psi = (C_n-z)^{-1}\phi$ it follows that

$$||B_n(C_n-z)^{-1}\phi|| \le c'||\phi||$$

and so $B_n(C_n-z)^{-1}$ is uniformly bounded. With this guide choose

$$M = \{1+(A+B)^+(A+B)\}^{1/2} = \{1+C^+C\}^{1/2} \quad .$$

Then B_nM^{-1} is uniformly bounded and $||(B_m-B_n)M^{-1}|| \to 0$.
Next we consider

$$MR_m(z)\psi = (1+C^+C)^{1/2}(C_m-z)^{-1}\psi \quad .$$

It is assumed that the C_m have a common dense domain $\underset{\sim}{D}$, which is the range of the resolvents. Since vectors of the form $(1+C^+C)^{-1/2}\phi$ are in the domain of all C_m, they are in the range of all $(C_m-z)^{-1}$. On these $MR_m(z)\psi$ is uniformly bounded. Since these results apply for Im $z \lessgtr 0$, this establishes the theorem. Note that in general the series solution (26) for (25) does not converge in the present case.

4. FIELD THEORY MODELS

A. General Remarks

Let us consider the regularized models discussed in Section 1 with Hamiltonians

$$\mathcal{H}_\kappa(g) = \mathcal{H}_0 + V_\kappa(g)$$

that made use of a momentum-space and configuration-space cutoff. In which of those cases is $\mathcal{H}_\kappa(g)$ self-adjoint?

For the regularized Fermi case the interaction term is a bounded operator. In fact all operators of the form

$$W_p = \int [\bar{\psi}_\kappa(\vec{x})\psi_\kappa(\vec{x})]^p g(\vec{x})\, d\vec{x}$$

are bounded and it is linear combinations that enter $V_\kappa(g)$. We have

$$\begin{aligned} ||W_p|| &= ||\int (\bar{\psi}_\kappa\psi_\kappa)^p g(\vec{x})\, d\vec{x}|| \\ &\leq \int ||\bar{\psi}_\kappa\psi_\kappa||^p |g(\vec{x})|\, d\vec{x} \\ &\leq d^p |\mathcal{K}|_2^{2p} \int |g(\vec{x})|\, d\vec{x} \quad , \end{aligned}$$

where d is the dimensionality of the γ matrices. Since $V_\kappa(g)$ is bounded and symmetric, and $\mathcal{H}_0$ is self-adjoint it follows from Rellich's theorem, for example, that $\mathcal{H}_\kappa(g)$ is self-adjoint. For the Yukawa theory we have

$$\begin{aligned} V_\kappa(g) &= \int :\bar{\psi}_\kappa(\vec{x})\psi_\kappa(\vec{x}): \phi_\kappa(\vec{x}) g(\vec{x})\, d\vec{x} = \\ &= \int [\bar{\psi}_\kappa(\vec{x})\psi_\kappa(\vec{x}) - c]\phi_\kappa(\vec{x}) g(\vec{x})\, d\vec{x} \end{aligned}$$

where $c = \langle 0|\bar{\psi}_\kappa(\vec{x})\psi_\kappa(\vec{x})|0\rangle$ is independent of $\vec{x}$ and satisfies $|c| \leq d|\mathcal{K}|_2^2$. Thus

$$||V_\kappa(g)\psi|| \leq 2d|\mathcal{K}|_2^2 \int |g(\vec{x}) \; ||\phi_\kappa(\vec{x})\psi||\, d\vec{x} \; .$$

If $\psi_{(n)}$ is a state with exactly n particles then

$$\sqrt{2}||\phi_\kappa(\vec{x})\psi_{(n)}|| = ||\{a(\chi) + a^\dagger(\chi)\}\psi_{(n)}||$$

$$\leq ||a(\chi)\psi_{(n)}||+||a^\dagger(\chi)\psi_{(n)}||$$

$$\leq 2|\chi|\sqrt{n+1}\ ||\psi_{(n)}|| \qquad (27)$$

where $|\chi|^2=\int|\chi(\vec{k})|^2 d\vec{k}/\omega(k)$. Hence for an arbitrary vector ψ

$$||\phi_\kappa(\vec{x})\frac{1}{N+r}\psi||^2 \leq 2|\chi|^2\sum \frac{n+1}{(n+r)^2}||\psi_{(n)}||^2$$

$$\leq \sup_n\{2|\chi|^2 \frac{n+1}{(n+r)^2}\}||\psi||^2 \equiv \varepsilon^2||\psi||^2 .$$

By choosing r large enough we can make ε arbitrarily small. The byproduct

$$||\phi_\kappa(\vec{x})\psi|| \leq \varepsilon||(N+r)\psi|| \leq \varepsilon r||\psi|| +$$

$$+ \varepsilon||N\psi||\leq \varepsilon r||\psi||+ (\varepsilon/m)||\mathcal{H}_0\psi||$$

leads to the inequality

$$||V_\kappa(g)\psi|| \leq a||\psi|| + b||\mathcal{H}_0\psi|| \quad ,$$

where we can always choose $b < 1$. Thus Rellich's theorem applies: $\mathcal{H}_\kappa(g)$ is self-adjoint and the resolvent admits a perturbation expansion. Extensive study of this model has been made by Lanford [20]. The Hamiltonian for the regularized ϕ^4 theory is likewise self-adjoint. This can be seen as a consequence of the Glimm-Jaffe theorem. From (27) we see that $\phi_\kappa(\vec{x})(N+1)^{-1/2}$ is bounded by $\sqrt{2}|\chi|$. In like manner one shows that $(N+1)^{p/2}\phi_\kappa(\vec{x})(N+1)^{-(1+p)/2}$ is bounded by $2^{(1+p)/2}|\chi|$. Thus $||\phi_\kappa^4(\vec{x})\psi|| \leq 2^5|\chi|^4||(N+1)^2\psi||$ Since $V_\kappa(g)$ can be represented as multiplication in some L^2 space it is clear that $V_\kappa(g)$ is essentially self-adjoint on $\underset{\sim}{D}(N^2)$ (for the vectors with at most a finite number of particles are analytic vectors for $\phi_\kappa(x)$). Likewise $\mathcal{H}_\kappa(g)$ is at least symmetric on the domain $\underset{\sim}{D}(\mathcal{H}_o)$ $\cap$ $\underset{\sim}{D}(N^2)$, which is evidently dense.

We make plausible the conditions which ensure that $\mathcal{H}_\kappa(g)$ is self-adjoint. Clearly $\mathcal{H}_\kappa(g)$ is semibounded since both $\mathcal{H}_o$ and $V_\kappa(g) = \int :\phi_\kappa^4(\vec{x}):g(\vec{x})d\vec{x}$, $g(\vec{x}) > 0$, are semibounded. The condition (21b) can be made plausible as follows. Let $V_\kappa(g)$ be replaced by λq^4, the potential of an anharmonic oscillator and $\mathcal{H}_o$ by $\frac{1}{2}(p^2+q^2)$. Then commutation of V with $\mathcal{H}_o^{1/2}$ tends to lower the power of the potential by one power, and two such commutations crudely yield the result q^2 . In fact the correct answer can be bounded by some multiple of $q^2 + c$. This expression can always be swamped by $\varepsilon\mathcal{H}_o^2 + b(\varepsilon)$, for ε arbitrarily small. This qualitative argument carries through more generally to the problem at hand [21]. In this way the conditions of the theorem can be established and $\mathcal{H}_\kappa(g)$ is self-adjoint on $\underset{\sim}{D}(\mathcal{H}_o) \cap \underset{\sim}{D}(V_\kappa(g))$. Hence in all cases the regularized quantum field theory exists.

What about removing the regularizations? In the Yukawa theory in a 2-dimensional space-time the addition of mass renormalization counter terms permits the removal of the ultraviolet cutoff in momentum space [22]. Although the resultant Hamiltonian is semi-bounded, it is not essentially self-adjoint so that its extension is not unique. On the other hand, for the $(\phi^4)_2$ theory, the quartic self-interaction in two space-time dimensions, the ultraviolet cutoff can be removed leaving a self-adjoint Hamiltonian. It is this example we turn to now.

B. A (ϕ_2^4) Theory

It is possible to approach this model in two ways, one through the regularized model discussed above and the other through a model which begins initially with only a spacial cutoff. Each method is particularly useful to discuss certain questions. The latter approach is particularly useful to build a local quantum theory as we shall show, and it is made possible by the fact that in one space dimension $V(g) = \int :\phi^4(x): g(x)dx$ is actually an operator for suitable functions $g(x)$. (We drop the vector notation since space has only one dimension.) To see this fact let $\psi \in \mathcal{D}(N^4)$ and consider

$$||\int :\phi^4(x): g(x)dx\psi||^2 = \int \langle 0|\phi(x)\phi(y)|0\rangle^4 g(x)g(y)dxdy||\psi||^2 + R$$

where R is finite provided that the first term is finite. Since $\langle 0|\phi(x)\phi(y)|0\rangle$ has only logarithmic singularities in one space dimension it suffices for $V(g)$ to be densely defined that $g(x) \in \mathcal{S}$. (In higher space dimensions there are pole singularities of order (s-1) which are not integrable when raised to a power 2 or larger.) It is reasonable that $V(g)$ is essentially self-adjoint on $\mathcal{D}(N^4)$, say, and that $\mathcal{H}(g) = \mathcal{H}_0 + V(g)$ is symmetric on $\mathcal{D}(\mathcal{H}_0) \cap \mathcal{D}(N^4)$, which is dense. Even though $V(g)$ is not semibounded (thanks to normal ordering and no momentum cutoff) it has been shown by Nelson [23], Glimm [24] and by Federbush [25] that $\mathcal{H}(g)$ is in fact semibounded and fulfills the condition (21a). The remaining condition (21b) is plausible on the basis of the single-degree-of-freedom analog and can be shown to hold true [1]. Thus the Glimm-Jaffe theorem applies and $\mathcal{H}(g) = \mathcal{H}_0 + V(g)$ is self-adjoint on $\mathcal{D}(\mathcal{H}_0) \cap \mathcal{D}(V(g))$. A neat bit of analysis which we have only touched upon!

Locality : Let us define

$$\mathcal{H}_n = \mathcal{H}_o + V(g_n)$$

where $g_n(x) = g(x/n)$ and $g \in \mathcal{D}$ is chosen to have the value λ for $|x| \leq 1$ and to vanish identically for $|x| \geq 2$. For n large $\mathcal{H}_n$ approximates a relativistic Hamiltonian in a region of space near the origin. Also it should propagate influence no faster than the speed of light, which we take as unity. Let us consider

$$\phi_n(x,t) = \exp\{i\,\mathcal{H}_n t\}\phi(x)\ \exp\{-i\,\mathcal{H}_n t\}\ . \quad (28a)$$

The maximum causal interval is given by $|x|+|t|$ and thus for $n>|x|+|t|$ it is plausible that $\phi_n(x,t)$ no longer varies with n and determines a local field

$$\phi(x,t) = \lim_{n\to\infty} \phi_n(x,t) \quad . \quad (28b)$$

This was first noted in perturbation theory by Guenin [26] and shown to hold by Segal [27] granted the assumption that $\mathcal{H}_n$ is self-adjoint.

To make these notions more precise let us introduce the local algebras

$$\mathcal{U}(O) = \{e^{i\phi(f)}e^{-i\pi(g)}\}''$$

for f and g restricted to have support only in O where O is an open set of R. The double prime denotes double commutant which in this case can be taken as the closure under weak limits. The algebra $\mathcal{U}(O')$, where O' is the complement of the closure of O, commutes with $\mathcal{U}(O)$. Thus the commutant $\mathcal{U}(O')' \supseteq \mathcal{U}(O)$. Let supp $g \subset O$, then $V(g)$ commutes with $\mathcal{U}(O'_\varepsilon)$, and $\exp\{iV(g)t\} \in \mathcal{U}(O_\varepsilon)$, where O_ε is O expanded by ε. Since ε is arbitrary, it follows that $e^{iV(g)t} \in \mathcal{U}(O)$.

To discuss the locality let us consider the Trotter product formula [28]

$$\exp\{it\mathcal{H}(g)\} = \lim_{n\to\infty}(\exp\{it\mathcal{H}_o/n\}\exp\{itV(g)/n\})^n$$

which holds since $\mathcal{H}(g)$ is self-adjoint. Since this limit is in the strong sense se have

$$\sigma_t(A) \equiv \exp\{it\mathcal{H}(g)\}A\exp\{-it\mathcal{H}(g)\} =$$

$$= \lim_{n\to\infty} (\exp\{i\mathcal{H}_o t/n\}\exp\{iV(g)t/n\})^n \times$$

$$\times A(\exp\{-iV(g)t/n\}\exp\{-i\mathcal{H}_o t/n\})^n \quad .$$

Let O be the region defined by $|x| < M$, $t=0$, and let $A \in \mathfrak{A}(O)$. We split $g(x)$ into two C^∞ functions, $g(x) = g_1(x)+g_2(x)$, where $\operatorname{supp} g_1(x) \subset O_\varepsilon$, and $\operatorname{supp} g_2(x) \cap O_{\frac{\varepsilon}{2}}$ is empty. Then

$$\exp\{iV(g)t/n\} = \exp\{iV(g_1)t/n\} \exp\{iV(g_2)t/n\} \quad .$$

Furthermore $\exp\{iV(g_1)t/n\} \in \mathfrak{A}(O_\varepsilon)$ and $\exp\{iV(g_2)t/n\}$ commutes with $\mathfrak{A}(O_{\varepsilon/4})$. Thus because of the causal propagation of $\mathcal{H}_o$

$$A_1(t) = \exp\{i\mathcal{H}_o t/n\}\exp\{iV(g)t/n\}A \exp\{-iV(g)t/n\} \times$$

$$\times \exp\{-i\mathcal{H}_o t/n\}$$

is contained within $\mathfrak{A}(O_{t/n+\varepsilon})$ and depends only on $g(x)$ in O_ε . In like manner

$$A_n(t) = \exp\{i\mathcal{H}_o t/n\}\exp\{iV(g)t/n\}A_{n-1}(t)\exp\{-iV(g)t/n\}\times$$

$$\times \exp\{-i\mathcal{H}_o t/n\}$$

is contained within $\mathfrak{A}(O_{t+n\varepsilon})$. Since ε is arbitrary it follows that $A_n(t) \in \mathfrak{A}(O_t)$ for each n and depends on $g(x)$ only in the closed region $\bar{O}_t$ where we choose $g(x)=\lambda$. As a strong limit $\sigma_t(A) = \lim_{n\to\infty} A_n(t)$ preserves these features. Note that $\sigma_t(A)$ commutes with any $B \in \mathfrak{A}(O')$ where

O' and O_t are disjoint. Hence the spacial cutoff has been removed and the resulting theory is local!

Physical Hilbert Space

We have argued above that the $(\phi^4)_2$ theory leads to a local algebraic theory. Although the time automorphism is locally implementable (by H(g)) it is not globally so since H(g) does not converge on Fock space. To introduce the physical Hilbert space we must first discuss states in algebraic quantum theories. A state ω is a linear functional $\omega(A)$ on the elements $A \in \mathcal{U} = U_O \mathcal{U}(O)$ which satisfies $\omega(I) = 1$, $\omega(A^\dagger) = \omega(A)^*$ and $\omega(A^\dagger A) \geq 0$. We can view the algebra $\mathcal{U}$ as a C^* algebra if we employ a suitable norm $\|.\|$ and complete the algebra with respect to that norm. As such $\mathcal{U}$ becomes a Banach space, and ω an element of the dual space $\mathcal{U}'$. The weak topology on $\mathcal{U}'$ is defined by a system of neighborhoods (of zero) given by

$$N(\varepsilon;A_1,\ldots,A_n) = \{\omega: |\omega(A_j)| < \varepsilon \ , \ j=1,\ldots,n\}$$

and in this topology $\mathcal{U}'$ is <u>compact</u>. This means that any (generalized) sequence ω_n has a convergent subsequence.

Hence consider the collection of states on $\mathcal{U}$ defined by

$$\omega_n(A) = (\Omega_n, A\Omega_n) \tag{29}$$

where Ω_n is the ground state of $\mathcal{H}_n$, i.e., for $g_n(x)=g(\frac{x}{n})$. Such ground states are known to exist [1]. Although Ω_n does not converge in $\mathcal{F}$, $\omega_n(A)$ has a convergence subsequence so that $\omega_n(A) \to \omega(A)$.

Two simple examples of such behaviour may prove helpful. Consider

$$\Omega_n \equiv \exp\{i\phi(f_n)\} \ \Omega_o$$

where $f_n \in \mathcal{S}$ is a sequence converging pointwise to unity,

$f_n(x) \to 1$. Now $(X,\Omega_n) \to 0$ for all $X \in \mathfrak{X}$ but consider

$$\omega_n(\exp\{i[\phi(f)-\pi(g)]\}) = (\Omega_n, \exp\{i[\phi(f)-\pi(g)]\}\Omega_n)$$

$$= (\Omega_o, \exp\{i[\phi(f)-\pi(g)]\}\Omega_o) \times$$

$$\times \exp\{-i(g,f_n)\} \quad .$$

Thus in the limit

$$\omega(\exp\{i[\phi(f)-\pi(g)]\}) =$$

$$= (\Omega_o, \exp\{i[\phi(f)-\pi(g)]\}\Omega_o) \exp\{-i\int g(x)dx\} \ ,$$

which is a perfectly legitimate state, and in fact one which corresponds to an inequivalent representation of the canonical commutation relations. Consider another example, this time for a single quantum mechanical degree of freedom

$$\omega_n(\exp\{i(pQ-qP)\}) = (\psi_n, \exp\{i(pQ-qP)\}\psi_n)$$

where in a Schrödinger representation $\psi_n=(\omega n/\pi)^{1/4} \times$ $\times\exp\{-(\omega n x^2/2)\}$. Then

$$\omega_n = \exp\{-1/4[(n\omega)^{-1}p^2+(n\omega)q^2]\} \ .$$

Thus

$$\omega(\exp\{i(pQ-qP)\}) = 0 \quad , \quad p \neq 0$$
$$= 1 \quad , \quad q \neq 0 \quad . \qquad (30)$$

Such a functional characterizes a state, allright, but not a regular state. A regular state by definition retains the continuity properties (11) required to recover the generators of the Weyl system of operators. Our interest evidently lies with the regular states. It is this feature which almost assuredly is lacking if one were to bindly consider constructions of the form (29) when Ω_n is the ground state of $\mathcal{H}_\kappa(g)$ for a ϕ^4 theory in 4-dimensional space time as the cutoffs are removed. Although we are guaranteed that the states have a convergent subsequence no regularity property of the limiting state is ensured. The value of this procedure in the $(\phi^4)_2$ theory is that the regularity of the resultant state can be determined. In particular Glimm and Jaffe [1], using the method of Jaffe and Powers [29] , show that

$$\omega_n(\phi(f)^2+\pi(f)^2) \leq |f|^2$$

for some suitably continuous norm on the test functions f(x), a norm which is independent of n. From this it follows that any weak limit of the states ω_n is regular.

Gelfand-Naimark-Segal Construction

Assume we are given a state $\omega(A)$ on the elements $A \in \mathcal{U}$. Then it is always possible to realize this on a (new) Hilbert space $\mathcal{H}$ in such a way that

$$\omega(A) \equiv \langle v_1, \pi(A)\, v_1\rangle$$

where $\pi(A)$ is a cyclic (but not necessarily irreducible) representation of the operators, and v_1 is an element of the representation space $\mathcal{H}$. This representation is uniquely determined by ω up to unitary equivalence.

One canonical functional realization is defined as follows.Let the elements $A \in \mathcal{U}$ themselves be the functional arguments, and introduce a total set of vectors for $\mathcal{H}_\omega$

$$v_B(A) \equiv \omega(A^\dagger B) \equiv \mathcal{K}(A;B)$$

for all $B \in \mathcal{U}$. Every vector in $\mathcal{H}_\omega$ is expressible in the form

$$v(A) = \Sigma c_j v_{B_j}(A) = \Sigma c_j\, \mathcal{K}(A;B_j)$$

where only those sequences are allowed for which the $(\text{norm})^2$

$$\langle v,v\rangle \equiv \Sigma\, c_k^* c_j \mathcal{K}(B_k;B_j)$$

is finite. With a second vector $v'(A) = \Sigma c_k' \mathcal{K}(A;B_k')$ the inner product reads

$$\langle v',v\rangle = \Sigma c_k^{*\prime} c_j \mathcal{K}(B_k';B_j)$$

We note that

$$\langle \mathcal{K}(\cdot;B),v\rangle = \Sigma\, c_j \mathcal{K}(B;B_j) = v(B)$$

and thus this realization is a Hilbert space with a reproducing kernel [30]. The representation of the operators in this space is given by

$$(\pi_\omega(C)v_B)(A) = v_{CB}(A) = v_B(C^\dagger A) \quad .$$

Thus if $v_1(A) \equiv \omega(A^\dagger)$, then

$$(\pi_\omega(B)v_1)(A) = v_B(A)$$

and

$$\mathcal{K}(B;C) = \langle \pi_\omega(B)v_1, \pi_\omega(C)v_1\rangle \quad .$$

If the algebra $\mathfrak{A}$ is separable in the norm topology and the state ω is norm continuous, $\omega(A_n) \to 0$ if $\|A_n\| \to 0$, then the representation space is separable. Note this is <u>not</u> the case in the example (30). In that example the representation space is conveniently taken as that of almost periodic functions $v(k)$ such that

$$\langle v',v\rangle = \lim_{K\to\infty} \frac{1}{2K} \int_{-K}^{K} v^{*\prime}(k)v(k)\,dk \quad ,$$

where $v_1(k)=1$. It is easily shown that this is a nonseparable Hilbert space.

Properties of the $(\phi^4)_2$ Model

We summarize some of the known facts regarding this model. Besides the local algebras $\mathcal{O}\!\!\mathcal{l}(O)$ one can study the local fields $\phi(x,t)$ defined by (28). One initially shows that

$$\phi(f) = \int \phi(x,t) f(x,t) dx\, dt$$

is a self-adjoint operator for real $f \in \mathcal{D}$ where $\mathcal{H}_n$ is chosen for the specific f, i.e., $n > |x| + |t|$ for all $(x,t) \in \mathrm{supp}\, f$. This field operator fulfills local commutativity, commuting for support regions which are space like separated. Given this field, one considers a sequence of operators $\phi(f_n)$, $f_n \in \mathcal{D}$ for which $f_n \to f \in \mathcal{S}$. Whether or not this sequence of self-adjoint operators converges to a self-adjoint operator $\phi(f)$, $f \in \mathcal{S}$, can be ascertained by the resolvent techniques discussed in Section 3. By this route Glimm and Jaffe have constructed a local field $\phi(f)$ defined for all $f \in \mathcal{S}$ which is identical to the original family when $f \in \mathcal{D}$. In much the same fashion

$$A(t) = \int \phi(x,t) f(x,t) dx \ ,$$

$$B(t) = \int \pi(x,t) f(x,t) dx \ ,$$

define self-adjoint operators, first for $f \in \mathcal{D}$ and then from these for $f \in \mathcal{S}$.

A space-time translation automorphism exists for the local algebra and the local fields such that for suitable g

$$\sigma_{\alpha,t}(A) = \exp\{-i\alpha\mathcal{P}\} \exp\{it\,\mathcal{H}(g)\}\, A \exp\{-it\mathcal{H}(g)\} \times$$
$$\times \exp\{i\,\alpha\mathcal{P}\} \ ,$$

with the space translation generator $\mathcal{P}$ defined on Fock space as in (6). As such $\mathcal{P}$ and $\mathcal{H}(g)$ do not commute.

In the physical representation there is a global Hamiltonian $\mathcal{H} \geq 0$,and space translation generator $\mathcal{P}$ which do commute and transform the physical fields properly. The existence of a well defined $\mathcal{H}$ hinges on the existence of unique normalizable ground states Ω_n for $\mathcal{H}_n$ with eigenvalue zero. In such a case

$$\omega_n(A(t)B) = (\Omega_n, A(t)B\Omega_n) =$$
$$= (\Omega_n, A\ \exp[-it\mathcal{H}_n]B\ \Omega_n) \quad ,$$

so that $\omega_n \to \omega$ where

$$\omega(A(t)B) = \langle v_1, \pi(A(t))\pi(B)v_1\rangle$$
$$= \langle v_1, \pi(A)\pi(e^{-it\mathcal{H}})\pi(B)v_1\rangle \quad .$$

Existence and spectral properties follow from this relation. Construction of a space translation invariant state ω can be carried out if the limiting operation ω_n includes, in addition, an ever widening average over space translations - in the limit this average leaves an invariant state.

Some important questions about this theory remain unanswered at this time. A boost generator has not been constructed so complete covariance has not been established. It is not known whether the physical vacuum is unique, or the related question, whether cluster decomposition holds. It is not known whether the physical representation of the canonical operators is irreducible or reducible. One can only hope that these questions will be clarified in the near future.

It is a pleasure to thank Professors Glimm and Jaffe for the opportunity to see their manuscripts prior to publication.

BIBLIOGRAPHY

1. J. GLIMM and A. JAFFE, A $\lambda(\phi^4)_2$ Field Theory Without Cutoffs I, Phys. Rev. 176, 1945(1968) ; A $\lambda(\phi^4)_2$ Field Theory Without Cutoffs II (to be published); A. JAFFE, Varenna Lectures 1968, Academic Press (to be published).
2. A. JAFFE, O. E. LANFORD III and A. S. WIGHTMAN, "A General Class of Cut-Off Model Fields " (to be publ.)
3. R. STREATER and A. S. WIGHTMAN, PCT, Spin & Statistics, and All That, W. A. BENJAMIN, Inc. (1964); L. STREIT, Bull. Am. Phys. Soc. 14, 86 (1969).
4. A.S.WIGHTMAN, 1964 Cargese Summer School Lectures, M. LEVY, editor, Gordon and Breach (1967).
5. R. HAAG and D. KASTLER, J. Math. Phys. 5, 848 (1964).
6. M. A. NAIMARK, Normed Rings, (translated by L. F. Boron), P. Noordhoff, Ltd. (1964), Chapter VI.
7. J. von NEUMANN, Math. Ann. 104, 570 (1931).
8. L. GARDING and A. S. WIGHTMAN, Proc. Nat. Acad. Sci . 40, 617 (1954); J. R. KLAUDER, J. McKENNA and E. J. WOODS, J. Math. Phys. 7, 822 (1966).
9. J. R. KLAUDER, J. Math. Phys. 6, 1666 (1965).
10. R. T. POWERS, Comm. Math. Phys. 4, 145 (1967).
11. K. SINHA and G. G. EMCH, Bull. Am. Phys. Soc. 14, 86 (1969).
12. H. ARAKI, J. Math. Phys. 1, 492 (1960).
13. J. R. KLAUDER, J. Math. Phys. 8, 2392 (1967).
14. H. GRIMMER, "Model Field Theories", University of Edinburg thesis (1968).
15. J. R. KLAUDER and E. C. G. SUDARSHAN, Fundamentals of Quantum Optics, W. A. Benjamin, Inc. (1968), Chapter 7.
16. F. RIESZ and B. Sz.-NAGY, Functional Analysis, Fredrick Ungar Publ.Comp. (1955).

17. E. NELSON, Annals of Math. (Princeton) 70, 572 (1959)
18. T. KATO, Perturbation Theory for Linear Operators, Springer Verlag (1966).
19. J. GLIMM and A. M. JAFFE, "Singular Perturbations of Self Adjoint Operators", Comm. Pure and Applied Math. (to be published).
20. O. E. LANFORD III, Construction of the Dynamics for a Cutoff Yukawa Theory, A. M. S. Memoir, to appear.
21. A. M. JAFFE, Dynamics of a Cutoff $\lambda\phi^4$ Field Theory, A. M. S. Memoir, to appear.
22. J. GLIMM, Comm. Math. Phys. 5, 343 (1967); 6,120 (1967).
23. E. NELSON, Mathematical Theory of Elementary Particles, R. Goodman and I. Segal, Editors, M.I.T.Press, 1966.
24. J. GLIMM, Comm. Math. Phys. 8, 12 (1968).
25. P. FEDERBUSH, J. Math. Phys. 10, 50 (1969).
26. M. GUENIN, Comm. Math. Phys. 3, 120 (1966).
27. I. SEGAL, Proc. Nat. Acad. Sci. USA 57, 1178 (1967).
28. H. F. TROTTER, Pacific Journal of Math. 8, 887 (1958).
29. A. JAFFE and R. POWERS, Comm. Math. Phys. 7, 218 (1968).
30. N. ARONSZAJN, Proc. Cambridge Phil. Soc. 39, 133(1943); Trans.Am. Math. Soc. 68, 337 (1950).

DUALITY IN STRONG INTERACTION PHYSICS†

By

M. JACOB
CERN - Geneva

1. INTRODUCTION

The purpose of this series of lectures is not to introduce a theory, but rather an approach to strong interaction physics. This approach appears at present as fruitful and exciting and many papers exploring the consequences of duality have recently been published or are being circulated in preprint form. Nevertheless, I should mention to begin with that the whole matter is still in a controversial form. There is not yet agreement on how far duality can be used, since no precise definition for it is yet available. As a consequence, results presented here as definitive success are still bitterly challenged there. I will therefore not even try to start by a tentative definition which will emerge only as we proceed. In any case, after Horn's series of lectures [1] on the use of finite energy sum rules, a very good idea of what it should be has already been gathered.

I shall start by listing a few facts which, refusing to call for mere coincidences, we will try to correlate in the most simple way, thus building a duality approach

---

† Lecture given at the VIII. Internationale Universitätswochen für Kernphysik, Schladming, Feb. 24 - March 8,1969.

to strong interaction physics where Regge behaviour and resonances are interlocked.

This will lead us to the prominent features of a simple bootstrap scheme based on rising Regge trajectories. This is not a full bootstrap programme. Nevertheless it is a simple way of reaching some general properties of two-body collision amplitudes which should, to a very good approximation, result from a more refined but still apparently too complicated approach, and this, independently of the details of so far poorly known inelastic mechanisms. We will then introduce the Veneziano model as providing a tentative answer to the solution looked for which meets many of the required properties. We will then see how it yields in fact very simple expressions for meson-meson scattering where current algebra constraints together with broken SU(3) relations can be indeed most simply incorporated. We will then close on a few remarks about its generalization to many particle production and reactions among particles with spin.

2. RISING TRAJECTORIES

The known particles show striking regularities when set on a Chew-Frautschi diagram. Several particles can now be safely associated with the same Regge trajectory. This is in particular the case for the three baryon trajectories which are drawn on Fig. 1, and which respectively include the Δ, the Y_1^* and the Λ [2], [3]. All three trajectories are amenable to a straight line approximation with a common slope α'. Another striking fact is the possibility of superposing on each of these trajectories another trajectory with opposite signature. Such coincident trajectories have in most cases different

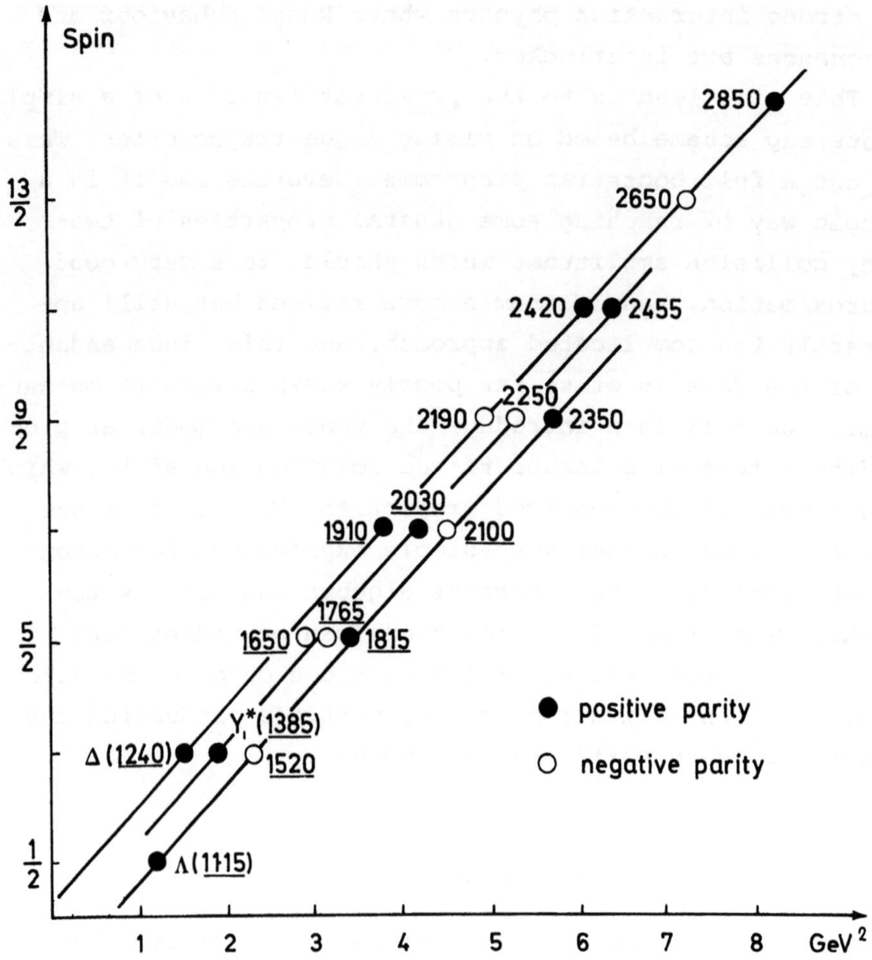

Figure 1 : Three baryon trajectories (Δ, Y_1^*, Λ) and their exchange degenerate partners. The negative signature trajectory which coincides with the Δ trajectory corresponds to isospin $\frac{1}{2}$. The underlined masses correspond to well-established spins and parities.

residue functions but their superposition will not be considered as a coincidence. As a result this will be associated with the absence of either direct or exchange forces in the channel where the corresponding resonances occur, a property which will in turn have to be interpreted in the realm of unitarity and crossing symmetry [3].

This degeneracy between trajectories of opposite signatures can be used to help drawing meson trajectories since only in the case of the ρ we already have two particles present: the $\rho(1^-)$ and the $g(3^-)$. The $f(2^+)$ indeed falls on the line joining the ρ and g points as if we had degenerate trajectories with opposite signatures. This assumption is supported by the similarity of the two trajectories determined for negative t values from the high energy scattering data. As a matter of fact, Regge analysis would favour slightly different values of the ρ and f (or P') trajectories in the scattering region. Nevertheless we will here refuse to consider their close similarity as a mere coincidence, or consider their degeneracy as a good approximation. The ω and the A_2 [4] have masses which are respectively very close to the masses of the ρ and the f. Drawing Regge trajectories through the corresponding points with the slope of the ρ trajectory, a procedure supported by the Regge analysis of high energy data, we get four almost coincident trajectories with isospin I=O and 1 and, for each isospin, the two signatures. Here again we probably oversimplify the picture since the masses involved show but an approximate equality and high energy Regge fits would prefer slightly different intercepts and behaviour. However, it is tempting to see here again more than a coincidence and to try to interpret such an apparent redundancy of parameters. We also note at this point that the equality in masses of the ω and ρ on the one hand, and of the f and A_2 on the other hand, follow from "ideal" nonet structure with a common mixing angle such that:

$$tg^2\Theta_V = tg^2\Theta_T = \frac{1}{2} \tag{1}$$

where V and T, respectively, stand for vector and tensor.

In this mixing scheme, we expect that the two other members of each nonet are equally spaced in mass squared. This is fairly well met by the facts. Furthermore, on the Chew-Frautschi diagram, the straight lines joining the $K^*890(1^-)$ to the $K^*1420(2^+)$ and the $\phi(1^-)$ to the $f'(2^+)$, as two coincident Regge trajectories with opposite signatures, have within a good approximation a common slope equal to the ρ slope and are equally spaced. This is displayed on Fig. 2 .

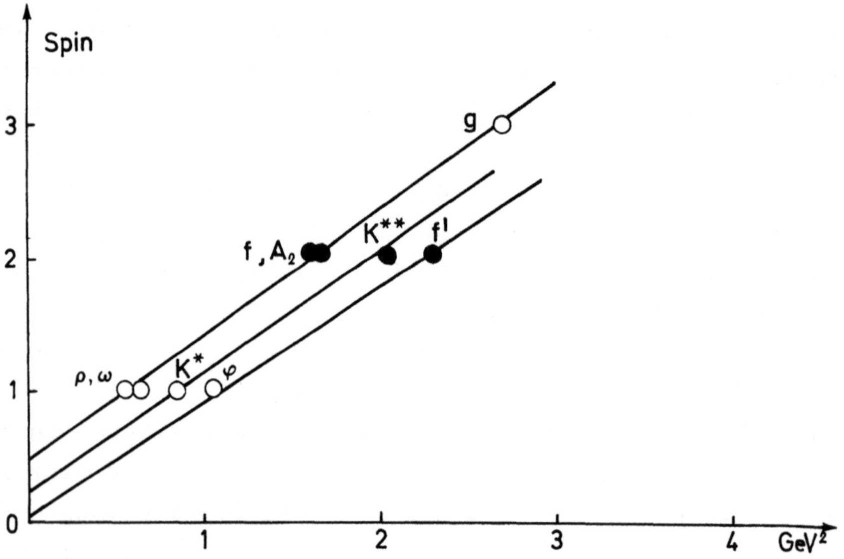

Figure 2 : Regge trajectories for the vector and tensor mesons. The black (white) dots correspond to positive (negative) parity.

This leads us to consider as an important fact the degeneracy of meson Regge trajectories of opposite signatures. We have so far considered only the natural parity trajectories and leave the others for later discussion. The natural parity trajectories are indeed the only relevant ones in meson-baryon scattering where at present these ideas can be submitted to experimental tests.

A third fact which will also not be considered as a coincidence is the equality between the slopes of the meson and baryon trajectories. All slopes are very similar, of the order of 1 GeV^{-2}, and may - within a good approximation - be considered as identical. This strongly suggests the possibility of a bootstrap between the different trajectories, the baryon resulting from the presence of forces associated with meson (and baryon) exchange in the crossed channels, and vice versa, self-consistency then imposing a common slope for all trajectories.

One must stress that such a bootstrap based on linearly rising trajectories may at present appear as too bold a goal. It is very likely that as energy increases, new structures will appear and that rising trajectories extrapolated from the yet sketchy data which we have (Figs. 1 and 2) should be considered as only a simple approximation [6]. Nevertheless, the basic philosophy underlying the bootstrap approach is that nearby singularities are enough to build themselves out in a self-consistent way, the effect of distant and unknown singularities being lumped together into an assumed simple asymptotic behaviour. We may therefore try to understand the low energy baryons lying on linear Regge trajectories in terms of the low energy mesons also lying on linear Regge trajectories and vice versa. We consider that the observed equality between the baryon and meson slopes is a strong hint of doing so. The observed degeneracy in quantum numbers is a further hint. Any resonance being associated with a particular set of internal quantum numbers, this calls for equal (or opposite) forces due to different sets of quantum numbers in the crossed channels. Baryons with different isospin would otherwise in general show degeneracy in mass. This is not met by the facts and imposes that some forces (and therefore here some meson trajectories) should show degeneracy in internal quantum numbers.

Our aim is therefore to build a bootstrap approach in which linear Regge trajectories in one channel result from the exchange of linear Regge trajectories in the crossed channels, thus linking together the direct channel resonance and crossed channel Regge pole descriptions of collision processes [7]. Whether Regge trajectories continue to rise indefinitely or eventually fall is not relevant here since our bootstrap approach postulates the existence of an approximate self-consistent solution involving only nearby singularities in all channels.

3. FINITE ENERGY SUM RULE BOOTSTRAP

The present basic difficulty with strong interaction physics is the impossibility to yet safely separate out a non-trivial but still simple process which, once understood, would give much insight for more complicated reactions. All processes are connected through unitarity and we cannot hope to understand in detail two-body collisions, about which we already have an impressive amount of data, without tackling at the same time more complicated production processes about which we still know very little.

In order to try to get anyway some understanding of two-body processes which we could readily test against experiment, we are forced to make guesses, writing down tentative expressions which incorporate many properties and carry but a very few adjustable parameters. In so doing, we hope that information yielded by unitarity and not yet exploited by lack of appropriate means, will eventually fix the parameters but not completely upset the present tentative form tried for collision amplitudes. This is in particular the case with a Regge amplitude used in the analysis of high s, low t results

$$F(s,t) = \sum_i \beta_i(t) \frac{\xi_i - \exp\{-i\pi\alpha_i(t)\}}{\sin\pi\alpha_i(t)} (s/s_o)^{\alpha_i(t)} \qquad (2)$$

where $\xi = \pm 1$ stands for signature ∓ 1. It is a priori not incompatible with unitarity, when a one-particle exchange expression:

$$F(s,t) \simeq \frac{B(t)}{t - M^2} P_J(\cos\Theta_t)$$

with $\cos\Theta_t = (u-s)/(4m^2-t)$, written for equal mass spinless scattering, would be.

Relation (2), however, does not explicitly satisfy unitarity and our hope in using it is that unitarity will eventually help fixing β and α but not fully upset (2) as a result of calling for dominating Regge cuts as opposed to dominant Regge poles. As far as we know, at present this hope is met by the facts [8].

In the present approach we will also have to make a guess in order to write a tentative expression for two-body reaction amplitudes but we will now try to incorporate Regge behaviour in all channels and this together with the presence of resonances, the resonances in one channel building up the Regge trajectories in the crossed channels and vice versa.

It is obvious that a finite set of resonances [9], which we may write in a zero width approximation as

$$F(s,\cos\ ) \simeq \sum_i \frac{B_i}{s-M_i^2} P_{J_i}(\cos\theta) \qquad (3)$$

will not give a Regge asymptotic behaviour, but rather a fixed pole behaviour at high s and that on the other hand a Regge amplitude such as (2) will not give s channel poles as usually associated with resonances. Imposing unitarity shifts the real axis poles, as they appear in the K matrix, say, to the unphysical sheets. Only an infinite set of s channel resonances can yield an asymptotic Regge behaviour at large s and only an infinite set of Regge trajectories may lead to a second sheet pole. These two approximations to two-body amplitudes, the first one (s channel resonances) being fruitful at low energies (< 2 GeV, say) and the second one (Regge poles) being fruitful at high energies (> 5 GeV, say) are far from equivalent when stopped short after a few terms, as legitimate in order to analyze either low or high energy experiments. Nevertheless, it has been realized that the two descriptions have a large overlap so that many features readily interpreted in terms of Regge pole exchange can be fruitfully analyzed as co-operative effects of direct channel resonances and conversely that properties most easily associated to the dominance of resonances can be interpreted as due to relations among Regge trajectories.

Relations (2) and (3) may then be considered at the same time as two possible ways to describe a reaction amplitude in the sense of asymptotic series. One is thus led to introduce a duality between the resonance and Regge pole description of hadron collisions [10]. Both expansions, when stopped short after a few terms correspond to two incomplete but complementary ways of describing a particular process. As a consequence, even at relatively low energies, the sum of all resonance contributions corresponding to a particular set of internal quantum numbers in the crossed channel should be well approximated in a semi-local sense [11] by the extrapolation of the Regge amplitude, with the same quantum numbers, determined from the analysis of high energy data. To the extent that such a property is met by the facts at energies where the amplitude is still very well known in terms of phase shift or multipole expansions, so that it can be safely continued to unphysical t values, the Regge trajectories can be determined from the low and intermediate energy data in terms of finite energy sum rules [1] (FESR):

$$\int_0^N \nu^n \{\mathrm{Im}\{F(\nu,t)\}d\nu = \sum_i \beta_i(t) \frac{N^{\alpha_i(t)+1+n}}{\alpha_i(t)+1+n} \qquad (4)$$

where $\nu = (s-u)/2m$.

Continuing to timelike t values, one eventually reaches right signature integer (or half-integer) points on the trajectory, the crossed channel particles being thus determined in terms of the direct channel ones [12]. The finite energy sum rules result from a combination of analyticity, which leads to the possible existence of superconvergence relations, Regge asymptotic behaviour, which warrants writing such superconvergence relations and l'Hopital's rule which, after a simple regrouping of terms [8] , directly leads to (4), as an asymptotic relation for large N. Provided that N can be taken low enough, this leads to a simple and powerful bootstrap scheme which has already met with several successes [10].

The main point of (4) is its linear character so that one may separate out the exchange of the quantum numbers of the vacuum, and in particular the problems associated with the complicated structure of the Pomeranchuk singularity, from the contributions of other well established simple Regge trajectories However, finite energy sum rules may lead to a bootstrap only to the extent that continuation to timelike t values is possible, which we shall do in an approximation with zero width resonances and real Regge trajectories. This means that we will thus have to neglect constraints imposed by unitarity on the ground that crossing symmetry is really important enough to yield a solution. This departs from standard N/D bootstrap, when one tries to keep both unitarity and crossing symmetry, but very poorly indeed however when it comes down to actual calculations.

As written explicitly in (4), the direct channel resonances "build up" the Regge trajectory in the crossed channel.

4. REGGE POLES AND ARGAND LOOPS

Conversely, Regge behaviour implies the presence of direct channel resonances or at least phase variations in partial wave amplitudes which could be attributed to the presence of resonances. Relation (2) imposes a fixed t structure to the differential cross section up to $\log(s/s_o)$ terms which we will first neglect. In a semiclassical picture this corresponds to a fixed impact parameter and therefore to an increase of "effective" angular momentum with energy given by

$$< J > \sim \sqrt{s} \tag{5}$$

The phase of F, as given by (2), is further fixed with t when each partial wave is associated with a particular Θ behaviour. Since Θ and t are related by:

$$t \sim \frac{s}{2}(1-\cos\Theta) \tag{6}$$

a phase fixed with t for all s implies phase variations in all partial waves. As can be readily checked in (2), each partial wave will increase and then decrease with energy, the corresponding maximum being reached for higher s values, the higher the angular momentum J [13]. Furthermore, with standard Regge parameters such as those of the ρ trajectory, the real part increases first and is then followed by the imaginary part. An anti-clockwise loop is thus eventually obtained on an Argand diagram, a feature which is generally associated with the presence of a resonance [14].

In order to analyze this in a more quantitative way, we can consider the partial wave projection of the phase factor in (2) [15]. It brings the most conspicuous t variation at relatively low energy, when $\log(s/s_o) \ll \pi$. Taking for simplicity a linear trajectory of slope α', this gives:

$$\int e^{-i\pi\alpha(t)} P_J(\cos\Theta)\, d\cos\Theta = e^{-i\pi(\alpha(o)-2q^2\alpha')} i^J j_J(-2q^2\alpha') \tag{7}$$

where we have introduced the spherical Bessel function:

$$j_J(z) = \int e^{iz\cos\Theta} P_J(\cos\Theta)\, d\cos\Theta$$

For each J value, the phase of the partial wave amplitude increases with $s\,(q^2 \sim s/4)$ and eventually reaches $\pi/2$ for $s=s_J$. For another partial wave, $J' = J + \Delta J$, the same phase will be reached but for a different value of s, $s_{J'}$ with - as readily seen from (7):

$$s_{J'} = s_J + \frac{\Delta J}{\alpha'} \tag{8}$$

One thus obtains from the Regge amplitude a phase behaviour for the partial wave amplitudes as if a series of direct channel resonances were present, these resonances being further spaced in energy as if they were on a rising trajectory with the same slope α' as the exchanged Regge trajectory.

This analysis is only very rough. Nevertheless, it carries the basic properties involved and is supported by a more detailed one taking all factors into account, but which has then to be carried out on a computer [15]. The Argand loops thus obtained do not yet correspond to direct channel resonances. The amplitude written as in relation (2), with a finite sum of Regge poles, does not have poles in s as normally associated with resonances [16]. A good approximation on the real axis (as an asymptotic series), it probably becomes very bad when continued to an unphysical sheet.

We will nevertheless assume that a partial wave phase increasing through $\pi/2$ (or $\pi/2 + n\pi$) obtained from a Regge amplitude indicates the presence of a direct channel resonance. The phase behaviour is taken as a common feature of (2) and (3). Only a more detailed Regge description of the amplitude, in terms of poles and cuts, would yield second sheet poles in s [8]. The Regge poles alone are however enough to locate these s poles. As a result exchange of linear Regge trajectories implies linear direct channel trajectories and vice versa, all trajectories having the same slope. Moreover, the presence (or absence) of resonances in one channel is to be associated with relations among the exchanged trajectories which summarize the forces involved. This is the full implication of the duality between the Regge pole and direct channel resonance descriptions of collision amplitudes as we will understand it here.

It should be stressed though that, if we assume that the conspicuous overlap between the two descriptions is meaningful, one should take seriously into account only big effects which should remain in a more refined analysis and disregard small ones which are most likely associated with the inaccurate character of either (2) or (3). In particular we will consider a Regge trajectory as definitively built of direct channel resonances only to the extent that different moments in (4) yield compatible Regge parameters. Conversely, we will take into account an Argand loop only if it does not lead to an excessively small inelasticity parameter. A tiny loop is very unstable against variation of the Regge parameters which we know but imperfectly, whereas a large and neat one is not. The recently much debated question of ancestors, Refs. [13], [17], is considered to offer a typical example of such spurious effects.

We are therefore aiming at a tentative expression for two-body collision amplitudes with direct and crossed channel resonance poles equally spaced in s and t and asymptotic Regge behaviour at high s and high t, the two properties being fully interlocked.

Relations (8) and (5) are not incompatible. Their simultaneous presence tells us that several lower rising trajectories have to be included in such a way that the contributions from these lower trajectories become so important as s increases that the "effective" $\langle J \rangle$ value is definitely smaller than the highest resonance angular momentum.

5. DUALITY AND POMERANCHUK SINGULARITY

The direct channel resonances being closely associated with exchanged Regge trajectories, it is necessary to leave aside the leading vacuum singularity with effective pole intercept 1 at t = 0. Since the positivity of cross sections requires that no trajectory with internal quantum numbers other than those of the vacuum can be degenerate with the leading vacuum singularity, direct channel resonances, if associated with Pomeranchon exchange, would show isospin degeneracy when this is not observed. The Pomeranchon should then not appreciably contribute to direct channel resonances when all other trajectories may do. This is indeed the case, following the previous analysis of partial wave amplitudes, if it is approximated as an almost flat Regge trajectory with $\alpha_P(0) = 1$; $\alpha'_P << 1\ \mathrm{GeV}^{-2}$.

The corresponding contribution to (2) is pure imaginary for all t values and its partial wave components will not have a phase varying with energy as obtained in (7). At least with $\alpha'_P << 1\ \mathrm{GeV}^{-2}$, its effect cannot be related to the observed resonance pattern.

Excluding the Pomeranchuk contribution from the bootstrap scope, one is therefore led to a scheme where only the resonances in one channel generate resonances in crossed channel, all resonances lying on approximately linear Regge trajectories. Appreciable forces are therefore to be considered only when resonances exist. They do not result from unitarity alone, whether resonances are present or not, as they do in ordinary N/D bootstrap. As a result, either direct or exchange forces may be absent from many channels where Pomeranchon exchange could have been a priori expected to generate these forces, even in the absence of resonances.

This separation of the Pomeranchon contribution, as possible in the linear bootstrap scheme offered by FESR, has several direct consequences [18]. Resonances being associated to the exchange of well established trajectories such as the ρ or the $f^{o}(P')$ trajectories with $\alpha(o) \sim \frac{1}{2}$ and $\alpha' \sim 1$ GeV^{-2}, the absence of resonances calls for exchange degeneracy between crossed channel trajectories. Two trajectories with opposite signatures have to coincide with the same α but further with opposite residues such that their sum would be purely real for all t values in the scattering region. Following (2) and (7), no resonances result. Conversely, as a result of the absence of resonances in a particular channel, one type of forces (direct or exchange) is missing and, as stressed already, resonances with both signature have to be present, thus leading to exchange degenerate trajectories.

If the standard Regge amplitude is real, the total cross section is proportional to the Pomeranchon contribution alone and should be constant already at relatively low energies instead of showing the standard fractional power Regge decrease [19]. Such properties are well met by the facts. As an example the $K^{+}p$ total cross section, a reaction with no direct channel resonance observed, is almost constant as compared to the decreasing $K^{-}p$ total cross section, a reaction where conspicuous resonances show up. Absence of resonances in $K^{+}p$ also calls for exchange degeneracy between the ρ and A_2 trajectories when both trajectories are indeed found to coincide (Fig. 2). Broken SU(3) nonet structure for exchanged trajectories requires the standard mixing angle $tg^2\theta = \frac{1}{2}$ [20], so that the singled out ϕ decouples.

In this scheme the Pomeranchon appears as the over-all effect of all the inelastic channels coupled to the particular elastic channel studied. A bootstrap solution is first obtained from the resonances alone in all channels. The cumulative effects obtained through unitarity from all production reactions are then globally added in each elastic channel as a Pomeranchukon contribution. As a result, the Pomeranchon should be considered only at relatively high energy when many channels get open. It should not be present below conspicuous inelastic thresholds, below 1 GeV, say, in $\pi\pi$ scattering.

6. EXOTIC STATES AND EXCHANGE DEGENERACY

The bootstrap scheme which is thus being built up is then based upon resonances as generating the only important forces. It lies, of course, on the assumed dominance of relatively nearby singularities which are then associated to rising Regge trajectories. In the framework of such a nearby singularity approach, one may then fully exclude from the scheme all types of possible resonance states which happen not to be found experimentally, even though they might well show up at higher energies. It is well known that all the presently unambiguous resonance states can be classified according to SU(3) octets and singlets (for mesons) and decuplets,octets and singulets (for baryons). In effect, all states which can be built out of a quark and an antiquark (for mesons) or three quarks (for baryons) [21]. This excludes in particular resonances with isospin or strangeness 2. Such states which cannot appear in the standard quark model are called "exotic" states and one then takes as a basic property that exotic channels do not generate appreciable forces. As a result, exchange degeneracy is expected in both crossed channels whenever an exotic channel is present.

As a particular example of exchange degeneracy, we

will consider here K^+ nucleon scattering, following a recent paper by Schmid [2]. Both the K^+p and K^+n channels are exotic ones and, as a matter of fact, no unambiguous resonances exist in these channels. The presence of an exotic channel implies then, according to our rules, exchange degeneracy in both crossed channels (Fig. 3): $K\bar{K} \to N\bar{N}$ (channel 2) and $\bar{K}N \to \bar{K}N$ (channel 3) .

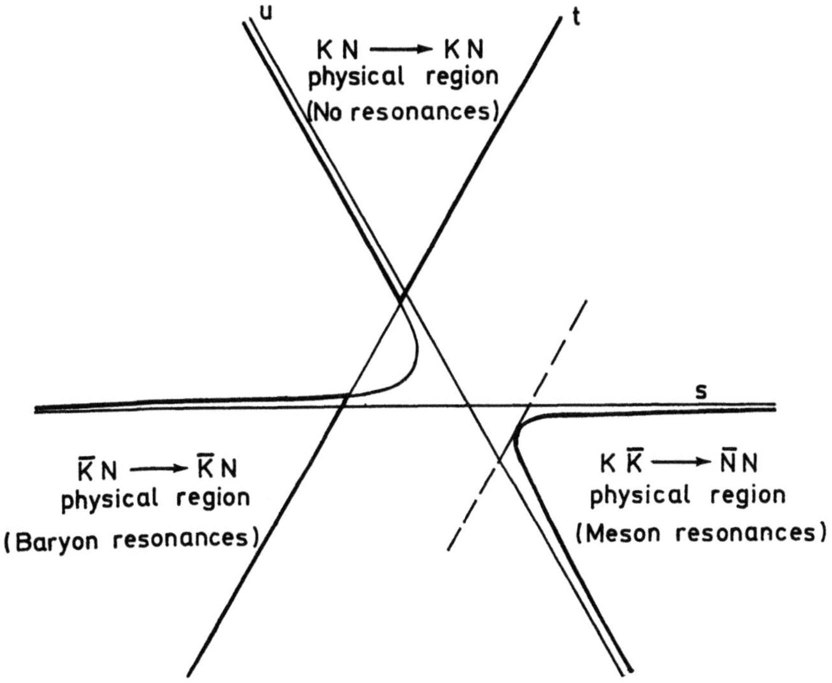

Figure 3 - Mandelstam diagram for KN→KN scattering. Absence of resonances in one channel (KN) corresponds to exchange degenerate resonances in both crossed channels.

In channel 2 one finds the ρ, ω, A_2 and f trajectories, the φ and f' being only very weakly coupled to the $N\bar{N}$ system. These four Regge trajectories sould show exchange degeneracy and since it must work for both K^+p and K^+n scattering, the isospin 1 and 0 trajectories have to be separately exchange degenerate

$$\alpha_\rho = \alpha_{A_2} \qquad \beta_\rho = -\beta_{A_2} \qquad (9)$$
$$\alpha_\omega = \alpha_f \qquad \beta_\omega = -\beta_f$$

When going from K^+N scattering to K^-N scattering, the ρ and ω couplings (odd under C) change sign, whereas the f and A_2 couplings (even under C) stay the same. Accordingly the Regge amplitude :

$$F(s,t) \simeq \sum_i \beta_i(t) \frac{\xi_i - e^{-i\pi\alpha_i(t)}}{\sin\pi\alpha_i(t)} (s/s_o)^{\alpha_i(t)}$$

which excludes the Pomeranchon, purely real for K^+N scattering becomes complex and Argand loops now occur in each partial wave, thus indicating the actual hyperon resonances. The Regge amplitude for charge exchange:

$$K^o + p \rightarrow K^+ + n$$

is, however, purely real with no Pomeranchon present. Figure 2 shows how the exchanged trajectories indeed appear to coincide. The absence of forces in the exotic $\pi^+\pi^+$ channel in ππ scattering further implies exchange degeneracy between the ρ and f trajectories so that the four trajectories coincide.

A similar degeneracy should further hold for channel 3. Indeed, as shown on Fig. 2, two exchange degenerate Y_o^* and Y_1^* trajectories can be drawn through known resonances, and the observed couplings are compatible with the equality of the residue functions [2]. It should be stressed, though, that this leaves aside several hyperon resonances which we cannot associate with exchange degenerate trajectories as one a priori should. Such as the $Y_o^*(1405,\frac{1}{2}^-)$; the $Y_o^*(1690,\frac{3}{2}^-)$, the $Y_o^*(1830,\frac{5}{2}^-)$; and the $Y_o^*(1660,\frac{1}{2}^-)$ on the one hand and the $\Sigma(1190,\frac{1}{2}^-)$ and $Y_1^*(1910,\frac{5}{2}^+)$ on the other. This leads us to two important points about FESR bootstrap. We are dealing with one particular set of channels connected through crossing and, if we draw conclusions from the absence of forces in the exotic K^+N channel, they have to affect only those states conspicuously coupled to the $\bar{K}N$ (channel 3) or $N\bar{N}$ (and $K\bar{K}$) systems (channel 2). We should not draw any conclusion for states relatively weakly coupled to those particular channels. This is indeed the case for all the resonances left apart which we already listed [22] except for the $Y_o^*(1405,\frac{1}{2}^-)$ and $Y_o^*(1660,\frac{1}{2}^-)$ states. As a first rule, we should then use duality arguments only among strongly coupled states and not trust them a priori with channels about which we know almost nothing (say, A_1 exchange in $\varepsilon\pi$ elastic scattering). The two leftover Y_o^* states have a very low spin for their relatively high mass. They therefore contribute but a negligible amount in "building up" the crossed channel Regge trajectories which, as already stressed, get their major contribution from those partial wave amplitudes which are prominent at high unphysical values of $\cos\Theta_s$:

$$\cos\Theta_s = 1 + \frac{t}{2q^2} \qquad (t \sim m_\rho^2)$$

They are the resonating peripheral partial waves. The low partial waves are of almost no effect on the leading Regge trajectories. Exchange degeneracy therefore tells us almost nothing about the low partial waves. Our bootstrap scheme relates peripheral partial waves to leading trajectories and leaves aside all secondary effects. One cannot use it a priori to make refined prediction such as polarization and we cannot conclude for instance that exchange degeneracy implies the absence of polarization effects in KN charge exchange [23]. At most can we say that the polarization due to interference between the high (real) and low (complex) partial waves should a priori show little structure.

A similar remark holds for NN charge exchange where the absence of dibaryon resonance also implies exchange degeneracy and therefore real peripheral waves.

In standard N/D bootstrap exchange degeneracy does not occur so generally as a result of the presence of coupled channels where both direct and exchange forces are present. In this simplified approach we obtain but a first guess of the whole amplitude which does not incorporate such effects. This, however, clearly shows some general constraints of the bootstrap solution which are to be at least approximately satisfied by the exact solutions and which can readily be reached.

To sum up, we aim at building a model where narrow resonances would be the only relevant objects, in scattering. This leads to many obvious oversimplifications but allows us to deal with two-body scattering apart. We guess directly at a solution which would satisfy crossing symmetry if not unitarity. We hope that this simple bootstrap approach, even though very approximate, will already teach us many things about resonance location and properties. The equality of slopes of all leading trajectories and present evidence for exchange degeneracy are strong hints that this is a fruitful track.

7. GENERALIZED INTERFERENCE MODELS

The duality model presented here is, of course, not the only way of getting resonances and Regge behaviour in all channels. As shown by Van Hove [25] an infinite sum of resonances can result in an asymptotic Regge behaviour. The scattering amplitude written as

$$F(s,t) = \Sigma^{\pm} \frac{b(J,t)}{M_J^2-t} P_J(\cos\Theta_t)$$

where the sum runs over even (+) or odd (-) non-negative integer values of J, can be rewritten in terms of a Sommerfeld-Watson transformation. Its high s behaviour is then dictated by the leading pole in J (the location of which depends on t) of the continued partial wave amplitude: $b(J,t) \times (M^2(J)-t)^{-1}$.

If the t channel resonances are associated with a rising Regge trajectory, $\alpha(M^2(t)) = J$, $M^2(\alpha(t)) = t$, the partial wave amplitude will have a pole at $J = \alpha(t)$ which will correspond to Regge behaviour $(-s)^{\alpha(t)}$ at high s. The contribution of an infinite set of t channel resonances can then be resummed as a t channel Regge pole.

It is possible to build explicit examples of amplitudes which have poles in t and behave as $(-s)^{\alpha(t)}$ as s goes to infinity. Such functions may be further constructed as to have cuts in s and t as imposed by unitarity in such a way that the poles are on the unphysical sheet. The residues of the pole in t are polynomial in s [26]. If it is further imposed that the function goes to zero faster than a power when t goes to infinity, as shown in Ref. [26], a sum of two such functions, where s and t are interchanged, will have resonance poles and Regge behaviour in both channels, even though no relation will exist between the resonances in one channel and the Regge poles in the crossed channel. Such an amplitude does not show duality and allows full dissociation between prominent resonances, such as the Δ, and asymptotic Regge behaviour [27]. This provides a generalized interference model. It is, however, much different from standard interference model adding somewhat casually resonance contributions to Regge pole contributions since the constraints imposed on either pieces of the amplitude lead in practice to the presence of a strong Regge background interfering with the resonance pole contributions. The residue function would therefore be in general very different from what one would assume considering standard Breit-Wigner formulae to describe the prominent resonances [27].

In these lectures we will consider that resonances and Regge behaviour are fully interlocked. As already stressed, the subject should however be considered as still controversial, both the dualist and the generalized interference approaches corresponding probably to too radical choices.

8. DUALITY DIAGRAMS

Isolating a particular set of two-body channels related through crossing, duality will not always act in such a way that the observed resonances in one channel can be considered as generated from the observed resonances in both crossed channels. Cancellations may occur in such a way that only one channel is relevant even if both do not carry exotic resonances. Consider as an example $K\pi$ scattering (Fig. 4a). Only isospin $\frac{1}{2}$ K^* resonances are to be considered in the s and u channels. They cannot mutually sustain themselves since a force with isospin $\frac{1}{2}$ in one channel will obviously generate at the same time effects in the isospin $\frac{1}{2}$ and $\frac{3}{2}$ states in the crossed channel. Selection of the isospin $\frac{1}{2}$ state can only result, in a bootstrap scheme based on nearby, and therefore non-exotic, singularities, from a particular relation between the I=0 and I=1 t channel forces obtained through ρ f exchange degeneracy. Conversely $I=\frac{1}{2}$ forces in the s and u channels will create both I=0 and 1 effects in the t channel.

Duality will therefore manifest itself between the s and t, u and t but not s and u channels. Such a property can be simply exhibited on duality diagrams where the SU(3) properties or quark content of each particles are clearly written out. Contraction between the SU(3) indices of the external particles, as obtained in terms of exchanged vector or tensor mesons in this case (or their corresponding Regge trajectories), is graphically represented as quark exchange (Figs. 4b and 4c, obtained from each other using s $\leftrightarrow$ u exchange). The coupling of a Regge trajectory to the external meson corresponds to

$$\begin{aligned} &\mathrm{Tr}(M[V,M]) && \text{F coupling, vector meson} \\ &\mathrm{Tr}(M\{T,M\}) && \text{D coupling, tensor meson} \end{aligned} \tag{10}$$

The pseudoscalar mesons here considered are represented by 3×3 matrices and the curly bracket stands for the anticommutator. We exclude a coupling of the type:

$$\mathrm{Tr}(T)\ \mathrm{Tr}(MM) \tag{11}$$

the absence of which can be seen to correspond to the vanishing of the $f'\pi\pi$ coupling once the f' and f are defined using the standard 35° nonet mixing angle; it is already forbidden by C invariance for vector mesons [28].

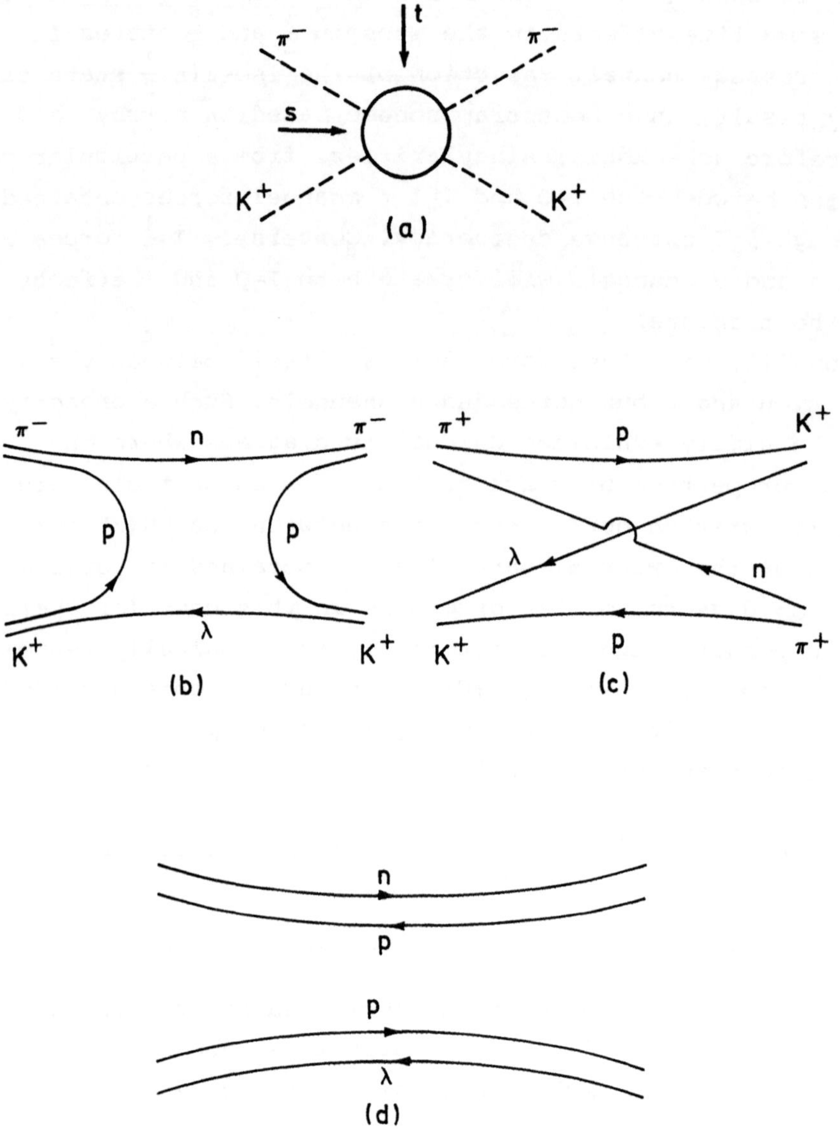

Figure 4 - Duality diagrams for Kπ scattering

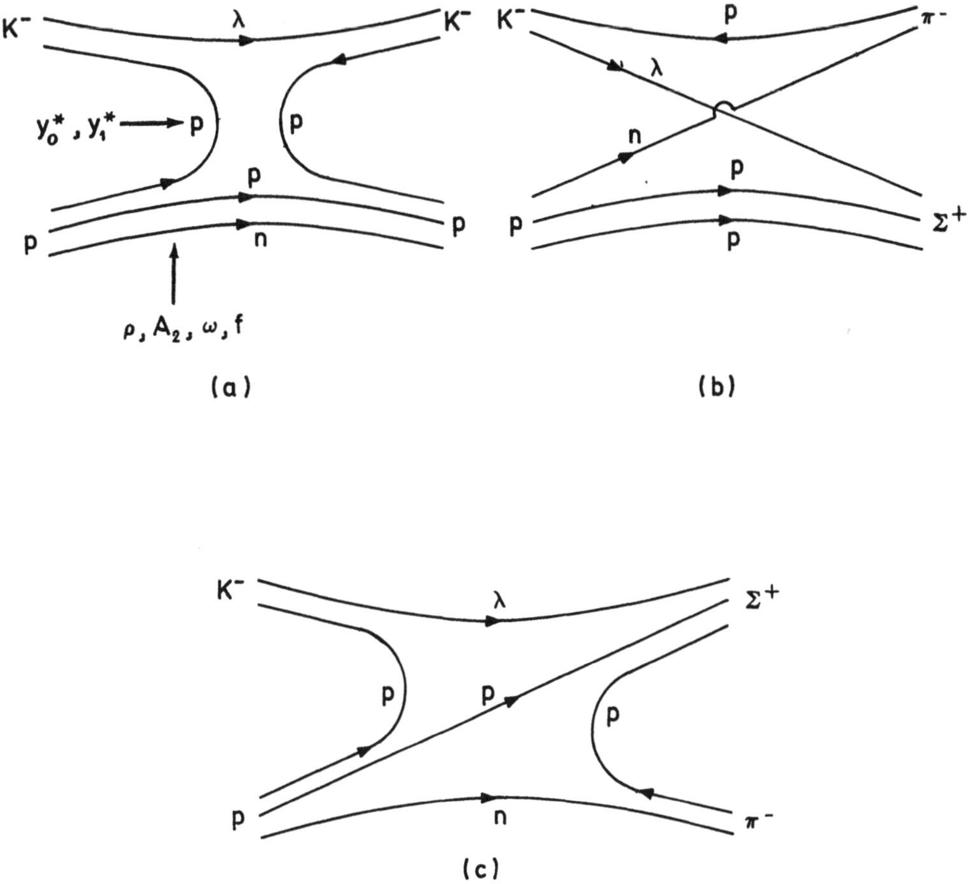

Figure 5 - Duality diagrams for $K^-p \to K^-p$ and $K^-p \to \Sigma^+\pi^-$ scattering .

With (10), and not (11), as possible couplings, the SU(3) indices of the two incoming particles are always connected through the exchanged meson indices. As a result, all graphs are connected as in Figs. 4b and 4c. Graphs like 4d, which would be associated with (11), are excluded. All possible diagrams being connected graphs, they can be "planar", as in 4b, or "not planar" as in 4c. The rule is then very simple [29],[30]. Duality is effective among channels associated with a planar graph. There is s,t duality (4b). The s channel K^* resonances in the s channel "build up" an imaginary part for high s which is to be associated with non exchange degenerate Regge exchange in the t channel. On the other hand, there is no s-u duality (4c). The $K^+\pi^+$ amplitude should be purely real at high s but for the Pomeranchon contribution. In other words, two channels connected by duality should have non-exotic quantum numbers and this is a fortiori the case when visualized in terms of quark-antiquark pairs exchanged in both channels, as in 4b, since a coupling like (11) is excluded. The rule is readily extended to meson-baryon scattering, each external baryon being represented in terms of three quarks. All graphs are again connected provided that the ϕ and f' are not coupled to the $N\bar{N}$ system, these two states being again defined together with the ω and f through standard nonet mixing : $tg^2\theta = \frac{1}{2}$ [31].

We have here merely stated the rule for the presence or absence of duality relation between two channels. Its derivation based on standard SU(3) and Regge pole analysis is given in Ref. [30]. Applications to several interesting cases and extension to many particle production are described in Ref. [29].

Coming back to $\bar{K}$ nucleon scattering as a further example we have s,t duality (Fig. 5a) and we would have u,t duality with K instead of $\bar{K}$. We do not have s,t duality for the reaction $K^-p \to \Sigma^+\pi^-$ (Fig. 5b), but u,t and s,u duality (Fig. 5c). This is particularly interesting. The resonances in the K^-p channel should not build up the exchange K^* trajectory. The t channel resonance having isospin $\frac{1}{2}$, this would have implied degeneracy between the I=0 and I=1, Y^* trajectories in order to exclude isospin $\frac{3}{2}$ resonances. This would go against experimental information (Fig. 1). On the other hand, the K^* trajectory is built from the N^* resonances in the u channel (u-t duality). This implies exchange degeneracy between decuplet and octet states and is indeed compatible with present information (Fig. 1). The $K^-p \to \Sigma^+\pi^-$ amplitude should be purely real at high s [30],[31] and small t values (no s-t duality). As already stressed, the rules obtained from duality diagrams imply more than SU(3) symmetry and the absence of exotic states. The decoupling of one of the nonet members from particles without strange quark here assumed will be necessary to incorporate standard broken SU(3) (SU(3) couplings but nonet (or octet) mass splitting) in the FESR bootstrap scheme. We will therefore use them to limit the number of possible terms in the amplitude.

9. VENEZIANO AMPLITUDE

After discussing the general ideas underlying a tentative linear bootstrap solution, we now apply them to the very simple and most interesting model proposed by Veneziano [32], which offers solutions exhibiting duality properties. We first consider for simplicity an amplitude with only s and t channel resonances ($\pi^+\pi^- \to \pi^+\pi^-$, say). The amplitude should show explicitly a string of resonances (poles) in both the s and t channels located on the linearly rising ρ and f Regge trajectories. The absence of force (resonance) in the $\pi^+\pi^+$ u channel implies exchange degeneracy of these two trajectories and resonances in both the s and t channels should then be spaced by one unit of spin, instead of two. The amplitude should also have s-t crossing symmetry, and asymptotic Regge behaviour in both channels. This is indeed the case for

$$V_{xy}(s,t) = -\lambda \frac{\Gamma(1-\alpha_x(s))\Gamma(1-\alpha_y(t))}{\Gamma(1-\alpha_x(s)-\alpha_y(t))} \tag{12}$$

where x and y label the trajectories exchanged in the s and t channel. Here both correspond to the ρ-f degenerate trajectories. We are working in a zero width approximation and the Regge trajectories are real and linear

$$\alpha(s) = \alpha(o) + \alpha' s \qquad \alpha' > 0 \tag{13}$$

Relation (12) shows poles in s and t associated with a zero or negative integer value of the gamma function arguments, that is, with $\alpha(o) \sim \frac{1}{2}$, for timelike s and t values. It has further no double pole. This is important since one should be able to associate each pole term in (12) to particle exchange. It means that the residue of each pole should be a polynomial in the crossed variable. When rewritten in terms of Legendre polynomials, it will correspond to a sum of parent and daughter contributions. A double pole would correspond to an infinite series expansion in the crossed variable leading to arbitrary high spin values (ancestors) which cannot be tolerated. The denominator in (12) indeed cancels all double poles and the residue of each pole is the ratio of two gamma functions which is readily rewritten as a polynomial in $\alpha(t)$ (for a pole in s), which, together with (13), gives a polynomial in t. The pole structure is illustrated in Fig. 6. For $\alpha(s) = J \geq 1$, we have together with the leading particle exchange with spin J a series of daughters with spins J-1,J-2,...,0. Relation (12) implies therefore an

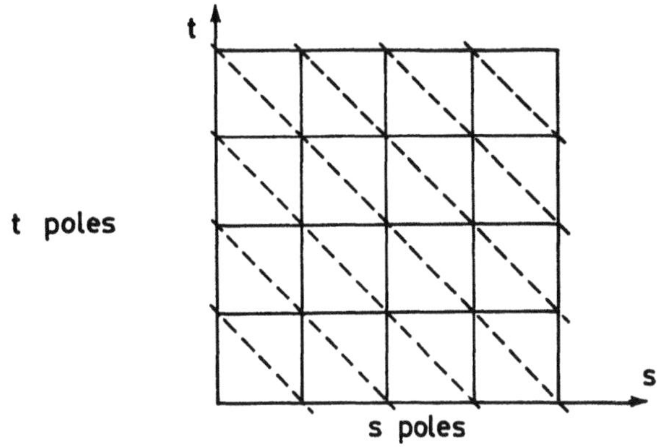

Figure 6 - s and t channel poles (solid lines). The dashed lines correspond to zeros (pole of the denominator) which eliminate double poles.

infinite series of equally spaced daughter trajectories (Fig. 7).

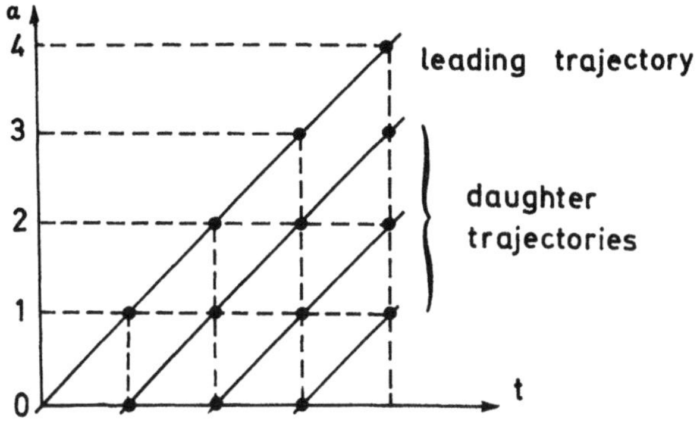

Figure 7 - Leading trajectory (two degenerate trajectories with opposite signatures) and daughter sequence.

For large values of the argument the ratio of two gamma functions is given by:

$$\frac{\Gamma(z+a)}{\Gamma(z+b)} = z^{a-b}(1+\frac{2}{2z}(a-b)(a+b-1)+O(z^{-2})) \qquad (14)$$

for $-\pi < \arg z < \pi$. Used in (12), with the appropriate limit procedure, this leads to asymptotic Regge behaviour:

$$\begin{aligned} V_{xy}(s,t) &\sim \Gamma(1-\alpha_y(t))(-\alpha's)^{\alpha_y(t)} \qquad s \to \infty \\ V_{xy}(s,t) &\sim \Gamma(1-\alpha_x(s))(-\alpha't)^{\alpha_x(s)} \qquad t \to \infty \end{aligned} \qquad (15)$$

V, as defined in (12), can be rewritten in terms of the Euler beta function

$$\begin{aligned} B(x,y) &= \int_0^1 u^{x-1}(1-u)^{y-1}du = \frac{\Gamma(x)\Gamma(y)}{\Gamma(x+y)} \\ V_{xy}(s,t) &= (1-\alpha_x(s)-\alpha_y(t))B(1-\alpha_x(s),1-\alpha_y(t)) \end{aligned} \qquad (16)$$

Relation (12), although extremely simple, cumulates many required properties for the solution we are guessing at. It has resonances and Regge behaviour and they are deeply interlocked as imposed by duality. Its basic drawback is of course its inconsistency with unitarity. In order to impose an analytic structure at least compatible with unitarity, we could depart from linear trajectories and introduce a branch point at the lowest threshold:

$$\begin{aligned} \alpha(t) &= \alpha(o) + \alpha't + \frac{t^2}{\pi}\int_{4m^2}^{\infty} \frac{\text{Im}\,\alpha'(t)}{t'^2(t'-t-i\varepsilon)}\,dt' \\ &\text{Im}\,\alpha(t) \neq 0 \text{ for } t > 4m^2 \end{aligned} \qquad (17)$$

This preserves crossing symmetry but introduces ancestors since the residue at one pole will no longer be a polynomial in the crossed variable. We could also impose unitarity in one channel, using the Baker-Blankenbecler method [24] or Padé approximations. This would, however, spoil crossing symmetry. Trying to guess a tentative solution for two-body scattering without using any definite information from many particle production amplitudes, we should not worry too much at not reaching an a priori perfect answer. We will use (12) on the ground that crossing symmetry is the main property and that it should be kept exactly, hoping that a bootstrap solution built, with zero width resonances, is not too far apart, as far as the general resonance pattern is concerned, from a solution which would take into account unitarity constraints which we cannot yet all incorporate. Relation (12) is built on $\rho(f)$ exchange and neglects the Pomeranchon contribution. As already mentioned, Pomeranchon exchange should be considered only at high energies where inelasticity becomes important. Relation (12) is then suitable for a "low" energy approximation to the $\pi\pi$ scattering amplitude. Nearby singularities are included as linear trajectories. The effect of distant singularities is lumped in the imposed Regge behaviour.

A term like:

$$V'_{xy}(s,t) = -\lambda' \frac{\Gamma(m-\alpha_x(s))\,\Gamma(n-\alpha_y(t))}{\Gamma(m+n+p-\alpha_x(s)-\alpha_y(t))} \tag{18}$$

has the same basic properties as (12). The first poles appear at $\alpha_x(s) = m$ and $\alpha_y(t) = n$ instead of 1 (we will take $m,n \geq 1$) and the asymptotic behaviours correspond, respectively to

$$s^{\alpha_y(t)-n-p} \quad \text{and} \quad t^{\alpha_x(s)-m-p}$$

with $n+p$ $(m+p) > 0$, this corresponds to daughter behaviour. Poles in s (t) will have polynomial residue in t (s) provided that $p \leq o$. We are therefore at liberty to add any such terms, called "satellite", to the initially considered solution (12), without modifying any of the desired properties we require. This is an important ambiguity in the bootstrap solution and such terms can be used to modify it at will at the daughter level without changing the leading trajectory contributions. Satellites can be added to eliminate unwanted daughter contributions and in particular odd daughters [33], if we want the solution to show trajectories spaced by two units of spin as corresponding to the minimum number of trajectories. Satellites will not affect the leading high energy behaviour directly obtained from (12).

In order to determine a signature factor, we consider an amplitude with specific s-u symmetry (t channel, I=0(1), $K\pi$ scattering, say, as discussed in detail later)

$$F^I = -\lambda^I (V_{xy}(t,s) + (-1)^I V_{xy}(t,u)) =$$

$$= -\lambda^I \left\{ \frac{\Gamma(1-\alpha_x(t))\Gamma(1-\alpha_y(s))}{\Gamma(1-\alpha_x(t)-\alpha_y(s))} + \right.$$

$$\left. + (-1)^I \frac{\Gamma(1-\alpha_x(t))\Gamma(1-\alpha_y(u))}{\Gamma(1-\alpha_x(t)-\alpha_y(u))} \right\} \tag{19}$$

We use the relation:

$$\Gamma(z)\,\Gamma(1-z) = \pi(\sin\pi z)^{-1} \tag{20}$$

to rewrite (19) as

$$F^I = -\frac{\pi\ \lambda^I}{\Gamma(\alpha_x(t))\sin\pi\alpha_x(t)} \left\{ (-1)^I \frac{\Gamma(1-\alpha_y(u))}{\Gamma(1-\alpha_x(t)-\alpha_y(u))} + \right.$$

$$\left. + \frac{\sin\pi(\alpha_x(t)+\alpha_y(s))}{\sin\pi\alpha_y(s)} \frac{\Gamma(\alpha_x(t)+\alpha_y(s))}{\Gamma(\alpha_y(s))} \right\} \sim \tag{21}$$

$$\sim -\frac{\pi\lambda^I}{\Gamma(\alpha_x(t))} \left\{ \frac{\cos\pi\alpha_x(t)+(-1)^I}{\sin\pi\alpha_x(t)} \left(\frac{s}{s_o}\right)^{\alpha_x(t)} + \right.$$

$$+ \text{cotg}\pi\alpha_y(s)\left(\frac{s}{s_o}\right)^{\alpha_x(t)} \Bigg\} \qquad (21)$$

when $|s| \to \infty$, $\arg s \to 0^+$.

The first term corresponds to the properly behaved real part with a signature factor associated with s-u symmetry. The second term corresponds to the absorptive part. It has a succession of poles on the real axis. At this state one may smooth them out, letting the imaginary part of α_y increase with s as it normally should. One then has $\text{cotg}\alpha_y(s) \to -i$ as $s \to \infty$ and obtains from (21) the standard Regge behaviour:

$$F^I \sim \beta^I(t)\,\frac{(-1)^I + \exp\{-i\pi\alpha_x(t)\}}{\sin\pi\alpha_x(t)}\left(\frac{s}{s_o}\right)^{\alpha_x(t)} \qquad (22)$$

Two important properties appear in this analysis. First s_o, as it is introduced in (21), is indeed α'^{-1}, the energy scale is therefore 1 GeV, as indicated through standard Regge pole analysis. Second, the residue function includes a factor $1/\Gamma(\alpha_x(t))$, which vanishes for zero or negative integer values of $\alpha_x(t)$. This corresponds to nonsense choosing with Gell-Mann mechanism [34], when writing a Veneziano term for each invariant amplitude in order to extend (22) to reactions with particles with spin (π nucleon scattering [35], for instance). Both the helicity flip and non-flip amplitude vanish when $\alpha(t) = 0$.

10. RESONANCE WIDTH IN THE VENEZIANO MODEL

In the neighbourhood of an s channel pole $\alpha(s) \sim J$ one may write $\Gamma(1-\alpha(s))$ as:

$$\Gamma(1-\alpha(s)) \sim \frac{(-1)^J}{(J-1)!\,\alpha'(s-M_J^2)}$$

with $\alpha(M_J^2) = J$. As a result, the pole residue in (12) is given by:

$$\lambda \frac{(2qq')^J}{(J-1)!} \alpha'^{J-1} \cos^J\Theta + O(\cos^{J-1}\Theta) \tag{23}$$

with $t = -2qq'(1-\cos\Theta)$, q and q' standing for the initial and final center of mass momenta.

The constant λ thus determines all pole residues on the rising trajectory and the coefficient of the leading power in $\cos\Theta$ is (in the spinless case) enough to determine the coupling constant and then the width of the leading trajectory resonances. This result is not affected by satellite terms.

Writing the $\rho\pi\pi$ coupling as: $G_\rho \varepsilon.(q_1+q_2)$, and therefore the ρ pole contribution in $\pi\pi$ scattering as: $((4qq')/M^2-s))G^2$ we get from (12) and (23):

$$G_\rho = \frac{\lambda}{2}$$

and

$$\Gamma_\rho = \frac{1}{12m_\rho^2} \frac{\lambda}{8\pi} (m_\rho^2 - 4m_\pi^2)^{3/2} \tag{24}$$

a width of 100 MeV corresponds to $\lambda = 50$. The ratio of the ρ and f elastic widths is readily obtained from (23). One finds:

$$\Gamma_\rho = 1.2\ \Gamma_f$$

The main point is that the range parameter which allows us to connect decay widths of resonances of different spins and therefore with decay coupling of different dimensions is the square root of the universal slope $\alpha' = 0.9\ \text{GeV}^{-2}$. Such a relation among the leading trajectory widths could be, for instance, checked through a detailed study of the $p\bar{p} \to \pi\pi$ reaction at low and intermediate energies.

The daughter width associated with a single term can be obtained from the expansion

$$B(x,y) = \sum_{J=0}^{\infty} (-1)^J \binom{y-1}{J} (x+J)^{-1}$$

or

$$V_{xy}(s,t) = \sum_{J=0}^{\infty} \binom{\alpha_y(t)+J+1}{J} \frac{1-\alpha_x(s)-\alpha_y(t)}{J+1-\alpha_x(s)} \qquad (25)$$

The ratio of the ρ and ε^0 width is then found to be 2/9, the ε^0 being the first ρ f daughter. This implies the existence of a wide (450 MeV) s wave resonance at the mass of the ρ [36]. At the same time, a p wave daughter is expected at the mass of the f, with a similar elastic width (9/10 ratio). This conclusion now relies, however, upon the absence of satellite terms. This ρ' resonance could, however, be already highly inelastic and therefore not spoil the isospin determination of the f which is known to be peripherally produced as the ρ' would also be.

As seen from (25), the daughter couplings will in general increase with the daughter order and, as the energy increases, daughters should then take a more and more important role in elastic scattering. The "effective" angular momentum will therefore increase much less than linearly with s (as $\sqrt{s}$ as a first approximation) to give a fixed forward peak up to logarithmic terms. The parent still plays a prominent part, however, when going to high unphysical t value and in particular when reaching the crossed channel trajectories.

11. PION-PION SCATTERING

We already wrote an amplitude for $\pi^+\pi^-$ scattering (12), taking into account the absence of force in the u channel. There is only one trajectory involved: the exchange degenerate ρ f trajectory. We exclude the f', assumed to have no coupling to $\pi\pi$ as a result of standard nonet mixing and a particular ratio of octet and singlet couplings. Had we kept the f' and considered it as separated from the f, we would have needed a new ρ' trajectory exchange degenerate with the f'. A broken SU(3) nonet structure therefore imposes the decoupling of the f' [20]. Considering

$$V_{\rho\rho}(s,t) = -\lambda \frac{\Gamma(1-\alpha_\rho(s))\Gamma(1-\alpha_\rho(t))}{\Gamma(1-\alpha_\rho(s)-\alpha_\rho(t))}$$

as the simplest tentative amplitude for $\pi^+\pi^-$ scattering, as opposed to more complicated terms involving satellites, the three s channel amplitudes can be written, imposing obvious symmetry relations, as:

$$\begin{aligned} S_0 &= \beta(V_{\rho\rho}(s,t) + V_{\rho\rho}(s,u)) + \gamma V_{\rho\rho}(t,u) \\ S_1 &= \alpha(V_{\rho\rho}(s,t) - V_{\rho\rho}(s,u)) \\ S_2 &= V_{\rho\rho}(t,u) \end{aligned} \tag{26}$$

The isospin crossing matrix between the s and t channel being

$$M_{ts} = \frac{1}{6}\begin{pmatrix} 2 & 6 & 10 \\ 2 & 3 & -5 \\ 2 & -3 & 1 \end{pmatrix}$$

with $T_i = (M_{ts})_{ij} S_j$.

One may exclude t poles in T_2, forbidden as exotic

resonances, provided that

$$\gamma = -\frac{1}{2} \qquad \beta = \frac{3}{2}\alpha$$

We should further impose that the pole at $\alpha(t) = 1$ in the T_o amplitude vanishes. This gives $\alpha = 1$. All $\pi\pi$ amplitudes can then be written in terms of one single constant λ [37]. We choose to write:

$$A(\pi^+\pi^- \to \pi^+\pi^-) = -\lambda\, V_{\rho\rho}(s,t) \tag{27}$$
$$A(\pi^+\pi^o \to \pi^+\pi^o) = -\frac{\lambda}{2}\,(V_{\rho\rho}(u,t) + V_{\rho\rho}(s,t) - V_{\rho\rho}(s,u))$$
$$A(\pi^o\pi^o \to \pi^o\pi^o) = -\frac{\lambda}{2}\,(V_{\rho\rho}(s,t) + V_{\rho\rho}(t,u) + V_{\rho\rho}(u,s))$$

An extremely important point about (12) is that the amplitude vanishes when $\alpha_\rho(s) + \alpha_\rho(t) = 1$, or with linear trajectories when $s+t = 2\mu^2$ with:

$$\alpha_\rho(\mu^2) = \frac{1}{2}$$

This coincides with the Adler condition [38] which requires the $\pi\pi$ amplitude to vanish when $s=t=u=m_\pi^2$, provided that $\mu^2 = m_\pi^2$. Such a condition is compatible with what is known about the ρ trajectory [37] and it is therefore tempting to write:

$$\alpha_\rho(t) = \frac{1}{2} + \alpha'(t-m_\pi^2)$$
$$\alpha' = \frac{1}{2(m_\rho^2-m_\pi^2)} = 0.88\ \text{GeV}^{-2} \tag{28}$$

thus having the Adler relation satisfied by (26) without having to call for satellite terms or a vanishing of λ with one of the external masses. Relation (26) can then be used to continue the $\pi\pi$ amplitude off the mass shell, considering s, t and u as independent variables. Its linear terms can further be identified with the Weinberg amplitude [39], since both amplitudes vanish at $s=t=u=m_\pi^2$ [40],[41]. The Weinberg amplitude reads:

$$F_{\beta\delta,\alpha\gamma} = -\frac{1}{f_\pi^2}\{\delta_{\alpha\gamma}\delta_{\beta\delta}(s-m_\pi^2)+\delta_{\alpha\beta}\delta_{\gamma\delta}(t-m_\pi^2)+\delta_{\alpha\delta}\delta_{\beta\gamma}(u-m_\pi^2)\}$$

with

$$S_0 = -\frac{1}{f_\pi^2}(3s+t+u-5m_\pi^2)$$

$$S_1 = -\frac{1}{f_\pi^2}(t-u)$$

$$S_2 = -\frac{1}{f_\pi^2}(t+u-2m_\pi^2) \qquad (29)$$

Writing a linear approximation to (12) around $s=t=u=m_\pi^2$ one gets $-\lambda\pi\alpha'(s+t-2m_\pi^2)$ and identifying (26) and (29) this gives:

$$\lambda = (\pi f_\pi^2\,\alpha')^{-1} \simeq 50\cdot \qquad (30)$$

thus yielding a ρ width of the order of 100 MeV, together with the Weinberg scattering lengths [39].

The structure of (29) follows from the Adler relation, current algebra and the absence of I=2 σ term. This last assumption is translated in our approach by the absence of force in the I=2 state which is due to the absence of resonance in this channel. We can therefore use current algebra through the low energy theorem; it provides to fix the value of λ . We already found:

$$G_\rho^2 = \frac{\lambda}{2} = \frac{m_\rho^2 - m_\pi^2}{\pi\, f_\pi^2} \qquad (31)$$

Following Sakurai, one may also assume ρ dominance and write the T_1 amplitude at t = 0 obtained from relations (29) as if it were dominated by ρ exchange. One obtains the KSFR* relation:

* Kawarabayashi, Suzuki, Fayyazuddin and Riazzuddin.

$$G_\rho^2 = \frac{m_\rho^2}{2f_\pi^2} \tag{32}$$

Relations (31) and (32) differ by a factor $2/\pi$, the ρ width obtained from the Veneziano formula being smaller than the one obtained from the KSFR relation. This reduction is needed in order to meet experiment.

The Veneziano amplitude is then very well suited to describe $\pi\pi$ scattering from low energy up to the first resonances. It can be checked that, as written, it has no ghost before deep daughter levels [42] and, as a result, barely meets Martin's bound based on positivity [43]. The first fixed t derivative with respect to s at $s=m^2$ is matched with the current algebra value. The second derivative of the amplitude, which can be calculated in terms of an unsubtracted dispersion relation involving the Pomeranchon contribution cannot, however, be identified with the value obtained from the Veneziano term. Nevertheless, the error made appears to be small.

The consistent picture for $\pi\pi$ s wave scattering thus obtained is based on small scattering length and large effective range (as imposed by the unphysical zero near threshold) as opposed to the large scattering length small effective range approximation initially tried.

As emphasized by Lovelace [37], the Veneziano amplitude can be used to describe three pion final state interaction and relation (27) is still valid (with different λ) when any of the pions is replaced by an isospin 1 0^- state. When comparing to experiment, it is however now necessary to include an imaginary part in α to replace physical region poles by resonances. One may, in particular, describe K and η decay assuming that graphs, as drawn on Fig. 8, are dominant. To the extent that one limits oneself to linear terms, the results are, however, identical to those which would result from the use of the Weinberg amplitude with the same dynamical assumptions. The Veneziano amplitude has more to say

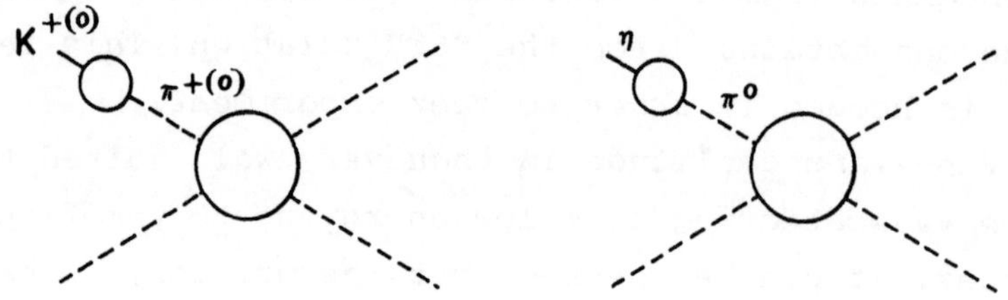

Figure 8 - K and η decay in the one-particle exchange approximation.

when phase space is big enough so that at least two resonances on the same trajectory can have important final state effects. This is the case of $\bar{p}n$ annihilation at rest [37] where the ε^0 and ρ' effects are correlated.

12. NONET STRUCTURE IN THE VENEZIANO AMPLITUDE

The $K\pi$ scattering amplitudes can be considered along the same lines [43]. From obvious symmetry reasons, one writes the $\pi\pi \to K\bar{K}$, t channel, amplitudes as:

$$
\begin{aligned}
T_0 &= \alpha'(V_{\rho K^*}(t,s) + V_{\rho K^*}(t,u)) \\
T_1 &= \beta'(V_{\rho K^*}(t,s) - V_{\rho K^*}(t,u))
\end{aligned}
\qquad (33)
$$

The (ρf) trajectory is exchanged in the t channel and the ($K^*(890)$, $K^{**}(1420)$) exchange degenerate trajectory is exchanged in both the s and u channels. Exchange degeneracy is imposed by the absence of resonance in the exotic $I=\frac{3}{2}$ channel. The s channel amplitudes are readily obtained in terms of the crossing matrix

$$M_{st} = \frac{1}{2}\begin{pmatrix} \sqrt{\frac{2}{3}} & 2 \\ \sqrt{\frac{2}{3}} & -1 \end{pmatrix}$$

and the absence of the $I=\frac{3}{2}$ s pole imposes $\alpha'=\sqrt{3/2}\,\beta'$. It also forbids the presence of a $V_{K^*K^*}(s,u)$ term in T_o which we already argued away. We get

$$s_{1/2} = \frac{\beta'}{2}\,(3V_{\rho K^*}(t,s) - V_{\rho K^*}(t,u)) \qquad (34)$$

The K^* pole residue in $s_{1/2}$ is proportional to $(3\beta')/2$, while the pole residue in T_1 is proportional to $2\beta'$. It then follows that

$$G^2_{K^*K\pi} = \frac{3}{4}\,G_{\rho\pi\pi}\,G_{\rho K\bar{K}} \qquad (35)$$

We also relate the K^{**} pole residue in $s_{1/2}$ to the f residue in T_o

$$G_{K^{**}K\pi} = \frac{\sqrt{3}}{2\sqrt{2}}\,G_{f\pi\pi}\,G_{fK\bar{K}} \qquad (36)$$

If we further relate the $\rho\pi\pi$ and $\rho K\bar{K}$ coupling through SU(3): $G_{\rho\pi\pi} = 2G_{\rho K\bar{K}}$, (35) corresponds to the relation we could have directly written using SU(3) :

$$G^2_{K^*K\pi} = \frac{3}{8}\,G^2_{\rho\pi\pi} \qquad (37)$$

while (36) corresponds to SU(3) coupling with standard nonet mixing. This follows from the fact that we have assumed the f' to decouple from $\pi\pi$ in order to be able to maintain the ff' mass difference without introducing a ρ'. We imposed symmetry between an s=0, I=1 and an $s=\frac{1}{2}$, $I=\frac{1}{2}$ object and it is therefore not too surprising that we obtain group structure constants of the simplest unitary group where these objects may be members of the same irreducible representation. It should be remarked that SU(3) breaking appears even at the first daughter level since only the parent trajectory coupling does not involve the external masses.

Normalizing (33) together with (26), with SU(3) symmetry, we get $\beta'=\frac{1}{2}$. The denominators in (33) will become infinite when the pion four momentum vanishes (Adler condition) if:

$$\alpha_\rho(m_\pi^2) + \alpha_{K^*}(m_K^2) = 1 \tag{38}$$

We will then impose $\alpha_{K^*}(m_K^2) = \frac{1}{2}$ in agreement with experiment (Fig. 2), so that (33) automatically satisfies the Adler condition. With common slope for the ρ and K^* trajectories, this implies the well satisfied relation

$$m_{K^{**}}^2 - m_f^2 = m_{K^*}^2 = m_\rho^2 = m_K^2 - m_\pi^2 \tag{39}$$

which is readily understood in terms of quark content (broken SU(6)×O(3) symmetry [21]).

The soft kaon limit corresponds to $s=m_\pi^2$, $t=m_K^2$ and, with linear and parallel trajectories, it also corresponds to a vanishing amplitude.

The K^* and the ρ f trajectories with both isospins can therefore sustain themselves through duality. The K^* and the f' could not and the f' must therefore decouple from the $\pi\pi$ channel if we require a difference in mass between f and f' and as a result different trajectories.

$\pi\eta$ scattering immediately raises a problem since the A_2 and f trajectories which can be, respectively, exchanged in the $\pi\eta$ and $\pi\pi \rightarrow \eta\eta$ channels have already been found to be degenerate. The Adler condition cannot be satisfied unless a satellite term is added since $\alpha_\rho(m_\eta^2) \neq \frac{1}{2}$ [41]. It would, however, readily hold with "ideal" nonet mixing for the pseudoscalar mesons, the π and the η having the same mass (no strange quark), thus departing from an SU(6)×O(3) classification. The X would then decouple from the A_2 and f trajectories since it would collect all the strange quarks. There seems to be, however, a very big difference in mixing schemes between the vector (and tensor) mesons and the pseudoscalar ones, which has to be considered as an important fact, the pseudoscalar mesons having an almost perfect octet structure. As a result, a solution tailored according to the mass pattern of the vector mesons cannot readily match all pseudoscalar meson channels and at least a satellite term has to be introduced, the coupling of which translates the difference between the vector and pseudoscalar meson mixing schemes. It should further be remarked that with the splitting of the A_2 also found in the $K\bar{K}$ channel [44] one cannot yet make any definite statement about exchange degeneracy involving the f and the A_2 trajectories when used in such reactions.

We finally consider KK or $K\bar{K}$ scattering [43]. The degenerate ρ, f, ω and A_2 trajectories can be exchanged in the s and u channels ($K\bar{K}$). These trajectories will be globally labelled according to the ρ. The ϕ(f') exchange degenerate trajectory can also be exchanged and is assumed to differ from the ρ (Fig. 2). No force (or resonance) is present in the t channel and this imposes exchange degeneracy for the crossed channel trajectories. For obvious symmetry reasons, one writes:

$$T_0 = \alpha''(V_{\rho\phi}(u,s) - V_{\rho\phi}(s,u))$$
$$T_1 = \alpha''(V_{\rho\phi}(u,s) + V_{\rho\phi}(s,u)) \quad (40)$$

A non-vanishing T_0 imposes ϕ coupling. We have already excluded an a priori possible $V_{\phi\phi}(s,u)$ term in T_1, since the isolated $I = 0$ ϕ trajectory cannot sustain itself. We also exclude a possible $V_{\rho\rho}(s,u)$ term since we take $\rho\omega$ fA_2 mass degeneracy and forbid $f'2\pi$ coupling. Such assumptions warrant the use of duality diagrams and a $V_{\rho\rho}(s,u)$ term cannot be obtained from a planar graph. (The $\lambda\bar{\lambda}$ quark pair has to go one way.)

The Adler condition will be readily satisfied for zero momentum kaon ($s=t=u=m_K^2$) if:

$$\alpha_\rho(m_K^2) + \alpha_\phi(m_K^2) = 1 \quad (41)$$

This implies:

$$\alpha'(m_\phi^2+m_\pi^2-2m_K^2) = \frac{1}{2}$$
$$\alpha' = (2(m_\rho^2 - m_\pi^2))^{-1} \quad (42)$$

which is well satisfied. The straight line ϕ trajectory is therefore written as:

$$\alpha_\phi(s) = \frac{1}{2} + \alpha'(s-2m_K^2+m_\pi^2) \quad (43)$$

with

$$\alpha_{K^*}(s) = \frac{1}{2} + \alpha'(s - m_K^2)$$
$$\alpha_\rho(s) = \frac{1}{2} + \alpha'(s - m_\pi^2)$$

Neglecting the pion mass as is legitimate here since $m_\omega^2 - m_\rho^2 > m_\pi^2$, we obtain three equally spaced trajectories, as readily expected in our "ideal" nonet scheme (Fig. 2).

The s channel amplitudes are obtained from the crossing matrix:

$$M_{st} = \frac{1}{2}\begin{pmatrix} -1 & 3 \\ 1 & 1 \end{pmatrix}$$

One imposes that there is no ϕ pole in the S_1 amplitude and get as a result $\beta'' = -\alpha''$. It follows that

$$S_o = \beta''(2V_{\rho\phi}(u,s) + V_{\rho\phi}(s,u))$$
$$S_1 = \beta'' V_{\rho\phi}(s,u) \quad (44)$$

One can then again relate the leading trajectory pole residues. The ρ and ω residues (s channel poles) are obviously equal, whereas the ϕ residue is equal to twice the ω residue (s channel poles in S_o). Similar relations hold for the tensor mesons. This reads:

$$g^2_{\rho K\bar{K}} = g^2_{\omega K\bar{K}} = \frac{1}{2}\, g^2_{\phi K\bar{K}}$$
$$g^2_{A_2 K\bar{K}} = g^2_{fK\bar{K}} = \frac{1}{2}\, g^2_{f' K\bar{K}} \quad (45)$$

These relations of course result from our excluding $V_{\rho\rho}$ and $V_{\phi\phi}$ terms in (40) as imposed by the nonet scheme. Relations (45) and SU(3) results with standard mixing angle $tg^2\Theta=\frac{1}{2}$ and with no $f'2\pi$ coupling.

Using SU(3) we can again normalize the $K\bar{K}$ amplitudes to (26). We get $\beta'' = -\frac{1}{2}$.

For $\pi\pi$, πK and KK scattering, where octet or nonet structure is irrelevant, one can then easily incorporate the broken SU(3) nonet scheme for the vector and tensor mesons generating the forces and resulting as resonances. In all three cases the simplest terms readily satisfy the constraints of PCAC and current algebra as the result of relations between the vector and pseudoscalar masses:

$$2(m^2_\rho - m^2_\pi) = 2(m^2_{K^*} - m^2_K) = \alpha'^{-1}$$

together with the relation:

$$m^2_\phi - m^2_{K^*} = m^2_{K^*} - m^2_\rho \quad .$$

13. GENERALIZATION OF THE VENEZIANO MODEL

These relations among masses may not be considered as a coincidence. One may attribute a dynamical "raison d'être" to the simplest Veneziano amplitude, thus written without satellite, and therefore gets as a result of the Adler condition, relations among trajectory intercepts which can be translated into mass formulae [45]:

One may attach a great importance to the fact that the simplest Veneziano amplitude written for $\pi\pi$ (and πK) scattering satisfies the Adler condition. As emphasized by Lovelace [37], this might be connected with chiral symmetry. Considered as a dynamical hypothesis [45], it leads to mass formulae. This cannot be true in general since there are several counter examples such as $\bar{p}n(0^-)\to 3\pi$, where the simplest amplitude differs from the one written for $\pi\pi$ scattering by only a numerical factor [37], and where the Adler condition is not satisfied. It might nevertheless hold in many cases in connection with a mass pattern which it is very interesting to compare against experiment. For this purpose one imposes that each term of the amplitude, written for the reaction:

$$\pi + A \to B + C$$

namely:

$$V_{xy}(s,t) = \frac{\Gamma(m-\alpha_x(s))\,\Gamma(n-\alpha_y(t))}{\Gamma(m+n+p-\alpha_x(s)-\alpha_y(t))} \qquad (46)$$

p, integer ≤ 0, should vanish when the momentum of the pion goes to zero ($s=m_A^2$; $t=m_B^2$; $u=m_C^2$); m and n are the lowest resonance spins which can contribute to this particular term.

It follows that

$$\alpha_x(m_A^2)+\alpha_y(m_B^2) = m+n+p+N = m+n+N_{AB}$$

where N is either zero or a positive integer.

The same condition imposed for the amplitudes where B=A (y=x) gives in either case

$$\alpha_x(m_A^2) = m + \frac{N_{AA}}{2}$$

or $$\alpha_y(m_B^2) = n + \frac{N_{BB}}{2} \tag{47}$$

If the required property holds for some A and B particles

$$N_{AA} + N_{BB} = 2\ N_{AB} \tag{48}$$

and both N_{AA} and N_{BB} must therefore be either even or odd integer. Since we already know that for $\pi\pi$ scattering, $N_{\pi\pi}$ is odd, the conclusion is that N_{AA} is odd whatever A is [45].

Particle A with spin s_A being itself on a linear Regge trajectory

$$m_A^2 = \frac{s_A-\alpha_A(o)}{\alpha_A'} \tag{49}$$

and

$$\alpha_x(o) + \frac{\alpha_x'}{\alpha_A'}(s_A-\alpha_A(o)) = m+ \frac{N_{AA}}{2} \tag{50}$$

With identical slopes [46], and s_A and m being obviously both integers or half-integers, one concludes that the intercepts of the A and X trajectories, associated to particles which can decay into each other by pion emission, differ by an odd integer. This result holds, provided that the A and X trajectories have opposite normality, defined as $P(-1)^J$ for mesons and $P(-1)^{J-(1/2)}$ for baryons [45]. The Adler condition is otherwise satisfied by definition and no conclusion follows.

The ρ and π; ρ and A_1; Δ and N; Y_1^* and Σ intercepts should then be spaced by half an integer (in effect $\frac{1}{2}$). This is very well satisfied. With the same slope for all trajectories, relations among intercepts can be rewritten as relations among masses. With $\alpha'=1/(2(m_\rho^2-m_\pi^2))$, the previously mentioned relations given for instance:

$$m_{A_1}^2 = 2m_\rho^2 - m_\pi^2$$
$$m_\Delta^2 = m_N^2 + m_\rho^2 - m_\pi^2 \tag{51}$$

They are both very nicely satisfied.

Generalizing these results to too many reactions, one however meets the already mentioned difficulty that the mass differences between the internal and external particles do not obey the same rules (nonet or octet structures of mesons). For instance one cannot satisfy, at the same time, the relations $m_f^2=3m_\rho^2-2m_\pi^2$ and $m_{A_2}^2=3m_\rho^2+m_\eta^2-3m_\pi^2$ with a standard nonet of tensor mesons and an almost pure octet of pseudoscalar mesons. At the same time we do not, a priori, know whether to write $m_{Y_1^*}^2 = m_\Sigma^2 + m_\rho^2 - m_\pi^2$ or $m_\Lambda^2 + m_\rho^2 - m_\pi^2$, a clash between decuplet and octet mass splittings. Applying duality arguments according to coupling strength, we choose the second relation. The correct result is half-way in between. There is not yet any general rule to tell when to impose the Adler condition on each of the leading Veneziano terms. In other words, we do not yet know whether the simplicity of the Veneziano amplitude comes from the mass pattern, or the other way round.

If satellites are needed, the Adler condition could be imposed in a different way. We know that the simplest Veneziano term has all desired properties for $\pi\pi$ scattering, but in other cases it is no longer true. Such complications readily occur when one tries to extend the model to reactions between particles with spin. In order to do so, one may write Veneziano expressions for each of the invariant amplitudes which satisfy the Mandelstam representation and which are introduced to describe the reaction.

Since these invariant amplitudes do not have the same asymptotic behaviour either for high s(u) or t values, each Veneziano term has to be tailored so as to meet the required properties. One may also use regularized helicity amplitudes where each Regge trajectory contributes asymptotically as $s^{\alpha(t)-\Lambda}$, where Λ is the corresponding helicity flip. A standard term such as:

$$V_{xy}(s,t) = \frac{\Gamma(m-\alpha_x(s))\,\Gamma(n-\alpha_y(t))}{\Gamma(m+n+p-\alpha_x(s)-\alpha_y(t))}$$

will have $s^{\alpha_y(t)-n-p}$ or $t^{\alpha_x(s)-m-p}$ asymptotic behaviours which are respectively obtained when s or t go to infinity. However, if the same trajectory is exchanged in both channels, the same asymptotic behaviour will automatically follow when the amplitude might have to correspond to different helicity flips in the two channels and therefore to different asymptotic behaviours. In order to maintain the required power in s or t, the standard Veneziano amplitude initially written down will then have to be multiplied by a polynomial in s or in t. For example :

$$(t-t_o)\frac{\Gamma(1-\alpha(s))\,\Gamma(1-\alpha(t))}{\Gamma(2-\alpha(s)-\alpha(t))}$$

will have $s^{\alpha(t)-1}$ and $t^{\alpha(s)}$ asymptotic behaviours. The trajectories being linear, it is readily written as the sum of two Veneziano terms. One has lost, however, the great simplicity initially met with $\pi\pi$ scattering.

Attempts have been made to describe as simply as possible πN and KN scattering [35],[47], writing Veneziano terms for the two invariant amplitudes. As is well known, linear trajectories lead to a difficult question with parity doubling.

We will rather discuss in some detail $\rho\pi$ scattering which brings into focus many interesting questions [48]. The amplitude is usually written as:

$$T_{\mu\nu} = AP_{\mu}P_{\nu}+B(P_{\mu}Q_{\nu}+P_{\nu}Q_{\mu})+C_1Q_{\mu}Q_{\nu}+C_2g_{\mu\nu} \tag{52}$$

where $P=\frac{1}{2}(p+p')$ and $Q=\frac{1}{2}(q+q')$; p and q (p' and q') stand for the pion and vector meson momenta in the initial (final) state. The four invariant amplitudes A, B, C_1 and C_2 are known to satisfy the Mandelstam representation. They are analytic functions of the variables $s=(p+q)^2$ and $t=(p'-p)^2$. The amplitudes are found to respectively have $s^{\alpha-2}$, $s^{\alpha-1}$, s^{α} and s^{α} asymptotic behaviour in s at fixed t, where α is the leading Regge trajectory, but the same t^{α} behaviour at large t and fixed s.

The absence of force in the exotic $\rho^+\pi^+$ and $\rho^+\rho^+ \to \pi^+\pi^+$ channels imposes exchange degeneracy in both channels. In the t channel ($\pi\pi \to \rho\rho$) we have the ρ,f exchange degenerate trajectories. In the s (or u) channel ($\pi\rho \to \pi\rho$), we have the ω, A_2 exchange degenerate trajectories on the one hand, and , a priori, the π and the A_1 trajectories on the other. It is tempting to use π A_1 exchange degeneracy from which the already obtained mass relation

$$m_{A_1}^2 = 2m_{\rho}^2 - m_{\pi}^2$$

readily follows. This, however, does not work. The π and the A_1 having the same isospin, their contribution at building up the crossed channel Regge trajectories, will not vanish in the isospin 2 state, if it is present in the isospin 0 and 1 states. We furthermore cannot get cancellations between the ω, A_2 and π, A_1 contributions, since the trajectories are half a unit apart. $\pi\rho$ duality therefore imposes that an unnatural parity trajectory with negative signature and isospin 0 should be exchange degenerate with the pion trajectory [49], and by the same token coincide with the A_1 trajectory, since we can consider the coincidence of the π and A_1 trajectories as an "experimental" fact. A 1^+ meson with isospin zero and G parity -1 should then be degenerate (or almost degenerate) in mass with the A_1. Since most of the evidence for the A_1 comes from diffraction dissociation experiments, such an isospin 0 state might have yet escaped unambiguous detection. In a similar way, one needs an isospin 0 trajectory exchange degenerate with the A_1, the first resonance of which would have spin 2^-.

In other words, if a forward peak is expected in the $\rho^+\pi^- \to \rho^+\pi^-$ reaction as a result of ρ f t-channel exchange, it should change into a backward peak for the reaction $\rho^-\pi^+ \to \rho^+\pi^-$ where no resonance occurs in the t channel. Charge conjugation applied on the initial state does not, however, modify the π, A_1 contribution in the s channel (same isospin) as it does for the ω, A_2 contribution (opposite isospin). We therefore cannot reach the same answer both ways (in terms of t or s channel exchanges) as implied through duality.

The C_2 amplitude can be written as

$$C_2(s,t) = \beta_4 \frac{\Gamma(1-\alpha(s))\Gamma(1-\alpha(t))}{\Gamma(1-\alpha(s) - \alpha(t))} + \gamma_4 \frac{\Gamma(1-\alpha_1(s))\Gamma(1-\alpha(t))}{\Gamma(1-\alpha_1(s) - \alpha(t))} \tag{53}$$

(we consider $\rho^+\pi^-$ scattering and therefore do not have u channel resonances). α represents the ρ (or f, ω, A_2) trajectory and α_1 the π trajectory (or the three other trajectories which should coincide with it). These two terms will be referred to as the $\rho\omega$ and $\rho\pi$ contributions. Since the pion pole does not contribute to C_2, the first resonance here included on α_1 has spin 1. C_2 thus written has the correct poles and asymptotic behaviours. C_2 is singular at $\alpha(s) = 1$. This corresponds to the ω pole. The pole residue is easily found to be

$$\lim_{s \to m_\omega^2} (s-m_\omega^2)\, C_2(s,t) = -\beta_4\left(\frac{\alpha(o)}{\alpha'} + t\right) \qquad (54)$$

This corresponds to a superposition of vector and scalar exchanges, but since nothing yet fixes the parity, we could also have a superposition of axial, vector and pseudoscalar exchanges. However, pseudoscalar exchange is excluded for C_2 and scalar exchange is forbidden by parity conservation. If ω exchange corresponds to vector exchange (and not to a mixture of vector and pseudovector, which would cancel the spin zero part), (54) would lead to some forbidden spin zero exchange unless an extra satellite term is added to (53) in order to cancel it [50]. This term affects (53) only at the first daughter level. It does not change the asymptotic behaviour and reads

$$\beta_4' \frac{\Gamma(1-\alpha(s))\Gamma(1-\alpha(t))}{\Gamma(2-\alpha(s)-\alpha(t))}$$

One finds : $\beta_4' = \beta_4/2$, considering that $\alpha' m_\pi^2 << 1$ and neglecting the $\rho\omega$ mass difference. The absence of satellite would impose the excluded condition $\alpha_\rho(o) \simeq 0$ [50]. Eliminating parity doublets therefore imposes satellites. Further terms have similarly to be added in order to eliminate unwanted terms at $\alpha(s) = 2,3,\ldots$. They will act at higher daughter levels. Veneziano amplitudes can therefore still be written and they relate in an unambiguous way the width of the resonances on the leading trajectories (ω, A_2,...). The simplicity of the $\pi\pi$ amplitudes is however lost!

The A amplitude in (52) behaves as $s^{\alpha(t)-2}$ at high s. We could write tentatively a Veneziano amplitude with such a behaviour for both the ω,ρ and the π,ρ contributions, namely:

$$A(s,t) = \beta_1 \frac{\Gamma(1-\alpha(s))\Gamma(2-\alpha(t))}{\Gamma(3-\alpha(s)-\alpha(t))} + \gamma_1 \frac{\Gamma(-\alpha_1(s))\Gamma(2-\alpha(t))}{\Gamma(2-\alpha_1(s) - \alpha(t))}$$

This is, however, not correct since either term leading separately to a superconvergent amplitude, a cancellation of s channel poles would be required. This is not possible since all residues, if correctly given by the Veneziano amplitude, have to have the same sign. One should therefore impose superconvergence only to the combination of the ω,ρ and π,ρ contributions, writing up to extra necessary satellite terms:

$$A(s,t) = \beta_1 \left\{ \frac{\Gamma(1-\alpha(s))\Gamma(2-\alpha(t))}{\Gamma(2-\alpha(s) - \alpha(t))} + (s-s_1)\alpha' \frac{\Gamma(-\alpha_1(s))\Gamma(2-\alpha(t))}{\Gamma(2 - \alpha_1(s) - \alpha(t))} \right\} \tag{55}$$

This expression has the required poles in s, no pole at $\alpha(t)=1$ and the correct asymptotic behaviours. The residues at the ω and π poles can be calculated, when they are, respectively, known to be $((G^2_{\omega\rho\pi})/8)(t-2m^2_\rho)$ and $G^2_{\rho\pi\pi}$.

One finds

$$(s_1 - m^2_\pi) = \frac{8G^2_{\rho\pi\pi}}{G^2_{\omega\rho\pi}}$$

when from saturation of superconvergence sum rules with the ω and π contributions alone, one gets [51]

$$G^2_{\rho\pi\pi} = \frac{1}{4}(m^2_\rho \; G^2_{\omega\rho\pi}) \tag{56}$$

Both results agree with $s_1 \sim 2m^2_\rho \sim \alpha'^{-1}$. Using (55) we should fix s_1 taking the experimental value for $G^2_{\rho\pi\pi}/G^2_{\omega\rho\pi}$. As already discussed we expect both an

I=1 and an I=0 contribution at the A_1 mass, and β_1 and s_1, thus determined, donot fix one of the $A_1\rho\pi$ coupling constants.

A generalization of the Veneziano formula to the N point function has been worked out [52], [53]. This is in effect obtained in the case of identical spin zero mesons. The Euler beta function:

$$B(x,y) = \int_0^1 u^x(1-u)^y \, du \qquad (57)$$

used in the standard, four-point function, Veneziano amplitude, has poles in x and y at negative integer values but no double poles. As seen on (57) a pole appears in x (or y) when u goes to zero (or to 1).
The aim is therefore to generalize (57) with as many independent variables as necessary to describe a production amplitude (5 for the five-point function, 9 for the six-point function,...), in such a way that certain multiple poles occur, but only some of them. As an example, the graph of Fig. 9 shows multiple poles in the variable

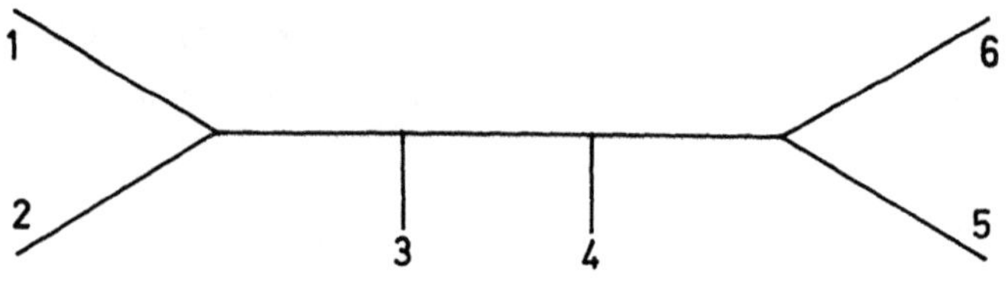

Figure 9 - A particular term involved in the six-point function

s_{12}, s_{123}, and s_{56} but a double pole in the variable s_{12} and s_{23} is not allowed. We refer to Refs. [52] and [53] for a detailed discussion of the construction proposed, which we do not attempt to sketch here. The amplitude is written as a multiple integral over as many variables as allowed coincident poles, while the integrand contains as many terms, all varying between 0 and 1 raised to the powers $(-1-\alpha(s_i))$, as there are different independent subenergy squared s_i. These terms are not independent and are related so that they cannot all vanish and be equal to 1 at the same time.

The four-point function:

$$B_4(x_1,x_2) = \int_0^1 du_1\, u_1^{x_1}\, u_2^{x_2}$$

with $u_2=1-u_1$, is thus generalized to a five-point function [52] as

$$B_5(x_1,x_2,x_3,x_4,x_5) = \int_0^1 du_1 \int_0^1 du_4\, \frac{1}{u_5}\, u_1^{x_1}\, u_2^{x_2}\, u_3^{x_3}\, u_4^{x_4}\, u_5^{x_5}$$

with $u_1=1-u_5u_2$, $u_2=1-u_1u_3, \ldots$. The six-point functions correspond to nine independent subenergies, but only three possible coincident poles (Fig. 9). As shown by Chan [53] the procedure can be extended to an arbitrary number of particles. The amplitude thus written has poles and Regge behaviour in all sub-channels, as required by duality. A very important point is that the residue of spin zero poles are found to be product of Veneziano amplitudes corresponding to lower multiplicity. Residues at a pole corresponding to integer values of $\alpha(s_i)$ give dual amplitude for reactions with particles with spin. As already mentioned, important complications occur in order to avoid parity doubling.

These constructions are very important since they allow to treat on a similar footing two-body collisions and multiparticle reactions, for which we need a common framework before unitarity can be fully used.

14. CONCLUSION

An attempt has been made to show some prominent facets of the approach to hadron physics followed under the name of duality. It is difficult to describe a method which claims at being somewhat general and nevertheless is at the same time bound to be only approximate as a result of too drastic steps made in dealing with unitarity. The resonance and Regge pole approximations to two-body amplitudes, which have been reconciled with powerful use of their overlap, still appear as different as they were initially. A generalized interference model compatible with the unitarity constraints, if not yet using them, is probably the next step. It should, however, be constructed in a dual way, as well as in a crossing symmetric way, interlocking the resonance recurrence in one channel with the asymptotic Regge behaviour in the same channel. Present checks of exchange degeneracy are enough to provide a very strong support to duality as thus imposed. One of the key points is, of course, the hitherto unobserved rising daughter trajectories. An immediate consequence is the postulated existence of a string of 1^- mesons separated in mass squared by 1.15 GeV^2. They should show up in e^+e^- collisions with the coming generation of storage rings. As already stressed, these resonances should be highly inelastic and could be searched for in multipion states, as well as in two-particle channels.

It is a pleasure to thank Drs. C. Lovelace and C. Schmid for many discussions. I also wish to express my deep gratitude to Professor T. D. Lee for his remarks and constructive criticisms while these lectures were being prepared.

REFERENCES AND FOOTNOTES

1. D. HORN, Lecture Notes, Schladming Winter School 1969.

2. V. BARGER, Communication at the Vienna Conference (1968);

   C. SCHMID, CERN preprint TH.960, to be published in Nuovo Cimento;

   A. H. ROSENFELD, Tables, UCRL-8030 (1969).

3. Continuation to complex values of the angular momentum has to be made separately for even and odd J values since $(-1)^J$ does not fulfil the conditions of the Carlson theorem. The absence of an absorptive part in either the s or u channel, say, will therefore lead to two coincident t channel Regge trajectories with opposite signatures.

4. We will not consider here in detail the implication of the splitting of the A_2 [5]. We will consider the A_2 as the nonet partner of the f. If the two A_2 peaks have to be associated with the same set of quantum numbers, we would then expect a similar splitting for all the nonet members.

5. W. KIENZLE, Philadelphia Conference on Meson Spectroscopy (1968);

   H. BENZ et al., Phys. Letters 28B, 233 (1968).

6. Writing a once subtracted dispersion relation for $\alpha(t)$, one sees that the observed widths of meson resonances cannot support the large observed slope.

   P. D. B. COLLINS, R. C. JOHNSON and E. SQUIRES, Phys. Letters 26B, 223 (1968).

   A linear trajectory may then be seen as the solution of a twice subtracted dispersion relation where the large contributions to the imaginary part are to be found only at energies much greater than 1 GeV ($s > \sigma$ 1 GeV^2) where quarks (say) show up as physical particles. We write therefore as a first approximation :

$$\alpha(t) \approx \alpha(o)+\alpha' t+ \frac{t^2}{\pi} \int_{\sigma}^{\infty} \frac{\mathrm{Im}\alpha(t')}{t'^2(t'-t-i\varepsilon)} dt'$$

neglecting branch points below σ.

7. S. MANDELSTAM, Phys. Rev. 166, 1539 (1968);
K. DIETZ, Lecture Notes, Schladming Winterschool 1969.

8. B. E. Y. SVENSSON, CERN Report 67-24 (1967);
L. VAN HOVE, CERN Lecture Notes (1968);
M. JACOB , Herceg Novi Lecture Notes (1968).

9. A sum of resonance contributions may be a good approximation to an inelastic (or charge exchange) amplitude. It is difficult to test in any definite way resonance saturation since there is much ambiguity at parametrizing resonances away from the poles, especially when they are highly inelastic.

10. G. F. CHEW, Comments Nucl. Particle Phys. 1, 121 (1967);
R. DOLEN, D. HORN and C. SCHMID, Phys. Rev. 166, 1768 (1968);
C. SCHMID, Phys. Rev. Letters 20, 628 (1968);
G. F. CHEW and A. PIGNOTTI, Phys. Rev. Letters 20, 1078 (1968).

11. The imaginary part (say) of a Regge amplitude cannot reproduce the wiggles associated with the direct channel resonances. Nevertheless, we expect at best the Regge amplitude to represent what we would obtain smearing out all resonance peaks. It should correspond to a good average even for several moments, which defines a semilocal approximation.

12. Continuation to positive t values located in the double spectral function region is, of course, not possible from a phase shift or a multipole expansion. Nevertheless, it is done on the ground that the zero width (real α) approximation is not too far from the actual case to still yield interesting results.

13. C. SCHMID, Phys. Rev. Letters 20, 689 (1968);

P. D. B. COLLINS, R. C. JOHNSON and E. J. SQUIRES, Phys. Letters 27B, 23 (1968);

C. SCHMID, CERN Preprint TH.958, to be published in Nuovo Cimento;

see also, Ref.[8], M. JACOB, Herceg Novi Lecture Notes (1968).

14. C. LOVELACE, Proceedings of the Heidelberg Conference (1967).

15. C. B. CHIU and A. KOTANSKI, Nuclear Physics 7, 615 (1968); 8, 553 (1969).

16. A second sheet pole, with dominant effect, is the simplest way to interpret the presence of an Argand loop. If further imposes the factorization property attached to a resonance. It is, however, not the only way of producing an Argand loop.

17. C. LOVELACE, Contribution to the Vienna Conference (1968).

18. H. HARARI, Phys. Rev. Letters 20, 1395 (1968); Contribution to the Vienna Conference (1968);

P. G. O. FREUND, Phys. Rev. Letters 20, 235 (1968).

19. D. MORRISON, Phys. Letters 22, 528 (1966).

20. C. B. CHIU and J. FINKELSTEIN, Phys. Letters 27B, 510 (1968).

21. R. H. DALITZ, Les Houches Lecture Notes (1965).

22. The decoupling of the Σ and $Y_1^*(1910)$ from the $\bar{K}N$ channel imposes a particular F/D ratio $\alpha = F/F+D=\frac{1}{2}$.

23. Standard absorption corrections using the Baker-Blankenbecler formalism, say [24], will leave the amplitude real if the initial Regge amplitude is real. To the contrary, the phases of the spin flip and non-spin flip amplitude become different when originally complex.

24. M. BAKER and BLANKENBECLER, Phys. Rev. 128,415(1962).

25. L. VAN HOVE, Phys. Letters 24B, 183 (1967);

L. DURAND III, Phys. Rev. 161, 1610 (1967).

26. R. JENGO, Phys. Letters 28B, 261 (1968); 28B, 606 (1969).

27. V. A. ALESSANDRINI, D. AMATI and E. J. SQUIRES, Phys. Letters 27B, 463 (1968).

28. V. BARGER and M. OLSSON, Phys. Rev. 146, 980 (1966).

29. H. HARARI, Phys. Rev. Letters 22, 562 (1969).

30. J. ROSNER, Tel Aviv University preprint (1968); G. P. CANNING, University of Oxford preprint (1969).

31. With standard nonet mixing the $\phi(\omega)$ and f'(f) have only strange (non-strange) quarks. This leads to the "ideal" nonet mass formula where two states (I=1 and I=O) are degenerate in mass and where the squared [21]. We postulate the absence of coupling between two states which have, one, only non-strange quarks and, the other, only strange quarks. The ϕ and f' couplings to non-strange mesons or baryons are indeed much weaker than the ω and f couplings.

32. G. VENEZIANO, Nuovo Cimento 57A, 190 (1968).

33. S. MANDELSTAM, Phys. Rev. Letters 21, 1724 (1968).

34. M. GELL-MANN, Geneva Conference (1962).

35. K. IGI, Phys. Letters 28B, 330 (1968).

36. E. MALAMUND and P. E.SCHLEIN, Phys. Rev. Letters 18, 1056 (1967);

C. LOVELACE, R. M. HEINZ and A. DONNACHIE, Phys. Letters 22, 322 (1966).

37. C. LOVELACE, Phys. Letters 28B, 265 (1968).

38. S. L. ADLER, Phys. Rev. 137, B1022 (1965).

39. S. WEINBERG, Phys. Rev. Letters 17, 616 (1966).

40. K. KAWARABAYASHI, S. KITAKADO and H. YABUKI, Phys. Letters 28B, 432 (1969);

C. LOVELACE, unpublished.

41. G. P. CANNING, Oxford preprint (1968);

J. ROSNER, Tel Aviv preprint (1968);

H. OLSON, London preprint (1968);

J. BAACKE et al., CERN preprint TH.983 (1968).

42. F. WAGNER, CERN preprint TH.978 (1969).

43. A. MARTIN, Nuovo Cimento 47A, 265 (1967).

44. A. MONTANET et al., CERN preprint (1969).

45. M. ADEMOLLO, G. VENEZIANO, and S. WEINBERG, Phys. Rev. Letters 22, 83 (1969).

46. The identity of the slopes is proved in Ref. [45]. Relation (30) can be also written for an external particle with the same spin S_A on the assumed parallel daughter. This implies that the ratio of the two slopes is an integer. Assuming that the Adler condition is satisfied the same way (through vanishing of each term) when A and X replace one another, the inverse ratio has also to be an integer and as a result $\alpha'_X = \alpha'_A$.

47. K. IGI and R. STORROW, to be published in Nuovo Cimento;

C. LOVELACE, to be published.

48. The analysis sketched here was done in collaboration with J. BAACKE, S. POKORSKY and C. SCHMID.

49. We disregard here the possibility that an isospin zero trajectory, at the pion daughter level, might "conspire" with the π A_1 trajectory in order to eliminate exotic resonances from the crossed channel. We choose to relate together only leading trajectories in either channels.

50. P. G. O. FREUND and E. SCHONBERG, Phys. Letters 28B, 600 (1969).

51. V. de ALFARO, S. FUBINI, G. FURLAN and C. ROSSETTI, Phys. Letters 21, 576 (1966);

F. GILMAN and H. HARARI, Phys. Rev. 165, 1803 (1968).

52. K. BARDAKCI and H. RUEGG, Phys. Letters 28B, 342 (1968).

53. CHAN HONG-MO, Phys. Letters 28B, 425 (1969);

CHAN HONG-MO and TSOU SHEUNG TSUN, Phys. Letters 28B, 485 (1969);

J. F. L. HOPKINSON and E. PLAHTE, Phys. Letters 28B, 489 (1969);

M. A. VIRASORO, Phys. Rev. Letters 22, 37 (1969);

J. G. GOEBEL and B. SAKITA, Phys. Rev. Letters 22, 257 (1969).

BOOTSTRAP OF INDEFINITELY RISING REGGE TRAJECTORIES†

By

K. DIETZ

Physikalisches Institut, Universität
Bonn, Germany

1. INTRODUCTORY REMARKS

The aim of these lectures is to discuss some of the more formal aspects of the concept of duality in strong interaction physics. As far as a detailed exposé of the underlying physical ideas and applications are concerned, we will vastly rely on the lectures by D. Horn and M. Jacob presented at this school and will not repeat them here.

We will mainly deal with questions which are pertinent to the bootstrap dynamics of infinitely rising Regge-trajectories. After a short outline of the general ideas, we describe (Section 3) the bootstrap of zero-width resonances which is shown to have rather serious defects (see also the Appendix). In Section 4 we turn to duality bootstraps of broad resonances. Under quite general circumstances we find that the Regge-trajectory and the Regge-residuum function are asymptotically determined in a unique manner. To counterpoint the idea of dual selfconsistency, we describe, in terms

---

† Lecture given at the VIII. Internationale Universitätswochen für Kernphysik, Schladming, Feb. 24 - March 8,1969.

of explicitly given functions, a generalized interference model which is fully crossing symmetric and gives almost no restrictions on rising trajectories (Section 5). In Section 6, finally,we extend the "dual" model of Section 4 to the case of full crossing symmetry and discuss, as an example, π-π scattering.

2. OUTLINE OF THE BOOTSTRAP IDEA

The theory of bootstrap [1] proposes a mechanism for the calculation of spectra of strongly interacting particles and is, in particular, devised to avoid the distinction between so-called elementary and composite particles, a distinction which frequently arises in some, more or less, defined way in other tentative models for theories of strong interactions. The fundamental idea is to try to construct a dynamical scheme in which all particles are bound states, their constituents being the particles themselves.

This vague concept can be translated, within the realm of S-matrix theory, in a construction algorithm. We start by considering the elastic scattering of two particles a and b

$$a + b \to a + b \tag{1}$$

The quantum numbers of a and b (spin, iso- or unitary spin, hypercharge, baryon number etc.) and their masses are assumed to be specified. A "self-generating" bound state or resonance is now to be determined by the following procedure:

Assume that a particle c with mass m_c and specified quantum numbers is exchanged in our scattering process and check whether the "exchange potential" is attractive. If so, determine its parameters (roughly speaking, its range $1/m_c$ and the coupling constants $g_{abc}^2, g_{bbc} \cdot g_{aac}$) such that at least one of its bound states has the same quantum numbers and parameters as the exchanged particle c. If such a solution exists, a new particle is born; if the self-consistent potential produces further bound-states c', exchange them and bind them self-consistently. At the same time production processes, as

$$\begin{aligned} a + b &\to a + c(c') \\ &\to b + c(c') \end{aligned} \tag{2}$$

have to be considered. New bound states will appear and so on. Finally, that at least is the hope and/or the belief, one will end up with channels in which the initial particles themselves appear as bound states and can be bootstrapped.

An obvious difficulty with the scheme described so far is the inclusion of multiparticle states in the direct and exchange channels. The approximate treatment, which considers only two-particle unitarity and single particle exchange, had, however, some qualitative success in explaining meson [2] and baryon[3] spectra.

On the other hand, since the concept of a Regge trajectory has emerged as a powerful tool for treating strong interaction problems, bootstrap equations should be formulated for the trajectories themselves such that they appear as the fundamental dynamical quantities. An approximate scheme of this kind has been developed by Cheng and Sharp [4] and actual calculations have been performed [5].

Mounting experimental evidence seems to suggest that Regge-trajectories might be indefinitely rising. A modification of the Cheng-Sharp equations which incorporates this suggestion has recently been proposed by Mandelstam [6]. Instead of going into any detail, we observe that the striking linearity of the experimentally found Regge-trajectories allows for the assumption that a strictly linear trajectory is a good approximation to the bootstrap problem we have in mind. Strict linearity of trajectories in turn is equivalent to an infinitesimally narrow resonance approximation (NRA) as is easily seen from the once subtracted dispersion relation for $\alpha(s)$

$$\alpha(s) = as + b + \frac{s^2}{\pi}\int ds' \frac{\mathrm{Im}\alpha(s')}{s'^2(s'-s)} \tag{3}$$

For the trajectory $\alpha(s)$ to be strictly linear, $\mathrm{Im}\alpha(s')$ has to vanish and the resonances on $\alpha(s)$ are (infinitesimally narrow).

In what follows a bootstrap scheme involving linear trajectories [6] will be described and a general outline of its consistency will be given.

The concept of an indefinitely rising Regge-trajectory $\alpha(s)$ is easily introduced into the above bootstrap algorithm by first summing over all resonances in the direct channel (which are supposed to lie on the self-generating trajectory $\alpha(s)$) and by performing a Sommerfeld-Watson transformation. In this way the Regge-trajectory is introduced in the usual way and analyticity in a larger s-t domain is achieved. Secondly as far as the exchange potentials are concerned, we follow Van Hove [7] and Durand [8] and observe that, under circumstances to be specified, the Regge asymptotic behaviour of a two-particle scattering amplitude determined by a certain trajectory $\alpha(s)$ can be reproduced by summing over the exchange graphs in which all the particles on the same trajectory are exchanged.

To be more specific, let us consider the case of the scattering of spin-less particles of equal mass m, a case which we will treat throughout these lectures unless it is specifically so indicated. Be p_a, p_b the incoming, p'_a, p'_b the outgoing four momenta. The total cms energy squared s and the momentum transfer t are then

$$s = (p_a + p_b)^2 \;, \qquad t = (p_a - p_b')^2$$

$$s + t + u = 4m^2 \equiv \Sigma \tag{4}$$

The cms scattering angles in the s- and t-channels are respectively

$$\cos\theta_s \equiv z_s = 1 + \frac{2t}{s-\Sigma} \quad \text{and} \quad \cos\theta_t \equiv z_t = 1 + \frac{2s}{t-\Sigma} \tag{5}$$

The t-channel exchange of a particle with spin j gives the contribution

$$A_j(s,t) = g_j^2 \, \frac{\sqrt{\pi}\,\Gamma(j+1)}{2^{j+1}\Gamma(j+\frac{3}{2})} \, (2j+1) \left(\frac{t-\Sigma}{M^2}\right)^j \frac{P_j(-z_t)}{m_j^2 - t} \tag{6}$$

to the scattering amplitudes (g_j is the coupling constant, m_j the mass of the exchanged particle, P_j is, of course, the Legendre polynomial of degree j, M is an arbitrary mass scale which we from now on choose to be unity). The factors in eq. (6) are arranged to give the asymptotic behaviour *

$$A_j(s,t) \underset{s\to\infty}{\sim} g_j^2 \, \frac{(-s)^j}{m_j^2 - t} \tag{7}$$

The forward scattering amplitude in the s-channel is then given, for large s, by the (in general divergent) series

$$A(s,o) \underset{s\to\infty}{\sim} \sum_{j=o}^{\infty} \frac{g_j^2}{m_j^2} \,(1 \pm (-1)^j)\, s^j \tag{8}$$

The u-channel contributions, assumed to be crossing symmetric to the t-channel, are taken into account and lead to the signature factor $(1\pm(-1)^j)$. Assuming a mass spectrum m_j and a set of coupling constants g_j the divergent

---

* The twiddle sign is defined by $f(s) \underset{s\to\infty}{\sim} g(s)$, if $\frac{f(s)}{g(s)} \underset{s\to\infty}{\longrightarrow} 1$.

series can be given sense by first summing for $|s| < c$, c sufficiently small, and then analytically continuing to large s [8]. For a linear mass spectrum $m_j^2 = m^2(j-\alpha(o))$ and at most exponentially decreasing g_j a Regge-behaviour of the scattering amplitudes is obtained

$$A(s) \underset{|s|\to\infty}{\sim} \text{const} \frac{1\pm e^{-i\alpha(o)}}{\sin\alpha(o)} (-s)^{\alpha(o)} \quad , \ \alpha(0)>-1 \qquad (9)$$

The result can be also extended to $t\neq 0$. Furthermore, a non-linear mass-spectrum will in general produce a high-energy behaviour corresponding to Regge cuts. With this information in mind we will now proceed in the formulation of bootstrap equations for linearly rising Regge trajectories.

In the bootstrap algorithm described above (two-particle) unitarity was an essential dynamical ingredient: the complex poles corresponding to bound-states or resonances of the unitarized exchange partial wave amplitudes had to be determined self-consistently to achieve a bootstrap. Now, by assuming strict linearity of Regge-trajectories, we already adopted an approximation to the unitarity equation. We are thus led to try to use crossing as the essential dynamical mechanism.

In an obvious generalization of our bootstrap algorithm we anticipate a self-consistency condition to arise from demanding that the sum over the exchange of resonances on a trajectory $\alpha(s)$ will reproduce, at least in some average sense, the sum over resonances on the same trajectory in the direct channel, direct and exchange channels being chosen to be crossing symmetric.

If the energy in the direct channel is taken high enough, the sum over resonance exchanges can be replaced by the asymptotic Regge-term as we just saw. The "average sense" will be provided by FESR which thus constitute the dynamical bootstrap equations. The latter simply follow from analyticity and the assumed asymptotic behaviour

$$c_t A^s(s,z_s) = A^t(t,z_t) \underset{s\to\infty}{\sim} \beta(t)(-s)^{\alpha(t)} , \qquad (10)$$

where $A^s(s,z_s)$ and $A^t(t,z_t)$ are the s- and t-channel amplitudes, c^t_s is the s-t isospin crossing matrix. From these assumptions one then derives

$$[A^s(v,z_s) = A^s(s,z_s); \quad v=s-u]$$

$$c \int^N dv\, v^n \operatorname{Im} \tilde{A}^s(v,z_s) \underset{N\to\infty}{\sim} \beta(t) \frac{N^{\alpha(t)+n+1}}{\alpha(t)+n+1} \qquad (11)$$

In these notes we need to consider only right-signature sum rules (see the lectures by D. Horn in these proceedings); c is again the crossing matrix.

Interpreted as a bootstrap equation by considering crossing symmetric s- and t-channels, the left hand side of eq. (11) represents the average sum over direct channel resonances, whereas the right hand side stands for the sum over the exchange of the same resonances.

If one differentiates eq. (11) formally with respect to N, one simply recovers the Regge form eq. (10). These expressions, then, are equivalent, whenever the order of the two limiting operations, $N\to\infty$ and differentiation with respect to N, is immaterial. However, the FESR is often useful because of the "smearing" effect it has on the rapid fluctuations of Im $A^s(s,z_s)$, which is, in particular, pertinent to the NRA, as will be seen in the next section.

3. GENERAL DISCUSSION OF THE BOOTSTRAP OF NARROW RESONANCES

The bootstrap of infinitely rising trajectories, as described in the preceding section, should arise from an answer to the following question: can one sum, in an average sense, a set of narrow resonances in the s-channel in such a way as to reproduce the Regge high-energy behaviour in the same channel (i.e. as $s\to\infty$)?

The average sum over s-channel narrow resonances is obtained by inserting the contribution of one Regge trajectory $\alpha(s)$ to the absorptive part (m_j is the mass of the j-th particle on the trajectory)

$$\mathrm{Im}\, A^s(s,z_s) = \pi\Sigma(2j+1)\beta(m_j^2)\; P_j(1+\frac{2t}{m_j^2-\Sigma})\delta(j-\alpha(s)) \quad (12)$$

into the FESR eq. (11) which then reads

$$\pi c \sum_{j=o}^{\alpha_I(N)} (2j+1)\beta(m_j^2)m_j^{2n}P_j(1+\frac{2t}{m_j^2-\Sigma}) \underset{N\to\infty}{\sim} \beta(t)\, \frac{N^{\alpha(t)+n+1}}{\alpha(t)+n+1} \quad (13)$$

The upper limit of the series, $\alpha_I(N)$, is the integral part of $\alpha(N)$ (in general, $\alpha_I(N)$ can only be even or odd depending on the signature, but this point can be easily allowed for). Taken literally, the asymptotic equality (13) is impossible, since the left hand side increases discontinuously with N, while the right hand side is smooth. However, if one demands that eq. (13) be satisfied only for those values of N for which, say, $\alpha(N)$ is an integer, with a smooth interpolating function between these points, then one no longer has a gross inconsistency. The narrow resonances have been effectively smoothed out.

It will now be shown that eq. (13) is in fact still inconsistent, because of the detailed functional dependences on t and N [9] . The task is to solve eq. (13) for

$\beta(t)$, given that the equation only holds at the discrete points $N=m_j^2$, j integral, and only in the limit $N\to\infty$. Instead of differentiating (13) with respect to N (as we discussed previously) which is no longer possible, one may take a finite difference:

$$\pi c(2j+3)\beta(m_{j+1}^2)P_{j+1}\left(1+\frac{2t}{m_{j+1}^2-\Sigma}\right)m_{j+1}^{2n} \underset{\substack{j\to\infty\\(N\to\infty)}}{\sim} \beta(t)\,\frac{(m_{j+1}^2)^{\alpha(t)+n+1}-(m_j^2)^{\alpha(t)+n+1}}{\alpha(t)+n+1}\,. \tag{14}$$

If one now uses the asymptotic form (valid for $t \neq 0$)

$$P_{j+1}\left(1+\frac{2t}{m_{j+1}^2-\Sigma}\right) \underset{j\to\infty}{\sim} \left(\frac{m_j}{4\pi j t^{1/2}}\right)^{1/2} \exp(2j\sqrt{t/m_j^2}) \tag{15}$$

one can rewrite eq. (14) as

$$\frac{c\beta(m_{j+1}^2)}{\beta(t)} \underset{j\to\infty}{\sim} \frac{t^{\frac{1}{4}}\left[(m_{j+1}^2)^{\alpha(t)+n+1}-(m_j^2)^{\alpha(t)+n+1}\right]}{(\pi j m_j)^{1/2}m_{j+1}^{2n}(\alpha(t)+n+1)\exp(2j\sqrt{t/m_j^2})} \tag{16}$$

Evidently, the right hand side of this expression should separate, as a function of j divided by a function of t, in the limit $j\to\infty$; but clearly it does not do so for any polynomially bounded mass spectrum m_j.

Hence, the sum of a tower of infinitesimally narrow resonances is not consistent with Regge behaviour in the same channel. Notice that the same argument works if the s- and t-channels are different (it suffices to change $\beta(t)$ to $\bar{\beta}(t)$, say, in eqs. (13) - (15)). The model fails, not because of the discontinuity inherent in the NRA (this was allowed for), but because of the fact that one Legendre polynomial is picked out at each energy. Moreover, the inclusion of a family of infinitely many Regge-trajectories, or even a finite number of families, does not vitiate the argument. Thus Regge behaviour must be produced either by the overlap of direct channel resonances (which then have finite width) or it must come from some background term which has nothing to do with the direct channel resonances.

Another way of stating our result is to say that the NRA cannot be strictly dual.

This failure is perhaps not too surprising since the NRA scheme is approximate anyhow. However, eq. (16) indicates, for example, that "solutions" for the couplings obtained by saturation at chosen values of the cut-off N and of the momentum transfer t, depend very strongly, in fact exponentially, on these parameters while they should be independent of them. The discussion of further implications resulting from the rather serious mismatch of the right- and left-hand sides of eq. (16), especially concerning the existence or non-existence of daughters, will be postponed to the Appendix.

Before ending this section, we will show, by giving a simple example [9][10] of a family of broad resonances which, summed in the s-channel, generates a Regge high-energy behaviour in the same channel.

Consider the s-u crossing symmetric form

$$A(s,t) = \pm\beta(s)\left[1+\frac{\alpha(t)\log(s-\Sigma)}{s-\Sigma}\right]^{\alpha(s)} + \\ +\beta(u)\left[1+\frac{\alpha(t)\log(u-\Sigma)}{u-\Sigma}\right]^{\alpha(u)} \tag{17}$$

As $s\to\infty$ above the unitarity cut and t is held fixed

$$A(s,t) \underset{s\to\infty}{\sim} \pm\beta(\infty)\exp\left[\alpha(t)(\log(-s)-i\pi)\ \frac{\alpha(s)}{s}\right] + \\ +\beta(-\infty)\exp\left[\alpha(t)\log(-s)\ \frac{\alpha(-s)}{-s}\right] \quad . \tag{18}$$

Suppose now that $\beta(s)$ tends to the same finite limit as $s\to\pm\infty$, and this is possible only if $\alpha(s)$ has an imaginary part (see eq. (3)) (i.e. the resonances on $\alpha(s)$ are broad) since $\alpha(s)$ includes a factor $\Gamma(-\alpha(s))$ required to give the resonance poles* .

If we now take $\alpha(s)$ to be**

$$\alpha(s) \underset{s\to\infty}{=} as + b + o(1) \tag{19}$$

then (18) becomes, <u>if a = 1</u>

$$A(s,t) \underset{s\to\infty}{\sim} \beta(\infty)(-s)^{\alpha(t)}(1 \pm e^{-i\pi\alpha(t)}) \tag{20}$$

which is the required Regge-asymptotic form. However, our ansatz cannot be made s-t crossing symmetric. This is because, letting $t\to\infty$ at fixed s in eq. (17):

$$A(s,t) \underset{t\to\infty}{\sim} \beta(s)\left[\frac{\log(s-\Sigma)}{s-\Sigma}\right]^{\alpha(s)} t^{\alpha(s)} \tag{21}$$

we observe a rather striking mismatch of the s- and t-channel residuum functions: for the s-channel residuum function $\hat{\beta}(t) = \beta(\infty)$ to be constant as required by eq.(20), we see from eq. (21) that, by demanding crossing symmetry, we have to have

$$\beta(s) = \beta(\infty)\left[\frac{s-\Sigma}{\log(s-\Sigma)}\right]^{\alpha(s)} \tag{22}$$

and thus $\beta(s) \equiv 0$.

It is by removing this mismatch among crossing symmetry channels, that we get, under rather general conditions, some sort of an asymptotic bootstrap as we will see in the next section.

Note, however, that in the case of non-crossing symmetric channels, which are dominated by the same trajectory (19), we already obtained from our example the con-

---

* The problem of ancestors will be regarded as inessential to the points this example is constructed to illustrate.

** The symbol o(1) stands for terms which vanish in the limit indicated.

dition a = 1, the physical interpretation of which is that the slope of the trajectory is given by the Regge-scale which we choose to be unity in the foregoing formulae.

4. CROSSING SYMMETRY AND INFINITELY RISING TRAJECTORIES OF BROAD RESONANCES

In the last section we found that bootstrap models involving a finite number of families of infinitesimally narrow-width trajectories can at best give a rather crude picture of a real crossing symmetric solution. Thus, if we want to get any further, we are forced to treat trajectories of broad resonances. Incorporating them into the scheme in any serious manner would, in general, mean that we have to tackle the problem of unitarity. A scheme of this kind has recently been proposed [11], but it goes beyond the scope of what we attempt to do here and will therefore not be discussed.

Another possibility is to pursue further implications of crossing symmetry for bootstraps of infinitely rising trajectories, unitarity being involved only in general terms, e.g. by respecting the Froissart bound or the fact that $\alpha(s)$ is a Herglotz function.The interesting question is then whether the restrictions on the trajectory $\alpha(s)$ obtained in this way will lead to a simplification of the unitarity condition. This would be desirable from a fundamental bootstrap point of view since the unitarity condition can be formulated only once the particle spectrum is anticipated. This last point is obviously not pertinent to the crossing condition which is thus favoured as a starting point for a bootstrap-type theory.

Again we ask the question whether a sum over now broad resonances on a certain trajectory, if forced to be crossing symmetric by FESR, leads to a restriction on the trajectory. For our purposes it is simpler to use directly the differential form (10) of the FESR; both are seemingly equivalent since with broad resonances no rapid fluctuations of the integrand in eq. (11), requiring an effective smearing, are to be expected.

An important point to be raised now, is what ansatz to take, to represent a sum over direct channel resonances. Usually, the expansion of a scattering amplitude into resonant partial waves, which is, of course, diverging outside the Lehmann ellipse, is extended to a larger domain by performing a Sommerfeld-Watson transformation. The well-known [12] contribution of a Regge trajectory is obtained. However, since we do not want to introduce any presupposed structure of daughters etc., we try to be as general as possible and write for the sum of resonances

$$A(s,t) = \beta(s)(\alpha(s) + g(t,s))^{\alpha(s)} \qquad (23)$$

The function g(t,s), this is our assumption, is allowed to be as general as the class of functions given by a double dispersion relation with a finite number of subtractions and with not infinitely often oscillating weight functions.

Clearly, the daughter series (17), or, for example, Khuri poles fall into this class. More general cases will be considered at the end of this section.

Our task is to find further conditions on g(t,s) such that A(s,t) has the following asymptotic behaviours*

$$A(s,t) \underset{s\to\infty}{\sim} \beta(t)\ (-s)^{\alpha(t)} \qquad (24)$$

$$A(s,t) \underset{t\to\infty}{\sim} \beta(s)\ (-t)^{\alpha(s)} \qquad (25)$$

* The case of full crossing will be discussed in Sect.6.

The signature factor and a factor $\Gamma(-\alpha(s))$ have tentatively been absorbed into β.

The trivial solution to these equations, namely $\beta(s)$=const and $\alpha(s)=c\log s$, is excluded since then $\alpha(s)\to+\infty$ for $s\to\pm\infty$ which is a flagrant violation of the Froissart bound, requiring $\alpha(s)\leq 1$ for $s\leq 0$. More generally, it can be shown that one must demand

$$\frac{\mathrm{Re}\,\alpha(s)}{\sqrt{s}} \underset{s\to\infty}{\longrightarrow} \infty \tag{26}$$

to avoid this contradiction.

To satisfy eq. (25) we must have

$$\frac{g(t,s)}{t} \underset{t\to\infty}{\longrightarrow} -1 \;. \tag{27}$$

The simplest situation arises if*

$$\frac{g(t,s)}{\alpha(s)} \underset{|s|\to\infty}{\longrightarrow} 0 \tag{28}$$

For large s, g(t,s) then exponentiates and we have

$$A(s,t) \underset{|s|\to\infty}{\sim} \beta(s)\alpha(s)^{\alpha(s)} \exp(g(t,s)) \tag{29}$$

and for this to agree with eq. (24), we must have

$$g(t,s) \underset{s\to\infty}{=} \alpha(t)\log(-s)+\log\beta(t)-\log[\beta(s)\alpha(s)^{\alpha(s)}]+ + o(1) \tag{30}$$

We conclude from this equation and from eq. (28)

$$\log\beta(s) \underset{s\to\infty}{\sim} -\alpha(s)\log\alpha(s) \tag{31}$$

a behaviour that has been suggested by Jones and Teplitz [13].

Consider now a double dispersion relation [14] for

* Because of our assumptions on g(t,s), taking the limit $|s|\to\infty$ requires no additional assumption.

g(t,s)

$$g(t,s) = P(s,t) + \frac{s^{m'}}{\pi} \int ds' \frac{\rho_s(s')}{s'^{m'}(s'-s)} + \tag{32}$$

$$+ \frac{t^{n'}}{\pi} \int dt' \frac{\rho_t(t')}{t'^{n'}(t'-t)} +$$

$$+ \frac{s^m t^n}{\pi^2} \int ds' \int dt' \frac{\rho_{st}(s',t')}{s'^{m'} t'^{n'} (t'-t)(s'-s)}$$

where P(s,t) is a subtraction polynomial. In order to reproduce the behaviour (30), it must be possible to express those parts X(s,t) of the double dispersion relation that depend on s only (i.e. the double integral, the s-single integral, and the part of the subtraction polynomial not independent of s in terms of the large-s expansion

$$X(s,t) \underset{s\to\infty}{=} \alpha(t) \log(-s) + f(t) + o(1) \tag{33}$$

where f(t) is some function of t only, which by the definition of X(s,t) has to result from the large s expansion of the double dispersion integral.

On the other hand, for fixed s and increasing t, g(t,s) must tend to -t, independently of s. This behaviour cannot come from the double dispersion integral which must give a s-dependent leading term in the large t expansion, since there is no function that has a constant Hilbert transform. Hence the -t behaviour of g(t,s) must come from the rest of eq. (32), i.e. the s-independent part of the subtraction polynomial and the single integral over t'. It follows then that f(t) in eq. (33) cannot increase as fast as t, for otherwise there would be an s-dependent term that interferes in the large t limit. This means that the large t behaviour of

$\log\beta(t)$ in eq. (30) will be given by the s-independent part of eq. (32) and hence must be

$$\log \beta(t) \underset{t\to\infty}{\sim} -t \qquad (34)$$

Eq. (31) then tells us that

$$\alpha(t) \underset{t\to\infty}{\sim} \frac{t}{\log t} \qquad (35)$$

This rather remarkable result, that $\alpha(s)$ and $\beta(s)$ are asymptotically determined, rests crucially on the assumption that the large t limit is uniform in s, which follows from our assumption that the double spectral function in (32) has at most a finite number of oscillations.

No attention has so far been paid to the extra dynamical branch points that $g(t,s)$ may have due to zeroes and the analytic properties of $B(s,t) \equiv \frac{A(s,t)}{\beta(s)}$:

$$g(t,s) = [B(s,t)]^{1/\alpha(s)} - \alpha(s) \underset{s\to\infty}{\sim} \log A(s,t) \quad (36)$$

Clearly, with our assumptions for $g(t,s)$, only a finite number of zeroes of $B(s,t)$ on, say, surfaces $s_i=f_i(t)$, can be allowed for. Furthermore, the functions $f_i(t)$ will in general have to be polynomially bounded if we want to remain in accord with our uniformity assumptions on $g(t,s)$. Whether or not a given type of zero can be included in our model (i.e. eq. (23) or eq. (36) together with the stated assumptions for eq. (32)) will, however, be a matter of detailed study of the presumed case and cannot be answered in general (see also the discussion below eq. (40)).

We still have to discuss what happens if eq. (28) were not true, that is if

$$\frac{g(t,s)}{\alpha(s)} \underset{s\to\infty}{\not\to} 0 \qquad (37)$$

Eq. (29) then has to be replaced by

$$A(s,t) \underset{s\to\infty}{\sim} \beta(s)\left[g(t,s)\right]^{\alpha(s)} \exp\left(\frac{\alpha^2(s)}{g(t,s)}\right) \tag{38}$$

If this to be consistent with the asymptotic behaviour (24), one must have

$$g(t,s) \underset{s\to\infty}{\sim} \exp\left[\frac{1}{\alpha(s)}\left(\alpha(t)\log(-s)+\log\beta(t)-\log\beta(s) - \frac{\alpha^3(s)}{g(t,s)}\right)\right] . \tag{39}$$

A detailed discussion shows that g(t,s) cannot be polynomially bounded in this case so that it falls outside the class of functions considered here.

We have thus shown, that our assumptions for g(t,s) together with asymptotic crossing symmetry allowed us to determine uniquely the asymptotic behaviour $\alpha(s) \underset{s\to\infty}{\sim} \frac{s}{\log s}$ for the dominating trajectory. That behaviour would, of course, be hard to distinguish from the experimentally suggested linearity of trajectories. In all the above considerations, s has been taken dimensionless, the scaling factor being in fact arbitrary. However, a most important point is that the slope of the trajectory is determined to be the same as this scaling factor. Hence there is only one parameter left to be determined by experiment. If the slope is taken to be 1 GeV^{-2}, as suggested by experiment, then the scaling factor should be 1 GeV^2.

Our result can be interpreted as an asymptotic bootstrap, in which the power law of the trajectory has been determined from very general considerations and in which the slope of the trajectory has been determined in terms of the energy scale.

The importance of our assumptions for g(t,s) in this asymptotic bootstrap can be accentuated by giving an example of a function g(t,s) which falls outside the class of functions considered above and gives more or less no restriction on $\alpha(s)$. The terms o(1) in eq. (30) can, for example, be represented by

$$- \log \left(1 + h(s)\, e^{f(t)}\right) \tag{40}$$

where h(s) and f(t) are analytic functions with

$$h(s) \underset{s\to\infty}{\rightarrow} 0 \qquad \text{and} \qquad f(t) \underset{t\to\infty}{\sim} t \quad . \tag{41}$$

Such a contribution to g(t,s) will vanish for $s\to\infty$ and t-fixed, in accord with eq. (30), and will, in the limit $t \to \infty$, s-fixed, contribute a term $-t$, in accord with eq. (27). The asymptotic t-behaviour has, therefore, no longer to be given by the log $\beta(t)$ - term in (30) and $\alpha(t)$ is to a large extent arbitrary. Clearly, this example goes beyond the class of functions that have been considered so far, since (40) tends to zero for large s and large t in a non-uniform way. Furthermore, (40) can in general be shown to correspond to an infinitely oscillating double spectral function because of the cyclic properties of the exponential.

This can be easily seen, for example, by writing eq. (40) as $-\log(h^{-1}(s)+e^{f(t)}) - \log h(s)$ and observing that, because of (41), there will be certainly values s_o, t_o such that the first logarithm will vanish. But then there will exist an infinite set t_n such that $f(t_n) = f(t_o) + 2n\pi i$ and, therefore, the term in question will also vanish at s_o, t_n. Hence, the term eq. (40) will oscillate infinitely many times at least around $-\log h(s_o)$ and our statement is obviated.

It seems that a behaviour of g(t,s) like (40) is intimately connected to a generalized interference model which we will discuss in the next section.

5. A GENERALIZED INTERFERENCE MODEL

In this section, we will show that crossing symmetry, analyticity and further general requirements on the scattering amplitude alone give almost no restriction on the parameters of indefinitely rising Regge trajectories. This was already indicated at the end of the last section; now we will provide explicit examples of scattering amplitudes $A(s,t,u)$ [15] which fulfil the following general requirements:

1. The amplitude $A(s,t,u)$ is symmetric in s,t,u;
2. for fixed t, it is analytic in the cut s-plane and is bounded by a polynomial in $|s|$;
3. for fixed c.m. angle in the s-channel, it is bounded by a polynomial in $|s|$.
4. $A(s,t,u)$ has resonance poles which lie on a polynomially bounded trajectory α, which is a Herglotz function with a right hand cut and which obeys $\mathrm{Re}\,\alpha(s) \to \pm\infty$, $s \to \pm\infty$.
5. The high-energy behaviour of $A(s,t,u)$ is of the Regge-type and is determined by the same trajectory.

The problem of crossing symmetry is facilitated if we split $A(s,t,u)$

$$A(s,t,u) = F(s,u;t) + F(u,t;s) + F(t,s;u) \tag{42}$$

For $A(s,t,u)$ to be symmetric in s,t,u, $F(s,u;t)$ has to be symmetric only in s and u.

Now we try the ansatz

$$F(s,u;t) \doteq \psi(-\alpha(t))\beta(t)\{(s_0-s)^{\phi(s,t)} + (u_0-u)^{\phi(u,t)}\} \tag{43}$$

where $s_0=t_0=u_0$ is e.g. an inelastic threshold. The logarithmic derivative $\psi(z)$ of the Γ-function

$$\psi(z) = \frac{\Gamma'(z)}{\Gamma(z)} \tag{44}$$

produces the resonance poles; $\beta(t)$ is a function with a right hand cut and is chosen to decrease exponentially for $|t| \to \infty$. An example is

$$\beta(t) = P(t)\exp(-(t_o-t)^{\gamma}) \tag{45}$$

with $0 < \gamma < \frac{1}{2}$ and $P(t)$ an arbitrary polynomial. The cuts in eqs. (43) and (45) are taken to run to the left of s_o, t_o, u_o .

The function $\phi(s,t)$ has to be constructed to give, obeying our assumptions, the large s or u Regge-behaviour as well as polynomial behaviour of the residua of the poles at integer $\alpha(t)$. An example is

$$\phi(s,t) = \alpha(t)\left\{1 - \frac{\log(-\alpha(t)+c)}{\psi(-\alpha(t))}\exp\left[-(s_o-s)^{\varepsilon}(t_o-t)^{\delta-\frac{1}{2}}\right]\right\}^{N} \tag{46}$$

where (c(t) will be discussed later)

$$0 < \varepsilon < \delta < \frac{1}{4}$$

and N is to be determined from the increase of the trajectory $\alpha(t)$: if $\alpha(t)$ is bounded by a polynomial of degree K, one has to have

$$N\left(\frac{1}{2}-\varepsilon-\delta\right) \geq K \tag{47}$$

for $\phi(s,t)$ to vanish for $|s|\to\infty$ and fixed z_s. Therefore*

$$\begin{aligned} &\phi(s,t) \Rightarrow \alpha(t) \quad , \; |s| \to \infty \qquad t \text{ fixed} \\ &\text{and} \\ &\phi(s,t) \to 0 \quad , \; |t| \to \infty \qquad s,u,z_s,z_t \text{ or } z_u \text{ fixed.} \end{aligned} \tag{48}$$

---

* The double arrow stands for exponential approximation, i.e. $y(t) \Rightarrow 0$ if there is an $\varepsilon > 0$ such that $y(t) \times \exp(|t|^{\varepsilon}) \to 0$ for $|t| \to \infty$.

Hence we conclude

$$F(s,u;t) \Rightarrow \psi(-\alpha(t))\beta(t)\{(s_o-s)^{\alpha(t)} + (u_o-u)^{\alpha(t)}\} \quad (49)$$

for $|s| \to \infty$, t fixed, and

$$F(s,u;t) \Rightarrow 0 \text{ for } |t| \to \infty, \quad s,u,z_s,z_t \text{ or } z_u \text{ fixed or}$$
$$\text{for } |s| \to \infty, \quad z_s,z_t \text{ or } z_u \text{ fixed} \quad (50)$$

so that we have

$$A(s,t,u) \Rightarrow \psi(-\alpha(t))\beta(t)\{(s_o-s)^{\alpha(t)} + (u_o-u)^{\alpha(t)}\} \quad (51)$$

for $|s| \to \infty$, t fixed and

$$A(s,t,u) \Rightarrow 0 \quad \text{for } |s| \to \infty, \quad z_s,z_t,z_u \text{ fixed .}$$

Furthermore, A(s,t,u) is holomorphic in the cut s and t planes with cuts starting at the physical thresholds of s,t,u - channels arising e.g. from the cuts of the trajectory α and logarithmic cuts starting at an inelastic threshold $s_o=t_o=u_o$. (The logarithm in (46) does not give rise to an additional branch cut since c(t) can always be arranged such that $\alpha(t)+c(t)$ will not vanish in the "physical" sheet and $\alpha(t) + c(t)$ has the asymptotic behaviour of $\alpha(t)$ because $\alpha(t)$ is a Herglotz function).

Conditions (2), (3) and (5) are therefore satisfied. Where the trajectory passes an integer $\alpha(t) = j$ (this will, for t above threshold, happen on the second sheet since we are considering broad resonances), A(s,t,u) will have a pole the residuum of which is a polynomial as is easily seen from (43) and (46). More precisely, for j=2n the polynomial will be of degree 2n, while for j=2n+1 it is of degree 2n. This is, of course, required by the underlying symmetry of the problem and means that at j=2n we have resonances starting at the parent trajectory, whereas for j=2n+1 we start with a resonance on the first daughter, spaced on unit angular momentum below. Furthermore, it is simple to show that no ghost problem is present in this example since all the resonance coupling constants are of the same sign.

This concludes the construction of our example.

6. EXTENSION OF THE MODEL OF SECTION 3 TO ASYMPTOTIC CROSSING SYMMETRY IN ALL CHANNELS

A fully crossing symmetric extension of the model of Section 4 is achieved, at least asymptotically by identifying the amplitude (23) with the function F(s,t;u) introduced in eq. (42). We obtain

$$\begin{aligned} A(s,t,u) = {} & \beta(s)(\alpha(s) + g(t,s))^{\alpha(s)} + \\ & + \beta(t)(\alpha(t) + g(u,t))^{\alpha(t)} + \\ & + \beta(u)(\alpha(u) + g(s,u))^{\alpha(u)} \end{aligned} \tag{52}$$

From eqs.(31) and (30) and the additional assumption that the term o(1) in eq. (30) is arranged to vanish for the simultaneous limits $s,t \to \pm\infty$ such that s/t=const (which is, of course, only a further specification of the assumptions already made) it is easily shown that

$$A(s,t,u) \underset{\substack{a\to\infty \\ b \text{ fixed}}}{\sim} \beta(b)(1+e^{-i\pi\alpha(b)})(-a)^{\alpha(b)} \tag{53}$$

where (a,b) are the six combinations (s,t),(s,u);(t,s), (t,u) and (u,s),(u,t).

For the sake of definiteness we assumed and will assume below the asymptotic behaviours (24) and (25) to hold for $|s|\to\infty$ and $|t|\to\infty$ respectively as is usually done in deriving FESR; eqs. (27), (30) and (31) then hold for $|t|\to\infty$; $|s|\to\infty$. More general cases can be treated; it would, however, be tedious to spell out the details.

Asymptotic crossing symmetry is thus achieved and the signature factor is provided. Of course, all our conclusions concerning the asymptotic bootstrap of $\alpha(s)$ and $\beta(s)$ still hold.

Furthermore, we see no reason, which could prevent us to adapt g(t,s) in such a manner as to achieve Mandelstam analyticity of A(s,t,u) (see also the discussion following eq. (36)).

Let us now check whether we have polynomial boundedness for $|s|\to\infty$ at $\cos\theta_s = z_s$ fixed; i.e. $t \underset{s\to\infty}{\sim} c_t s$ and $u \underset{s\to\infty}{\sim} c_u s$ with $-1 \le c_t, c_u \le 0$. A short calculation shows

$$A(s,t,u) \underset{\substack{|s|\to\infty \\ z_s \text{ fixed}}}{\sim} (1-c_t\log(-c_t))^{\alpha(s)} + \\ + \left(1+\frac{\log(-c_u)}{c_u}\right)^{\alpha(u)} + \\ + \left(1+\frac{c_u}{c_t}\log\left(-\frac{c_t}{c_u}\right)\right)^{\alpha(t)} \begin{cases} \to 0 & \text{for } 0 \neq c_t, c_u \neq 1 \\ \to \text{const} & \text{for } c_t, c_u = 0 \text{ or } 1 \end{cases} \tag{54}$$

In the same way, polynomial boundedness in $|s|$ for fixed t etc. is guaranteed.

Resonance poles at, say $\alpha(s) = j$ are provided by $\beta(s)$ which was to contain a factor $\Gamma(-\alpha(s))$. However, for this pole to correspond to a resonance, g(t,s) has to be a linear function in t for $\alpha(s) = j$. Of course, we know g(t,s) only asymptotically, but even there this condition can be arranged by writing, for example,

$$g(t,s) = \alpha(t)\,\frac{\log^2(-s)}{\log\log s\;\psi(-\alpha(s))} + \log\beta(t) + \\ + \frac{1}{\psi(-\alpha(s))}\, o(\log(-\alpha(s)) \tag{55}$$

where ψ is again the logarithmic derivative of the Γ-function. This $g(t,s)$ will have the asymptotic behaviour eq. (30) for s on the physical sheet, since the zeros occur on the second sheet. If we now require eq. (34) to hold for all t, i.e. $\log \beta(t) = -t$, we arrive at $[\beta(s) = \hat{\beta}(s)\Gamma(-\alpha(s))]$

$$A(s,t,u) \approx \hat{\beta}(s)\ \Gamma(-\alpha(s))(-t)^{j}\ , \quad \alpha(s) \approx j \qquad (56)$$

and have thus constructed an example, within the realm of our assumptions for $g(t,s)$, which finally matches conditions (4) and (5).

It must, however, be stressed, that the behaviour near resonances, which after all concerns finite energies, cannot be answered from asymptotic considerations alone. No claim will therefore be made, that the daughter structure, implied by (56), is predicted from our model. For this to be gradually answered, more terms have to be kept in the asymptotic expansions (24) and (25). The same statements hold for the equation $\log\beta(t)=-t$, used in deriving (56), which can be improved upon, at least for finite t, again by considering more terms in the asymptotic expansion. All what we want to say is that the resonance-Regge-asymptotics structure can, in general, be easily incorporated in our model.

Another point can already be seen from our asymptotic considerations; namely the mismatch of signature factors in eqs. (56) and (53). More precisely, $\beta(s)$ or $\hat{\beta}(s)$ in eq. (56) must contain a positive signature factor (a single amplitude, $A(s,t,u)$, being (asymptotically) fully crossing symmetric with an asymptotic behaviour (53), can only have even-spin resonances and will, within the known quantum numbers, describe the scattering of self-conjugate particles). Hence in eq. (53), there appears a signature factor squared which is not compatible with the usual Regge-picture. A bootstrap of a Regge-trajectory of self-conjugate bosons, therefore, does not seem to be possible without giving an explicit mechanism to eliminate the "odd daughters".

A simple way out is to introduce isospin and to take the expression

$$F(s,t;u) = \beta(s)\ (\alpha(s)+g(t,s))^{\alpha(s)} \qquad (57)$$

to describe an exchange degenerate trajectory in the sense described above eq. (23). (The residue function $\beta(s)$ then contains a factor $\Gamma(-\alpha(s))$, but not a signature factor). Take $\pi - \pi$ sacttering as an example, $\alpha(s)$ is now the exchange degenerate ρ-f trajectory. We write, following Lovelace [16], for the s-channel amplitudes in the three isostates I=0,1,2

$$A_0(s,t,u) = \frac{3}{2}[F(s,t;u)+F(s,u;t)]- \frac{1}{2}\ F(t,u;s)$$

$$A_1(s,t,u) = F(s,t;u) - F(s,u;t)$$

$$A_2(s,t,u) = F(t,u;s) \quad . \qquad (58)$$

From eq. (56), we immediately see, that A_0 has only even-spin resonances, A_1 only odd-spin, whereas A_2 has no resonances. It furthermore, has the correct asymptotic behaviour, as is easily checked with the help of the isospin crossing matrix.

$$A_I(s,t,u) = C_{II'}\ A_{I'}(t,s,u)$$

$$C_{II'} = \begin{pmatrix} \frac{1}{3} & 1 & \frac{5}{6} \\ \frac{1}{3} & \frac{1}{2} & -\frac{5}{6} \\ \frac{1}{3} & -\frac{1}{2} & \frac{1}{6} \end{pmatrix} \qquad (59)$$

For details we refer the reader to the lectures by M. Jacob in these proceedings.

Thus, as our final result, we established that an exchange degenerate (ρ-f) trajectory can sustain itself in π-π scattering; it is asymptotically determined by asymptotic crossing symmetry and the assumptions on g(t,s), formulated in Section 3, in a unique manner : $\alpha(s)$ is almost linear, i.e. $\alpha(s) \underset{s\to\infty}{\sim} \frac{s}{\log s}$, and its slope is given by the Regge scaling parameter.

7. CONCLUSIONS

In the foregoing chapters, we studied general properties of bootstraps of indefinitely rising Regge-trajectories. In particular, a scheme involving NRA to simplify the unitarity condition and FESR to provide for crossing symmetry and for an averaging of NRA was shown to have rather important deficiencies. A detailed discussion of bootstraps of this type shows that especially the results concerning daughters have to be considered with caution (see the Appendix). On the other hand these deficiencies do not immediately apply to the model proposed by Veneziano [17], since his form does not, strictly speaking, fulfil FESR either.

Our conclusion from this section was that zero-width resonances cannot be strictly dual, even after an effective smoothing provided by FESR. A dual resonance - Regge asymptotics structure seems, at best, to be able to sustain itself by an appreciable overlap of broad resonances, even at not too high cut-off parameters. The following sections were, therefore, devoted to the question whether crossing symmetry and boundedness assumptions on the scattering amplitude, both for fixed momentum transfer and fixed c.m. scattering angle, lead to restrictions on infinitely rising Regge-trajectories of broad resonances. It turned out, that the possibility of an (asymptotic) bootstrap rests crucially on how a resonance, or rather the Sommerfeld-Watson transformed sum over resonances, is assumed to behave: with certain uniformity assumptions on such a sum, the dominant trajectory $\alpha(s)$ and the residuum function $\beta(s)$ were asymptotically determined, without those assumptions, almost no restriction on these quantities exists. Thus, an indication, at least, is given in what sense the "Horn Schmid duality" [18] depends on the definition of what a resonance is.

It is of interest to note, that the asymptotic behaviour of $\beta(t)$ predicted by our "dual" model of broad resonances is precisely the same as the one predicted in the Veneziano model namely that it will <u>increase</u> with increasing space-like momentum transfers - t, at least from a certain "asymptotic" value on. This behaviour of $\beta(t)$ seems to be intimately connected with the viability of a self-consistent infinitely rising trajectory and provides for the possibility of an experimental test of this idea. We furthermore predicted that the dominant self-consistent trajectory is (asymptotically) linear up to a factor 1/logs, which in turn demands broad resonances by the analyticity of $\alpha(s)$. In accord with the Veneziano model the slope of the trajectory was given by the Regge scaling parameter. No attempt has been made to explore unitarity beyond its implying the Froissart bound and the positivity of the imaginary part of $\alpha(s)$. In this approach unitarity will enter (if it does not prevent us from making our uniformity assumptions!) only as a second step to the bootstrap calculations.

In the last chapter, we also showed that the simplest way of getting a fully crossing symmetric amplitude, within our model, is to introduce isospin and an exchange degenerate trajectory, e.g. considering π-π scattering. As in other models, the Pomeranchon contribution plays an unknown role (a way out could perhaps be a

Pomeranchuk trajectory which is not infinitely rising, or a cut contribution. In both cases our deductions would not apply and, furthermore, the Pomeranchon contribution would not interfere with our conclusions).

As a final remark, we would like to point out, that "crossing bootstraps" seem to be superior to the previously considered N/D-type "unitarity bootstraps" since unitarity can be formulated only once the particle spectrum is known. The latter, however, is supposed to be obtained from the bootstrap itself. Obviously, this logical difficulty is not inherent in the crossing relation, which, and this was our main point, is particularly simple to exploit if Regge trajectories are infinitely rising.

APPENDIX

In this section we discuss in some detail FESR-bootstraps in the NRA. Particular emphasis is given to an illustration of the general statements of Section 3 : we try to point out which of the results of NRA-FESR bootstrap calculations might, to some degree, be independent of the crude approximations involved and which are not.

Certainly, the considerations of Section 3 do not exclude the possibility that the NRA approximation to the FESR yields solutions, which approximate experimental findings quite well, if the cut-off is chosen at a certain value N_o and if the momentum transfer is restricted to a region around some t_o. However, t_o and N_o have then to be considered as a posteriori parameters which can be eliminated only by having recourse to a much more sophisticated theory.

A very elegant reaction for the analysis of the $I^{GC}=1^{+-}$

negative signature ρ-trajectory is [19] [20]

$$\pi + \pi \rightarrow \pi + \omega \tag{A.1}$$

which is fully crossing symmetric and is described by only one invariant amplitude. The latter we take to be the t-channel helicity amplitude $f^t_{oo\lambda o}(t,z_t)$, $\lambda=\pm1$.

Reggeizing

$$f^t_{oo\lambda o}(t,z_t) = \Sigma_j (2j+1) f^j_{oo\lambda o}(t)\; d^j_{\lambda o}(z_t) \tag{A.2}$$

by means of a Sommerfeld-Watson-Mandelstam transformation, the dominant trajectory α(t) gives the contribution [21]

$$f^t_{oo\lambda o}(t,z_t) = \beta(t)\,(\alpha(t)+\tfrac{1}{2})\,(1-e^{-i\pi\alpha(t)})\,e^{-\alpha(t)-1}_{-1,o}(-z_t) \tag{A.3}$$

In the NRA, the residuum function β(t) is [6], up to the threshold factor, the sense-nonsense factors at α=0 and its mirror point at α=-1 and the factor $K(t)=q_t\, q_t'\, \sqrt{t}$ which eliminates a kinematical branch point present because of unequal-mass kinematics (q_t, q_t' are the cms momenta in the initial and final states in the t-channel) an entire function E(t):

$$\beta(t) = \left(\frac{2q_t q_t'}{v_1}\right)^{\alpha(t)} \frac{\sqrt{t}\sqrt{\alpha\ (\alpha+1)}}{\Gamma(\alpha+3/2)}\, E(t) \tag{A.4}$$

(the factor $\Gamma(\alpha+3/2)^{-1}$ is extracted to kill negative half-integer poles).

Inserting into the first right-signature sum-rule (n=1, in eq. (11)) we obtain

$$\pi \sum_j^{j(N)} (2j+1)(2m_j^2+t-\Sigma)\, P_j'\left(\frac{m_j^2+2t-\Sigma}{4q_s q_s'(m_j^2)}\right)\left(\frac{2q_s q_s'(m_j^2)}{v_1}\right)^{j-1} \frac{E(m_j^2)}{\Gamma(j+3/2)}$$

$$\underset{N\to\infty}{\sim} \frac{4\sqrt{\pi}\; N}{(\alpha(t)+1)\,\Gamma(\alpha(t))}\, E(t) \left(\frac{N}{v_1}\right)^{\alpha(t)} \tag{A.5}$$

Taking the asymptotic equality to be approximate already above the first resonance on the trajectory we obtain

$$\frac{E(t)}{E(m_j^2)} \approx \frac{(\alpha(t)+1)\Gamma(\alpha(t))(2m_1^2+t-\Sigma)}{N} \left(\frac{v_1}{N}\right)^{\alpha(t)} \qquad (A.6)$$

We now assume a linear trajectory $\alpha(t) = at+b$ and evaluate at $\alpha(t_o) = 0$, i.e. $t_o = -\frac{b}{a}$ and at $\alpha=1$, i.e. $t_1 = \frac{1-b}{a}$. Using that E(t) is entire, we find

$$\frac{2-3b}{a} = \Sigma = m_\omega^2 + 3m_\pi^2 \qquad (A.7)$$

and

$$\frac{N^2 a}{v_1} = 2 \qquad (A.8)$$

The first of these equations expresses, of course, the fact that the zero of the right-hand side of eq. (A.5) at $\alpha=0$ is, for the chosen value of N, to be given by the ρ-resonance contribution. Such a feature might well be independent of the NRA and the chosen value of N (see the discussion of the dip mechanism in the $B^{(-)}$-amplitude of charge-exchange π-N scattering and its relation to FESR given by D. Horn, op.cit.). In fact, demanding that the ρ-trajectory passes the ρ-mass: $m_1^2 = m_\rho^2 = 0.58\ \mathrm{GeV}^2$ one gets from (A.7)

$$\alpha(t) = 0.45 + 0.9\ t \qquad (A.9)$$

which agrees amazingly well with the experimental information we have.

To draw further conclusions we rewrite eq. (A.6), using (A.7) and (A.8), as

$$\frac{E(t)}{E(m_1^2)} = \frac{\Gamma(\alpha(t)+2)}{2} \left(\frac{Na}{2}\right)^{\alpha(t)-1} \qquad (A.10)$$

If we now insist, that E(t) be approximately constant in a region $0 \leq \alpha(t) \leq 1$, Na can be determined to be

$$Na \approx 1 \qquad (A.11)$$

or, from (A.8), $a\, v_1 \approx \frac{1}{2}$. This is exactly the result that we obtained in Section 4 from much more general considerations, namely that the slope of the self-consistent trajectory is given by the Regge scale.

We now will shortly discuss what happens if we increase N and include more resonances on the left-hand side of eq. (A.5). In principle, we already solved that problem in Section 3 , in praxis this problem has been considered by the authors of ref. [19]. The next step, the inclusion of a 3^- resonance, reproduces in a very impressive manner the results of the first step. If, however, one includes 5^- and higher resonances, the resonance side of eq. (A.5) can no longer keep up with the Regge-side. This is the mismatch which we already expected from our general considerations. The authors of ref. [19] point out that a self-consistent solution for cut-off N beyond a 3^- - resonance can be constructed if daughters parallel to the ρ-trajectory are introduced. Again, we know that this is only an a posteriori argument for the existence of parallel daughters *, since we showed that there can be no solution to FESR involving only a finite number of towers of parallel daughters. That is to say, that by increasing N we will meet a point where a tower of parallel daughters, established at lower N, can no longer match the Regge-side of FESR; a new tower would have to be introduced, and so on. Hence, to establish a physically relevant solution, we would have to know, from some other arguments, what to take for the cut-off N. Otherwise, such tower solutions reflect only the shortcomings of the NRA.

* Daughters have, of course, to be present in general because of the unequal mass kinematics. Their particle structure might, however, be different from the one predicted from that kind of bootstrap.

REFERENCES

1. See e.g. R. E. CUTKOSKY, "Self-Consistent Dynamical Models", in Particle Symmetries, Brandeis Lectures 1965, Gordon and Breach Science Publishers, Inc. New York 1966, for a comprehensive outline of classical bootstraps and original references.
2. R. H. CAPPS, Purdue University preprint 1968; CHAN, H. M., P. de CELLES and J. PATON, Phys. Rev. Lett. 10, 312 (1963); CHAN, H. M., K. DIETZ, and C. WILKIN, Nuovo Cim. 34, 250 (1964), and further literature cited there.
CHAN H. M. and C. WILKIN, Ann. of Phys. 39, 300 (1966).
3. P. CARRUTHERS, and M. N. NIETO, Phys. Rev. Lett. 18, 297 (1967); P. CARRUTHERS, Phys. Rev. 154, 1399 (1967).
4. H. CHENG and D. SHARP, Ann. of Phys. 22, 481 (1963) and Phys. Rev. 132, 1854 (1963); S. C. FRAUTSCHI, P. KAUS, and F. ZACHARIASEN, Phys. Rev. 133B, 1607 (1964).
5. M. J. ABBE, P. KAUS, P. NATH and Y. N. SRIVASTAVA, Phys. Rev. 141, 1513 (1966).
6. S. MANDELSTAM., Phys. Rev. 166, 1539 (1968).
7. L. Van HOVE, Phys. Lett. 24B, 183 (1967).
8. L. DURAND III, Phys. Rev. 161, 1610 (1967).
9. D. ATKINSON, K. DIETZ and J. HONERKAMP, Zeitschrift f. Phys. 216, 281 (1968).
10. C. GOEBEL, Phys. Rev. Lett. 21, 383 (1968), N. N. KHURI, UCRL 18246.
11. G. F. CHEW and A. PIGNOTTI, UCRL 18275, G. F. CHEW, Proceedings of the Conference on High Energy Physics, Vienna 1968.
12. For a review of different representations see e.g. P. D. B. COLLINS and E. J. SQUIRES, "Regge Poles in Particle Physics", Springer Tracts, Vol.45, Berlin, Heidelberg, New York, 1968, Chapter III.

13. C. E. JONES and V. L. TEPLITZ, Phys. Rev. Lett. 19, 135 (1967).
14. D. ATKINSON, and K. DIETZ, Rutherford preprint RPP/A/42, to be published in Physical Review.
15. J. KUPSCH, "Generalized Interference Model with Indefinitely Rising Regge Trajectories", University of Bonn, preprint 1968, J. BOGUTA, "Crossing Symmetric, Regge Behaved Amplitude For A General Trajectory", University of Bonn preprint 1968, R. JENGO, CERN preprint TH 988.
16. C. LOVELACE, CERN preprint, TH 950 .
17. G. VENEZIANO, Nuovo Cim. 52, 190 (1968).
18. For an account of the underlying ideas and original references, see D. Horn, Proceedings of this conference.
19. M. ADEMOLLO, H. R. RUBINSTEIN, G. VENEZIANO, and M. A. VIRASORO, "Bootstrap of Meson Trajectories from Superconvergence", Weizmann Institute preprint (1968) and further papers referred to there.
20. D. J. GROSS, Phys. Rev. Lett. 19, 1303 (1967).
21. For reggeisation of helicity amplitudes see e.g. W. DRECHSLER, Nuovo Cim. 53A, 115 (1968). We follow his notations.

OFF-MASS-SHELL APPROACH TO BARYON RESONANCES†

By

Masao SUGAWARA

Department of Physics, Purdue University
Lafayette, Indiana

I. INTRODUCTION

The off-mass-shell approach to baryon resonances may be considered as a new formulation of bootstrap idea in which the dispersion theory and the current-commutator algebra [1] are incorporated in a systematic and unambiguous manner. We study in this approach the off-mass-shell continuation of the matrix element in which a baryon resonance decays into another baryon resonance with the emission of a single pion, kaon, or η-meson. We study, in particular, the dispersion relation satisfied by the off-mass-shell decay amplitude with respect to the external mass of the baryon resonance. We assume no subtraction in this dispersion relation. We assume, furthermore, that the dispersion relation can be saturated in terms of the baryon resonances, which are assumed in this approach to behave more or less like stable particles. Because of this bootstrap assumption, the above dispersion relation reduces to a set of self-consistency conditions satisfied by the physical decay amplitudes among the baryon resonances.

---

† Lecture given at the VIII. Internationale Universitäts-wochen für Kernphysik, Schladming, Feb. 24 - March 8,1969.

However, the above set of conditions alone does not appear to be useful. The important point in the off-mass-shell approach is that the current-commutator algebra [1] allows us to derive another set of self-consistency conditions. This set of conditions is based upon the once-subtracted dispersion relation satisfied by the off-mass-shell decay amplitude with the subtraction constant determined by the current-commutator algebra. For this purpose, we define the continuation with respect to the boson mass shell in terms of the divergence of the corresponding axial-vector current. The off-mass-shell continuation regarding the baryon resonance is defined in terms of the baryon field if the spin of the baryon resonance is $\frac{1}{2}$. If the spin is higher, the off-mass-shell continuation is defined in terms of the high-spin baryon field which is essentially a product of vector and axial-vector currents and the (spin-$\frac{1}{2}$) baryon field. These currents are assumed to satisfy the current-commutator algebra at equal times. The above high-spin field satisfies the equal-time commutator with the time-component of the axial-vector current which is the same as that satisfied by the (spin-$\frac{1}{2}$) baryon field except for the additional Schwinger term. This allows us to determine the subtraction constant and, therefore, to derive the second set of self-consistency conditions.

The off-mass-shell approach to baryon resonances consists of studying the baryon resonances in terms of the above sets of off-mass-shell self-consistency conditions combined. These conditions imply non-trivial consequences concerning the baryon resonances. The most non-trivial of all is that the self-consistency conditions require a certain minimum set of baryon resonances, which appears not too far from what we observe in the experiments. This minimum set of baryon resonances consists of parity doublets which are the pairs of two resonances with the same quantum numbers including spin but with the opposite parities. The above minimum set of baryon resonances includes also parity quartets each of which consists of four baryon resonances which are, more or less, two parity doublets combined. The off-mass-shell approach suggests that the baryon resonances with spin equal to or greater than presumably $\frac{5}{2}$ appear as parity doublets and the baryon resonances with the lower spins appear mostly as parity quartets.

The self-consistency conditions also determine all the decay amplitudes for the baryon resonances to decay among themselves. Thus the off-mass-shell approach allows us to analyze the decays of baryon resonances beyond what can be done by the conventional phase-shift analysis. The self-consistency conditions can also determine the coupling constants of baryon resonances with pion, which could not be determined otherwise. These clearly demonstrate the usefulness of this off-mass-shell approach in understanding the baryon resonances.

In Section II, we discuss the off-mass-shell continuation of the decay matrix element in which a baryon resonance decays into another baryon resonance with the emission of a single pion, kaon or η-meson.

In Section III, we construct the high-spin baryon field which is a product of vector and axial-vector currents and the (spin $\frac{1}{2}$) baryon field and satisfies with the time-component of the axial-vector current the equal-time commutator which is the same as that satisfied by the (spin $\frac{1}{2}$) baryon field except for the additional Schwinger term.

In Section IV, we derive the off-mass-shell self-consistency conditions.

In Section V, we discuss the parity doublet, the minimum set of baryon resonances that can satisfy the self-consistency conditions. In particular, we discuss the parity-doublet mass formula and the determination of the type of the axial-vector current to be used in the off-mass-shell approach.

In Section VI, we discuss the parity quartet, the smallest set of baryon resonances that the members of a parity doublet can decay into.

In Section VII, we summarize some of the main features of the off-mass-shell approach, together with some of the major conclusions that can be made based upon the analyses of the observed baryon resonances.

II. OFF-MASS-SHELL CONTINUATION

We consider a pionic decay of a baryon resonance, denoted by B_k $(k=k,\ell,\ldots)$, with spin $\frac{3}{2}$ and four-momentum p, into a baryon resonance, denoted by B_a $(a=a,b,\ldots)$, with spin 1/2 and four momentum q. We denote their masses by m_k and m_a, and their parities by $-\varepsilon_k$ and ε_a, respectively, where ε's are either +1 or -1. By reducing the pion in the above decay matrix element, we can express it as

$$\sqrt{q_o p_o / m_a m_k}\ \langle B_a(q)\,|\,j_\pi(0)\,|\,B_k(p)\rangle =$$

$$= (\bar{u}(q)\Gamma_a \gamma_5 q_\mu\ \Gamma_k\ u_\mu(p))\ F_{ak}(-(q-p)^2) \qquad (1)$$

where $j_\pi(x)$ is the source of the pion field, Γ_a and Γ_k are 1 or $i\gamma_5$ according as ε_a and ε_k being +1 or -1, $u_\mu(p)$ is a free Rarita-Schwinger spinor, and $F_{ak}(-(q-p)^2)$ is the decay amplitude. The right-hand side of (1) is the most general expression for the matrix element on the left-hand side of (1).

In order to take advantage of the current commutator algebra [1], we define the above off-mass-shell continuation with respect to pion in terms of the divergence of the corresponding axial-vector current. Thus, we choose $j_\pi(x)$ in (1) as

$$j_\pi(x) = \frac{(q-p)^2 + m_\pi^2}{c_\pi} \partial_\mu A_\mu(x) \tag{2}$$

where the constant c_π is given by

$$\sqrt{2k_o} \langle 0|\partial_\mu A_\mu(0)|\pi(k)\rangle = c_\pi \tag{3}$$

This choice of c_π insures that the decay matrix element (1) with $j_\pi(x)$ given by (2) reduces to the physical decay matrix element on the pion mass shell. Furthermore, if the axial-vector current in (2) and (3) is the weak axial vector current, the observed pion life-time determines c_π as

$$c_\pi = 0.94 \; m_\pi^3 \quad . \tag{4}$$

The extension to the decay which involves a kaon or η-meson is self-evident.

The decay matrix element (1) can be continued off the mass shell regarding B_a or B_k in terms of the baryon field $\Psi(x)$ or the Rarita-Schwinger field $\Psi_\mu(x)$, respectively.

These fields have the matrix elements given by

$$\sqrt{q_o/m_a}\langle 0|\Psi(0)|B_a(q)\rangle = c_a\Gamma_a \; u(q) \tag{5}$$

$$\sqrt{p_o/m_k}\langle 0|\Psi_\mu(0)|B_k(p)\rangle = c_k\Gamma_k \; u_\mu(p) \tag{6}$$

We require $\Psi_\mu(x)$ to satisfy

$$\gamma_\mu \Psi_\mu(x) = 0 \tag{7}$$

The above condition can always be satisfied since the combination, $\Psi_\mu - \frac{1}{4}\gamma_\mu(\gamma_\lambda\Psi_\lambda)$, satisfies the condition (7) identically. The off-mass-shell continuation of (1) regarding B_k can be defined in terms of the above $\Psi_\mu(x)$ and c_k by

$$\frac{i}{c_k}\int\sqrt{q_o/m_a}\,\langle B_a(q)|[j_\pi(0),\,\bar{\Psi}_\mu(x)]|0\rangle\theta(-x_o)\,\times \quad (8)$$

$$\times\;[(i\gamma.p+x_k)\delta_{\mu\nu}+i\alpha p_\mu\gamma_\nu]\Gamma_k u_\nu(p)\;e^{ipx}\,dx =$$

$$= (\bar{u}(q)\Gamma_a\gamma_5[q_\nu F_{ak}+p_\nu F_{ak}^{(1)}+\,i\gamma_\nu F_{ak}^{(2)}\;+$$

$$+(q_\nu F'_{ak}+p_\nu F_{ak}^{(1)'}+i\gamma_\nu F_{ak}^{(2)'})(i\gamma.p+x_k)]\Gamma_k u_\nu(p)$$

where

$$x_k \;=\; \varepsilon_k\, m_k \quad , \qquad (9)$$

$\theta(x_o)$ is unity for $x_o > 0$ and zero for $x_o < 0$, α is an arbitrary parameter, and the off-mass-shell decay amplitudes on the right-hand side depend on two variables, $-(q-p)^2$ and $-p^2$. The right-hand side of (8) is also the most general expression for the left-hand side of (8). In (8), the baryon four-momentum p is off the mass shell ($p^2+m_k^2 \neq 0$) and, therefore, $u_\nu(p)$ does not satisfy the usual free equations of motion.

We can show that the left-hand side of (8) reduces to the matrix element in (1) as p becomes physical. To show this, we insert an expansion $\sum_n|n\rangle\langle n|=1$ between $j_\pi(0)$ and $\bar{\Psi}_\mu(x)$ on the left-hand side of (8). If we then carry out the x-integration, the first term of the commutator in (8) gives rise to

$$\int\theta(-x_o)\;\exp\{i(p-p_n)x\}dx = \frac{i(2\pi)^3\delta(\underset{\sim}{p}-\underset{\sim}{p}_n)}{(p_o-p_{n,o})+i\varepsilon} \quad (10)$$

where p_n is the four-momentum of $|n\rangle$ in the above expansion. The energy-denominator in (10) implies that only the term $|B_k\rangle\langle B_k|$ in the above expansion gives rise to a pole as p becomes physical. It is easy to see that the second term of the commutator in (8) does not yield any pole as long as p and q are off the pion mass shell.

Since the kinematical factor that appears on the left-hand side of (8) vanishes identically as p becomes physical, only the term $|B_k\rangle\langle B_k|$ in the above expansion contributes to the limit of the left-hand side of (8) as p becomes physical. A simple manipulation shows that this limit is exactly the matrix element in (1). We also note in this manipulation that the term proportional to the arbitrary parameter α makes no contribution to the above limit. Thus, the left-hand side of (8) reduces to the matrix element in (1), independently of α. The off-mass-shell decay amplitudes are defined on the right-hand side of (8) in such a way that only F_{ak} remains as p becomes physical. Therefore, the off-mass-shell decay amplitude F_{ak} in (8) reduces on the mass shell to the decay amplitude in (1).

The definition (8) of the off-mass-shell continuation with respect to B_k, a spin - $\frac{3}{2}$ resonance, can be extended to the cases of other spins. The details are referred to the original work[2],[3],[4],[5]. In particular, a detailed study [4] is made on the arbitrary parameters (such as α in (8)) that can appear in the off-mass-shell continuation regarding the high-spin baryon resonance. We cann show [4],[5] that these parameters do not affect the self-consistency conditions that follow from the off-mass-shell decay amplitude which reduces to the physical decay amplitude on the mass shell (such as F_{ak} in (8)), but affect the additional conditions that follow from the other amplitudes (such as the F's other than F_{ak} in (8)). We use, in the off-mass-shell approach, only the self-consistency conditions that are independent of the above arbitrary parameters.

We assume that all the F's defined by (8) are analytic in $-(q-p)^2$ and $-p^2$ in the sense of the conventional dispersion theory. Thus, the imaginary parts of these F's with respect to $-p^2$ are all expressed, except for the kinematical factors, in terms of the matrix elements $\langle B_a(q)|j_\pi(0)|n\rangle$ and $\langle n|\bar{\Psi}_\mu(0)|0\rangle$, where $|n\rangle$ is the intermediate state. These matrix elements do not vanish identically for all the baryon resonances with the quantum numbers of B_k, spin 1/2 and 3/2, and both parities. However, the matrix element $\langle n|\bar{\Psi}_\mu(0)|0\rangle$ vanishes identically for the baryon resonances with the higher spins. The contributions from the spin - 3/2 resonances are expressed in terms of the physical decay amplitude defined by (1) and the resonance parameter defined by (6). The contributions from the spin - 1/2 resonances are expressed in terms of the physical decay amplitude among the spin - 1/2 resonances which is defined by

$$\sqrt{q_o p_o/m_a m_b}\,\langle B_a(q)|j_\pi(0)|B_b(p)\rangle = (\bar{u}(q)\Gamma_a i\gamma_5\Gamma_b u(p)) \times \\ \times\; F_{ab}(-(q-p)^2) \qquad (11)$$

where B_b is another spin - 1/2 baryon resonance, and the resonance parameter defined by

$$\sqrt{q_o/m_a}\,\langle 0|\Psi_\mu(0)|B_a(q)\rangle = \left(\frac{4iq_\mu}{3x_a} + \frac{1}{3}\gamma_\mu\right) c_a'\,\Gamma_a\,u(q) \qquad (12)$$

or equivalently

$$\sqrt{q_o/m_a}\,\langle 0|\partial_\mu\Psi_\mu(0)|B_a(q)\rangle = x_a\,c_a'\,\Gamma_a\,u(q) \qquad (13)$$

where we have assumed that B_a has the quantum numbers of B_k except for the spin and parity. These considerations explain the reason why the off-mass-shell self-consistency conditions include all the baryon resonances with spins up to, but not higher than, a given spin. As long as no relation is assumed between $\Psi(x)$ and $\partial_\mu\Psi_\mu(x)$, the parameter c_a' defined by (13) is not related to c_a defined by (5).

The c's defined by (5), (6), and (12) or (13) are the normalization constants. Thus, if a given field yields the c's which are comparable in magnitude for all the baryon resonances with the same quantum numbers including the spin but not the parity, this particular field can become approximately the renormalized field simultaneously for all the above baryon resonances. This is certainly very attractive and seems to imply that this field can describe all these baryon resonances equally well. Therefore, we require in the off-mass-shell approach that the above c's are in fact comparable in magnitude. We can express the reason for this requirement also as follows. If some of the c's are substantially larger than the others, the baryon resonances with the larger c's necessarily dominate in the self-consistency conditions such as (34) and (36). This means that the self-consistency conditions can no longer describe the baryon resonances equally well. Thus, it is in fact necessary to impose the above requirement in the off-mass-shell approach. However, there seems no a priori way to find the numerical figures for the bounds on the acceptable ratios among the c's. The detailed analyses [6],[7] of the observed nucleon resonances suggest that we may allow the c's to differ roughly by a factor of two but presumably not as much as by a factor of three.

III. HIGH-SPIN BARYON FIELD

One of the most attractive features of the off-mass-shell approach is that the high-spin baryon field that describes the high-spin baryon resonances is constructed as a product of vector and axial-vector currents and the (spin -1/2) baryon field. In particular, the high-spin baryon field is constructed in such a way that its equal-time commutator with the time-component of the axial-vector current is readily identifiable in terms of the current-commutator algebra [1]. The use of such a high-spin baryon field enables us to evaluate the limit of the left-hand side of (8) as p approaches q. We can use this limit as the subtraction constant in the once-subtracted dispersion relation satisfied by the off-mass-shell decay amplitude, which then reduces to the second set of the self-consistency conditions mentioned in Section I. The off-mass-shell self-consistency conditions do not appear to be useful unless the above set of conditions is combined with the set of conditions which follows from the unsubtracted dispersion relations satisfied by the off-mass-shell decay amplitude. In other words, the usefulness of the self-consistency conditions depends crucially on the high-spin baryon field of the type described above.

We denote vector and axial-vector currents by $V_{\mu,a}^{b}$ and $A_{\mu,a}^{b}$, respectively, and a nonet of baryon fields by Ψ_{a}^{b}, all in the usual tensor notation. The current-commutator algebra [1] can be satisfied by expressing these currents in terms of Ψ_{a}^{b} as

$$V_{\mu,a}{}^{b} = i(\bar{\Psi}_{c}{}^{b}\gamma_{\mu}\Psi_{a}{}^{c} - \bar{\Psi}_{a}{}^{c}\gamma_{\mu}\Psi_{c}{}^{b}) \; , \qquad (14)$$

$$A_{\mu,a}{}^{b} = i(\bar{\Psi}_{c}{}^{b}\gamma_{\mu}\gamma_{5}\Psi_{a}{}^{c} - \bar{\Psi}_{a}{}^{c}\gamma_{\mu}\gamma_{5}\Psi_{c}{}^{b}) \; , \qquad (15)$$

$$A'_{\mu,a}{}^{b} = i(\bar{\Psi}_{c}{}^{b}\gamma_{\mu}\gamma_{5}\Psi_{a}{}^{c} + \bar{\Psi}_{a}{}^{c}\gamma_{\mu}\gamma_{5}\Psi_{c}{}^{b}) \; , \qquad (16)$$

where the last two expressions stand for the axial-vector currents of f and d type, respectively. The expressions (15) and (16) imply that the baryon fields satisfy with the time-component of the axial-vector current the equal-time commutator,

$$\delta(x_o)[\Psi_a{}^b(x), A_{o,r}{}^s(0)] = \delta(x)\gamma_5\{\delta_{as}\Psi_r{}^b(0) \mp \delta_{br}\Psi_a{}^s(0)\} \tag{17}$$

where the upper and the lower signs on the right-hand side refer to the axial-vector currents of f- and d-type, respectively. The current-commutator algebra is satisfied by the sets $(V_{\mu,a}{}^b, A_{\mu,a}{}^b)$ and $(V_{\mu,a}{}^b, A'_{\mu,a}{}^b)$, but cannot be satisfied if $A_{\mu,a}{}^b$ and $A'_{\mu,a}{}^b$ are mixed. However, if we define the vector current of d-type by

$$V'_{\mu,a}{}^b = i(\bar{\Psi}_c{}^b \gamma_\mu \Psi_a{}^c + \bar{\Psi}_a{}^c \gamma_\mu \Psi_c{}^b) , \tag{18}$$

the current-commutator algebra can be satisfied by the set which consists of $\frac{1}{2}(V_{\mu,a}{}^b + V'_{\mu,a}{}^b)$ and $\frac{1}{2}(A_{\mu,a}{}^b + A'_{\mu,a}{}^b)$ and also by the set $\frac{1}{2}(V_{\mu,a}{}^b - V'_{\mu,a}{}^b)$ and $\frac{1}{2}(A_{\mu,a}{}^b - A'_{\mu,a}{}^b)$. For these sets of currents, the baryon fields satisfy

$$\delta(x_o)[\Psi_a{}^b(x), \tfrac{1}{2}(A_{o,r}{}^s(0) + A'_{o,r}{}^s(0))] = \\ = \delta(x)\gamma_5\, \delta_{as}\, \Psi_r{}^b(0) \tag{19}$$

$$\delta(x_o)[\Psi_a{}^b(x), \tfrac{1}{2}(A_{o,r}{}^s(0) - A'_{o,r}{}^s(0)] = \\ = -\delta(x)\gamma_5\, \delta_{br}\, \Psi_a{}^s(0) . \tag{20}$$

It is easy to show that the above four sets of currents are all the combinations of the currents given by (14), (15), (16), and (18) that satisfy the current-commutator algebra.

When we refer to the nucleon field and the isovector current, the equal-time commutators (17) and (19) all reduce to the commutator assumed in the analyses of the pionic decays of the observed nucleon resonances [6],[7]. Thus, the axial-vector current of f-, d-, or $\frac{1}{2}$(f+d)-type is consistent with the experiments on nucleon resonances. However, the equal-time commutator (20) does not reduce to the above commutator. Therefore, the experiments on nucleon resonances certainly rule out the axial-vector current of $\frac{1}{2}$(f-d)-type.

For the axial-vector current of f-, d-, or $\frac{1}{2}$(f+d)-type, we can construct the high-spin baryon field which satisfies with the time-component of the axial-vector current the equal-time commutator which is the same as (17) or (19) except for the additional Schwinger term. For the axial-vector current of f-type, we consider the nonet of spin - 3/2 baryon fields given by

$$\Psi_{\mu,a}^{\ \ b} = (V_{\mu,c}^{\ \ b} + \gamma_5 A_{\mu,c}^{\ \ b})\,\Psi_a^{\ c} \quad \text{or} \quad (V_{\mu,a}^{\ \ c} + \gamma_5\, A_{\mu,a}^{\ \ c})\,\Psi_c^{\ b} \tag{21}$$

Using the equal-time commutator (17) with the upper sign assumed and also the current-commutator algebra including the space-time commutator, we can show by direct evaluation that the spin -3/2 baryon fields (21) satisfy (17) with the upper sign assumed, Ψ's replaced by Ψ_μ's and also the additional Schwinger term. For the axial-vector current of d-type, we can show similarly that the nonet given by

$$\Psi_{\mu,a}^{\ \ b} = (V_{\mu,c}^{\ \ b} - \gamma_5\, A'^{\ b}_{\mu,c})\,\Psi_a^{\ c} \quad \text{or} \quad (V_{\mu,a}^{\ \ c} + \gamma_5\, A'^{\ c}_{\mu,a})\,\Psi_c^{\ b} \tag{22}$$

satisfies (17) with the lower sign assumed, Ψ's replaced by Ψ_μ's and the additional Schwinger term. For the axial-vector current of $\frac{1}{2}$(f+d)-type, the nonet given by

$$\Psi_{\mu,a}^{\ \ b} = (V_{\mu,c}^{\ \ b} - V'^{\ b}_{\mu,c})\,\Psi_a^{\ c} \quad \text{or} \quad \gamma_5\,(A_{\mu,c}^{\ \ b} - A'^{\ b}_{\mu,c})\,\Psi_a^{\ c} \tag{23}$$

satisfies (19) with Ψ's replaced by Ψ_μ's and the additional Schwinger term. We add that the currents that are used in (23) are of $\frac{1}{2}$(f-d)-type which commute at equal times with the currents of $\frac{1}{2}$(f+d)-type.

It is casy to see that the spin -3/2 fields (21), (22), and (23) can be made to satisfy the condition (7) without modifying the equal-time commutators (17) and (19). It is also obvious that the above construction can be extended to the baryon field with an arbitrarily high spin. It is essential in the above construction that the currents satisfy the current-commutator algebra [1]. Therefore, we do not expect that such a construction is possible except for the cases of the axial-vector current of f-, d-, or $\frac{1}{2}$(f+d)-type and also the case of $\frac{1}{2}$(f-d)-type which is, however, ruled out by experiments.

It is necessary in the off-mass-shell approach that the high-spin baryon field satisfies with the time-component of the axial-vector current the equal-time commutator which is readily identifiable. This equal-time commutator does not have to be the same as that satisfied by the spin 1/2 baryon field, though the high-spin baryon fields discussed here all satisfy the same equal-time commutator as that satisfied by the spin - 1/2 baryon field. The important point is that it does not appear to be possible to construct any other high-spin baryon field that satisfies the equal-time commutator which is readily identifiable.

IV. OFF-MASS-SHELL SELF-CONSISTENCY CONDITIONS

We consider the dispersion relation satisfied by the off-mass-shell decay amplitude F_{ak} defined by (8) with respect to $-p^2$. It follows from the definition (8) that the resonance contributions to Im F_{ak} are given by

$$\text{Im } F_{ak}(-(q-p)^2,-p^2) = \pi\sum_{\ell}\delta(p^2+m_\ell^2)(m_k^2-m_\ell^2)F_{a\ell}(-(q-p)^2)\frac{c_\ell}{c_k} \tag{24}$$

where the sum is over the spin - 3/2 resonances. It is easy to see why the spin -1/2 resonances and also the arbitrary parameter α in (8) do not contribute to the above imaginary part. Thus, if we saturate in terms of the baryon resonances the unsubtracted dispersion relation satisfied by $F_{ak}(-(q-p)^2,-p^2)$ with respect to $-p^2$, we obtain

$$\sum_k F_{ak}(-(q-p)^2)c_k = 0 \tag{25}$$

where the sum is over the spin -3/2 resonances. Similarly, the off-mass-shell continuation of (1) with respect to B_a yields [3]

$$\sum_a c_a F_{ak}(-(q-p)^2) = 0 \tag{26}$$

where the sum is with respect to the spin -1/2 resonances. The conditions (25) and (26) constitute the set of off-mass-shell self-consistency conditions that follow from the unsubtracted dispersion relation. This set of conditions alone does not appear to be useful, unless the additional set of conditions based upon the once-subtracted dispersion relation is combined.

In order to derive the additional set of conditions, we study the limit of the definition (8) as p approaches q. The soft-pion technique [8] enables us to evaluate this limit of the left-hand side of (8), as long as we use the pion source given by (2) and the equal-time commutator between $\Psi_\mu(x)$ and $A_0(x)$ is given by (17) or (19). When the baryon resonances are all the nucleon resonances with the charge of proton and $A_0(x)$ is the neutral isovector current, the above commutator becomes

$$\delta(x_0)[\Psi_\mu(x), A_0(0)] = \frac{1}{\sqrt{2}}\delta(x)\gamma_5\Psi_\mu(x) \tag{27}$$

except for the additional Schwinger term, which, however,

does not contribute to the above limit. In this case, the above soft-pion limit consists of the pole term which is proportional to the matrix elements $<B_a(q)|j_\pi(0)|B_a(q)>$ and $<B_a(q)|\bar{\Psi}_\mu(0)|0>$ and also the equal-time commutator term which contains only the matrix element $<B_a(q)|\bar{\Psi}_\mu(0)|0>$. Thus, the soft-pion limit of the left-hand side of (8) is expressed, aside from the kinematical factors, in terms of c_a' defined by (12) or (13) and $F_{aa}(0)$, the pion coupling constant of B_a, defined by (11).

The above soft-pion limit of the left-hand side of (8) determines the soft-pion limit of the off-mass-shell decay amplitudes that appear on the right-hand side of (8). Using the free equation of motion satisfied by $\bar{u}(q)$ (but not the one by $u_\nu(p)$), we can rewrite the right-hand side of (8) as

$$\bar{u}(q)\Gamma_a\gamma_5[q_\nu\{F_{ak}+F_{ak}^{(1)}-2F_{ak}^{(2)\prime}+(x_a+x_k)(F_{ak}'+F_{ak}^{(1)\prime})\} +$$

$$+ i\gamma_\nu\{F_{ak}^{(2)}+(x_k-x_a)F_{ak}^{(2)\prime}\}+ \ldots]\Gamma_k u_\nu(p) \qquad (28)$$

where the dots stand for the terms which are proportional to p-q. Because of this factor, p-q, the terms denoted by the dots in (28) become proportional to the residues of the amplitudes at the pole at p=q. Since all these residues can be evaluated directly from the definition (8), the terms denoted by the dots in (28) do not yield any new information. Thus, the soft-pion limit of the left-hand side of (8) determines the soft-pion limits of the two combinations of the off-mass-shell decay amplitudes that appear inside the two curly brackets in (28). A straightforward evaluation shows that

$$F_{ak} + F_{ak}^{(1)} - 2F_{ak}^{(2)\prime} + (x_a+x_k)(F_{ak}'+F_{ak}^{(1)\prime}) \to \qquad (29)$$

$$\xrightarrow[p\to q]{} \frac{2(x_a-2x_k)}{3x_a}\,\frac{F_{aa}(0)c'_a}{2x_a c_k} + \frac{2(x_a+2x_k)}{3x_a}\,\frac{m_\pi^2 c'_a}{\sqrt{2}\,c_\pi c_k}\,, \tag{29}$$

where the last term is due to the equal-time commutator (27). It is also straightforward to show that the soft-pion limit of the left-hand side of (8) contains the terms which are exactly the same as the soft-pion limit of the terms denoted by the dots in (28), as is should.

We remark that the soft-pion limit in (29) is independent of the arbitrary parameter α in (8). In fact, it is not difficult to see how this follows. Much more non-trivial to see is that the imaginary part of the combination of the off-mass-shell decay amplitudes in (29) is also independent of α, since most of these amplitudes actually depend upon α . The detailed studies made in the original work [4], [5] show that the soft-pion technique [8] determines the soft-pion limits of more than one combination of the off-mass-shell decay amplitudes when the spins of baryon resonances change in the decay. However, the off-mass-shell decay amplitude that reduces on the mass shell to the physical decay amplitude (such as F_{ak} in (8)) appears only in one of these combinations. We can show [4],[5]in general that the soft-pion limit and also the imaginary part of only this particular combination of the off-mass-shell decay amplitudes are independent of the arbitrary parameters that can appear in the definition of the off-mass-shell continuation. Therefore, we do not consider in the off-mass-shell approach the rest of the combinations of the amplitudes the soft-pion limits of which are determined. The self-consistency conditions that follow from these other combinations of the amplitudes are discussed in detail in the original work [4]. We also add that no soft-pion limit can be determined [3],[4] with respect to the off-mass-shell continuation of the decay matrix element regarding the baryon resonance whose spin is lower than the spin of the other baryon resonance. This is simply because both sides of the definition of the off-mass-shell continuation (such as (8)) vanish identically at the soft-pion point because of kinematical factors.

In order to utilize the soft-pion limit in (29), we consider the amplitude given by

$$F_{ak} + \frac{m_k^2+p^2}{m_k^2-m_a^2}\,[F_{ak}^{(1)}-2F_{ak}^{(2)}{}'+(x_a+x_k)(F'_{ak}+F_{ak}^{(1)}{}')] \qquad (30)$$

The above amplitude reduces on the mass shell to the physical decay amplitude defined by (1) and tends, as p approaches q, to the same soft -pion limit as that in (29). Thus, we can write the once-subtracted dispersion relation for the amplitude (30) with $-(q-p)^2 = 0$ and the subtraction constant chosen as the soft-pion limit in (29). By saturating the above dispersion relation in terms of the baryon resonances, we find [3]

$$\sum_\ell \left[1-\frac{(x_k-x_\ell)(x_a-x_\ell)}{3x_\ell^2}\right]F_{a\ell}(0)\ c_\ell = \qquad (31)$$

$$= \sum_b \frac{2(x_b-2x_k)}{3x_b}\,\frac{F_{ab}(0)c'_b}{x_a+x_b} + \frac{2(x_a+2x_k)}{3x_a}\,\frac{m_\pi^2\ c'_a}{\sqrt{2}\ c_\pi}\ ,$$

where the sum over ℓ includes all the spin -3/2 resonances and the sum over b includes all the spin -1/2 resonances.

As is discussed in the subsequent section, the off-mass-shell self-consistency conditions cannot be satisfied unless there exists at least a pair of two resonances with the same quantum numbers including the spin but with the opposite parities. If there are two or more spin -3/2 resonances, the conditions (31) split into two sets of conditions because of the dependence of the conditions (31) on x_k. If we introduce the dimensionless amplitudes by

$$f_{ab} = \frac{\sqrt{2}\, c_\pi}{m_\pi^2} \frac{F_{ab}(0)}{x_a + x_b} , \tag{32}$$

$$f_{ak} = \frac{\sqrt{2}\, c_\pi}{m_\pi^2} F_{ak}(0) . \tag{33}$$

The conditions (25), (26) and (31) can be expressed as

$$\sum_a c_a f_{ak} = 0 , \qquad \sum_k f_{ak} c_k = 0 ,$$

$$\sum_k f_{ak} y_k c_k = 2y_a (\sum_b f_{ab} c_b' + c_a') ,$$

$$\sum_k f_{ak} y_k^2 c_k = 2y_a^2 (\sum_b f_{ab} c_b' + c_a') + \tag{34}$$

$$+ 4y_a (\sum_b f_{ab} y_b c_b' - y_a c_a') ,$$

where

$$y_k = \frac{1}{x_k} = \frac{1}{\varepsilon_k m_k} \tag{35}$$

etc., the sums over a and b include all the spin -1/2 resonances and the sum over k is with respect to all the spin -3/2 resonances. The conditions (34) are all the off-mass-shell self-consistency conditions that can be derived for this decay because no soft-pion limit can be determined [3] with respect to the off-mass-shell continuation of the matrix element (1) regarding B_a .

The derivation of the off-mass-shell self-consistency conditions for the other decays is referred to the original work [2],[3],[5],[9]. The decay amplitudes among the spin -1/2 baryon resonances satisfy [2]

$$\sum_b f_{ab} c_b = c_a , \quad \sum_b f_{ab} x_b c_b = -x_a c_a \tag{36}$$

when the spin -1/2 baryon field also satisfies the equal-time commutator (27), with $\Psi_\mu(x)$ replaced by $\Psi(x)$. We can show [5], in particular, that the conditions (36) are satisfied by the decay amplitudes among the baryon resonances with an arbitrary spin. In order to explain this further, we refer to a pionic decay of B_k into B_ℓ, another spin -3/2 baryon resonance. The decay matrix element can be expressed as

$$\sqrt{q_o p_o/m_\ell m_k}\ \langle B_\ell(q)|j_\pi(0)|B_k(p)\rangle =$$

$$= \bar{u}_\mu(q)\Gamma_\ell i\gamma_5[\delta_{\mu\nu}F_{\ell k}(-(q-p)^2) + p_\mu q_\nu G_{\ell k}(-(q-p)^2)]\times$$

$$\times\ \Gamma_k\ u_\nu(p) \quad , \tag{37}$$

where two decay amplitudes appear corresponding to the two partial waves involved in this decay. If we define the dimensionless amplitude by

$$f_{\ell k} = \frac{\sqrt{2}\ c_\pi}{m_\pi^2}\ \frac{F_{\ell k}(0)}{x_\ell + x_k} \quad , \tag{38}$$

this amplitude satisfies the conditions (36) with the indices replaced. No soft-pion limit can be determined for $G_{\ell k}$ in (37) that involves the higher partial wave. Thus, $G_{\ell k}$ satisfies [5] only

$$\sum_k G_{\ell k}(-(q-p)^2)c_k \quad = \quad 0 \tag{39}$$

which is based upon the unsubtracted dispersion relation. In other words, the off-mass-shell approach provides a sufficient amount of conditions on $F_{\ell k}$ but not necessarily on $G_{\ell k}$. However, it is hoped that $G_{\ell k}$ is in fact a minor amplitude and can, therefore, be ignored in the decay matrix element in (37). The extension to the spin-5/2 resonances is referred to the original work [4],[5]

The inhomogeneous terms (or the terms which do not include f's) in the self-consistency conditions such as (34) and (36) are all due to the equal-time commutator (27). The equal-time commutators (17) and (19) all reduce to (27) in the case of the pionic decays of nucleon resonances. However, the commutators (17) and (19) differ in the cases of the pionic decays of strange baryon resonances and also the decays of baryon resonances with the emission of a kaon or η-meson. However, the self-consistency conditions for these decays also assume the forms such as (34) and (36) except that the inhomogeneous terms are changed according as the equal-time commutators (17) and (19). The complete set of off-mass-shell self-consistency conditions is given in the original work [9] for the decays of all the baryon resonances with the emission of a single pion, kaon, or η-meson for the axial-vector current of f-, d-, or $\frac{1}{2}$(f+d)-type.

The decay amplitudes that appear in the self-consistency conditions such as (34) and (36) all refer to the point where the external boson mass vanishes. The analyses [6],[7],[10] of the observed baryon resonances imply that the off-boson-mass-shell correction does not appear to be appreciable in the pionic decays, though this may not necessarily be the case in the decays which involve kaon and η-mesons. However, independently of whether or not the off-boson-mass-shell correction is small we expect that all the self-consistency conditions are satisfied as long as they are well satisfied in the pionic decays. The reason for this expectation is merely that the self-consistency conditions are the dispersion relations with respect to the external mass of the baryon resonance saturated in terms of the baryon resonances.

V. PARITY-DOUBLET MASS FORMULA

If we eliminate f's in the conditions (36), we find

$$\sum_a x_a c_a^2 = 0 \tag{40}$$

The c's defined by (5), (6), and (12) or (13) are all real as long as the baryon resonances behave like a nucleon under time-reversal. The condition (40) then cannot be satisfied by the resonances with the same parity alone. In other words, the self-consistency conditions (36) yield no solution unless there exists at least a pair of two resonances with the same quantum numbers including the spin but with the opposite parities. We call such a pair a parity doublet. For a parity doublet (B_a, B_b), the solution to the self-consistency conditions (36) is uniquely determined in terms of the masses as

$$\begin{aligned} c_a &= 1 \quad , \qquad c_b = \sqrt{m_a/m_b} \sim 1 \; , \\ f_{aa} &= -f_{bb} = \frac{m_b - m_a}{m_b + m_a} \sim \text{small} \\ f_{ab} &= \frac{2\sqrt{m_a m_b}}{m_a + m_b} \sim 1 \quad , \end{aligned} \tag{41}$$

where we have normalized c_a to unity and chosen c_b to be positive. The approximate equalities in (41) are due to the fact that the masses of baryon resonances are roughly the same in magnitude. The solution (41) is satisfactory since $c_a \sim c_b$, which is the requirement imposed in the off-mass-shell approach.

In order to appreciate the solution (41), we point out that the known pion-nucleon coupling constant corresponds to

$$f_{NN} = 1.18 \quad , \tag{42}$$

according to the normalization of the amplitude by (32). Thus, the solution (41) implies that the coupling constants with pion of the members of a parity doublet are much smaller than the known pion-nucleon coupling constant. In order to appreciate the decay amplitude in (41), we find that the decay rate of N(1466) into a pion and a nucleon would be ∿1200 MeV, versus 139 MeV according to the experiments [11], if the decay amplitude there were ∿1. This means that the solution (41) implies an intense decay amplitude between the members of a parity doublet. Therefore, we expect a parity doublet to appear with a mass difference which is not large compared with the pion mass since, otherwise, the internal decay between them could become too intense to agree with experiments.This is,in fact,in general agreement with the experiments [11]. We refer a further discussion to the original work [5].

The pionic decays of Λ- and Σ-resonances are of special interest in this context. We denote them by Λ_a, Λ_b,... and by Σ_α, Σ_β ,..., all with the same spin . We denote the amplitudes for the decays $\Lambda_a \to \Sigma^o_\alpha + \pi^o$ and $\Sigma^+_\alpha \to \Sigma^+_\beta + \pi^o$ by $f_{a\alpha}$ and $f_{\alpha\beta}$, respectively, which are normalized also by (32) or (38) except for the indices. We add that the pionic decays are forbidden among the Λ's. In the case of the axial-vector current of $\frac{1}{2}$(f+d)-type, the above amplitudes satisfy [9]

$$\sum_\beta f_{\alpha\beta}\, c_\beta = c_\alpha\,, \qquad \sum_\beta f_{\alpha\beta} x_\beta c_\beta = -\, x_\alpha c_\alpha\,, \tag{43}$$

$$\sum_\alpha f_{a\alpha}\, c_\alpha = c_a\,, \qquad \sum_\alpha f_{a\alpha} x_\alpha c_\alpha = -x_a\, c_a\,, \tag{44}$$

$$\sum_a c_a\, f_{a\alpha} = c_\alpha\,, \qquad \sum_a x_a c_\alpha f_{a\alpha} = -x_\alpha\, c_\alpha\,, \tag{45}$$

where c_α is the resonance parameter of Σ_α with respect to Ψ_Σ in the sense of (5). However, c_a is the resonance parameter of Λ_a defined, not with respect to ψ_Λ , but in terms of

$$\psi_1 = \frac{1}{\sqrt{3}}(\psi_\Lambda + \sqrt{2}\psi_{\Lambda'}) = \frac{1}{\sqrt{2}}(\psi_1^1 + \psi_2^2) \tag{46}$$

where $\psi_{\Lambda'}$ is the ninth member of the baryon nonet Ψ_a^b that appears in the currents (16) and (18) of d-type. Since both ψ_Λ and $\psi_{\Lambda'}$ have the quantum numbers of Λ_a, any linear combination of ψ_Λ and $\psi_{\Lambda'}$ can be used to define the off-mass-shell continuation of the decay matrix element regarding Λ_a. This ambiguity can be desolved by requiring that the resulting self-consistency conditions such as (44) and (45) assume the simplest form and, therefore, become the most self-constrained. This requirement determines [9] the combination as that given by (46). The self-consistency conditions (43), (44) and (45) can then easily be derived by analyzing the equal-time commutator (19) for the decays concerned. We refer to the cases of f- and d-type later.

If we eliminate f's in the conditions (43), (44), and (45), we find

$$\sum_\alpha x_\alpha c_\alpha^2 = \sum_a x_a c_a^2 = 0 , \tag{47}$$

$$\sum_\alpha c_\alpha^2 = \sum_a c_a^2 , \quad \sum_\alpha x_\alpha^2 c_\alpha^2 = \sum_a x_a^2 c_a^2 . \tag{48}$$

The conditions (47) imply that the self-consistency conditions yield no solution unless both Λ- and Σ-resonances appear at least as parity doublets. However, in the case when they actually appear as parity doublets (Λ_a, Λ_b) and (Σ_α,Σ_β) , the conditions (47) and (48) constitute four conditions for three c's since one of the c's can always be normalized to unity. Thus, a mass formula follows for the above parity doublets. In order to derive this parity-doublet mass formula, we first solve $\sum_a x_a c_a^2 = 0$

and $\sum_a c_a^2 = \sum_\alpha c_\alpha^2$ as

$$c_a^2 = \frac{x_b}{x_b - x_a}(\sum_\alpha c_\alpha^2) \ , \qquad c_b^2 = \frac{x_a}{x_a - x_b}(\sum_\alpha c_\alpha^2) \ , \tag{49}$$

the right-hand sides of which are always positive. If we substitute (49) into the second condition in (48), we find

$$-x_a x_b (\sum_\alpha c_\alpha^2) = \sum_\alpha x_\alpha^2 \ c_\alpha^2 \ . \tag{50}$$

If we combine (50) with $\sum_\alpha x_\alpha c_\alpha^2 = 0$, we find the mass formula,

$$x_\alpha x_\beta = x_a x_b \quad , \text{ i.e., } m_\alpha m_\beta = m_a m_b \ . \tag{51}$$

In the above derivations, we use the fact that $x_a \neq x_b$ and $x_\alpha \neq x_\beta$, which states that no two resonances with the same quantum numbers can have simultaneously the same mass and parity. We can show, furthermore, that the self-consistency conditions (43), (44), and (45) yield a satisfactory solution if the above resonances appear as parity doublets with the masses which satisfy the above mass formula. It is straightforward to find that this solution consists of the c's and the f's for the Σ-doublet which are the same as (41) except for the indices and also

$$c_a = \frac{\sqrt{m_b(m_\alpha + m_\beta)}}{\sqrt{(m_a + m_b) m_\beta}} \sim 1 \ , \qquad c_b = \frac{\sqrt{m_a(m_\alpha + m_\beta)}}{\sqrt{(m_a + m_b) m_\beta}} \sim 1 \ ,$$

$$f_{a\alpha} = \frac{m_\beta - m_a}{m_\alpha + m_\beta} c_a \sim \text{small, or} \quad = \frac{m_\beta + m_a}{m_\alpha + m_\beta} c_a \sim 1 \ , \tag{52}$$

depending upon whether Λ_a and Σ_α have the same parity or the opposite parities, respectively. Thus, the solution (52) implies that Λ_a and Σ_α can also form a parity doublet, if they have the opposite parities, in the sense that they have an intense decay amplitude between them.

The available experiments [11] indicate that Λ- and Σ -resonances with spin $\frac{5}{2}$ appear exactly as parity doublets with the masses

$$m_\alpha = 1765^- \ , \quad m_\beta = 1910^+$$
$$m_a = 1815^+ \ , \quad m_b = 1830^- \ , \qquad (53)$$

in the units of MeV, where + and - stand for the parities. The above masses satisfy the mass formula (51) very well since they yield

$$\frac{m_\alpha m_\beta}{m_a m_b} = 1.015 \ . \qquad (54)$$

The largest mass difference among the masses in (53) is $m_\beta - m_\alpha = 145$ MeV which is only barely larger than the pion mass. Thus, the above four resonances can very well be parity doublets. Three Λ- and Σ-resonances with spin $\frac{7}{2}$ have been observed [11] with the masses

$$m_\alpha = \ldots. , \ m_\beta = 2030^+ \ ,$$
$$m_a = 1860^+ , m_b = 2100^- \ , \qquad (55)$$

in the units of MeV, where m_α is the mass of $\Sigma(\frac{7}{2}^-)$ that can complete the spin -7/2 parity doublets. This mass is predicted as

$$m_\alpha = 1924 \text{ MeV} \qquad (56)$$

according to the mass formula (51). With the masses (55) and (56), the largest mass difference that corresponds to an intense decay amplitude is $m_\beta - m_\alpha = 106$ MeV. This is smaller than the pion mass. Thus, the above four resonances may in fact be parity doublets. We also note that the mass (56) is quite close to m_β in (53). Thus this new $\Sigma(\frac{7^-}{2})$ is expected to appear very close to the observed $\Sigma(\frac{5^+}{2})$. This may explain the reason why this new $\Sigma(\frac{7^-}{2})$ has not yet been observed. It seems worthwhile to examine experimentally whether the observed $\Sigma(\frac{5^+}{2})$ consists of two resonances, $\Sigma(\frac{5^+}{2})$ and $\Sigma(\frac{7^-}{2})$, the latter being the new resonance.

If the axial-vector current is of f-type, $f_{\alpha\beta}$ satisfies the conditions (43) with the right-hand sides doubled and $f_{a\alpha}$ satisfies the conditions (44) and (45) with the right-hand sides replaced by zero [9]. Thus, the conditions (47) and (48) and, therefore, the mass formula (51) do not follow in this case from the above conditions. However, the complete set of off-mass-shell self-consistency conditions contains many other sets of conditions which can give rise to the conditions such as (47) and (48) when the amplitudes are eliminated. We can show [9] in fact, for the axial-vector currents of both f- and d-type, that the conditions (47) and (48) follow not only for Λ- and Σ-resonances but also for nucleon- and Ξ - resonances, which we denote by N_i, N_j,... and Ξ_ξ, Ξ_η ,... , respectively, all with the same spin. Therefore, in the cases of f- and d-type, we find the parity-doublet mass formula which reads

$$m_i m_j = m_\alpha m_\beta = m_a m_b = m_\xi m_\eta \tag{57}$$

On the other hand, we find [9], for the axial-vector current of $\frac{1}{2}$(f+d)-type, that the conditions (47) and (48) are satisfied by Λ-, Σ-, and Ξ-resonances but nucleon resonances satisfy only the conditions (47). Thus, in the case of $\frac{1}{2}$(f+d)-type, the parity-doublet mass formula becomes

$$m_\alpha m_\beta = m_a m_b = m_\xi m_\eta \tag{58}$$

which does not include the nucleon parity doublet.

This difference is due to a characteristic difference in the self-consistency conditions satisfied by the amplitudes for the kaon decays [9].

The available experiments [11] on Ξ-resonances are too meager to test the above parity-doublet mass formulae. The nucleon resonances with spin 5/2 are observed [11] with the masses

$$m_i = 1678^- , \quad m_j = 1692^+ , \tag{59}$$

which are, however, too small compared with the masses in (53) to satisfy the mass formula (57). Of course, this may mean that the nucleon resonances with spin 5/2 appear in a much higher multiplet than a parity doublet. On the other hand, the nucleon resonances with spin 7/2 are observed [11] with the masses

$$m_i \sim 1983^+, \quad m_j = 2190^- . \tag{60}$$

The above masses are now too large compared with those in (55) and (56) to satisfy (57). The mass formula (57) cannot be satisfied unless the Λ- and Σ- resonances with spin 7/2 form higher multiplets than parity doublets.

Thus, unless there exist considerably more resonances than those which are known at present, we must conclude that the mass formula (57) is violated but the mass formula (58) is consistent with the available experiments [11]. This determines the type of the axial-vector current that is to be used in the off-mass-shell approach as $\frac{1}{2}$(f+d)-type.

The same conclusion can be made based upon the analysis [10] of the observed Λ- and Σ-resonances with spin $\frac{1}{2}$ and $\frac{3}{2}$ in terms of the self-consistency conditions (43), (44) and (45). These observed resonances satisfy the above conditions quite well [10]. However, for the axial-vector current of f-type, $f_{\alpha\beta}$ satisfies the conditions (43) with the right-hand sides doubled. For the axial-vector current of d-type, $f_{a\alpha}$ satisfies the conditions (44) and (45) with all the right-hand sides also doubled [9]. This effectively enhances either $f_{\alpha\beta}$ or $f_{a\alpha}$ virtually by the same factor. According to the analysis [10] of the observed resonances, the difficulties encountered are always associated with too intense, but not too weak, decay amplitudes. Thus, the above change makes the self-consistency conditions much harder to satisfy. In fact, we can conclude [10] that the axial-vector current of f- or d-type is to be ruled out unless there exist considerably more resonances than those which are known at present [11].

VI. PARITY QUARTET

We discuss the self-consistency conditions (34) for the decay in which a spin - 3/2 resonance decays into a spin - 1/2 resonance. The right-hand sides of (34) include the decay amplitudes among the spin - 1/2 resonances which satisfy the self-consistency conditions (36). We note that these right-hand sides of (34) vanish identically if

$$c'_a \propto x_a c_a \quad . \tag{61}$$

According to the definitions (5) and (13) of c_a and c'_a, respectively, c'_a given by (61) is different from, but has the sign of, what is implied by $\partial_\mu \psi_\mu(x) \propto \psi(x)$. Anyhow, if c_a represents a satisfactory solution to the self-consistency conditions, then c'_a given by (61) is also satisfactory because the masses of baryon resonances are roughly the same in magnitude. We, therefore, see that the self-consistency conditions (34) possess a satisfactory solution which consists of (61) and also $f_{ak} = 0$. We can make the same observation in the case of the decay in which a spin -5/2 resonance decays into a spin -1/2 resonance [4] and also in all the decays in which the spins of baryon resonances change by one unit [12]. In other words, all these self-consistency conditions yield satisfactory solutions in which all the amplitudes vanish iden tically for the decays in which the spins of baryon resonances change. We remark that the amplitudes for the decays among the baryon resonances with a given spin cannot vanish identically since they satisfy the self-consistency conditions of the type of (36).

In order to discuss under what condition the self-consistency conditions (34) possess a solution in which f_{ak} does not vanish identically, we rewrite (34) as

$$\sum_a c_a f_{ak} = 0 \quad , \quad \sum_k f_{ak} c_k = 0 \tag{62}$$

$$\sum_k f_{ak} y_k c_k = g_a \quad , \quad \sum_k f_{ak} y_k^2 c_k = h_a \quad ,$$

where g_a and h_a stand for the right-hand sides of the third and the fourth conditions in (34), respectively. Because of the first condition in (62) and also the conditions (36) satisfied by f_{ab}, we can show that g_a and h_a satisfy

$$\sum_a c_a g_a = 0 \; , \; \sum_a x_a c_a g_a = c \; , \quad \sum_a x_a^2 c_a g_a = 0 \; , \tag{63}$$

$$\sum_a c_a h_a = 0 \; , \; \sum_a x_a c_a h_a = 0 \; , \quad \sum_a x_a^2 c_a h_a = - c \; , \tag{64}$$

where

$$c = 4 \sum_a c_a c'_a \tag{65}$$

First, we discuss the case when the spin -1/2 resonances appear as a parity doublet (B_a, B_b). Even when the masses of B_a and B_b are degenerate, the first two conditions in (64) are linearly independent and, therefore, imply that $h_a=0$. The last condition in (64) then implies $c=0$ and, subsequently, $g_a=0$ because of the conditions (63). With $g_a=h_a=0$, the last three conditions in (62) imply $f_{ak}=0$, unless the spin -3/2 resonances appear at least as a quartet, because these three conditions are all linearly independent even when some of the masses are degenerate. Thus, the decays are forbidden between a spin -1/2 doublet and the spin -3/2 resonances unless the latter form at least a quartet of resonances.

We then consider the case when the spin -3/2 resonances appear as a parity doublet (B_k, B_ℓ). We can solve the second and the third conditions in (62) as

$$f_{ak}c_k = \frac{g_a}{y_k - y_\ell} \quad , \quad f_{a\ell}c_\ell = \frac{g_a}{y_\ell - y_k} \quad . \tag{66}$$

The last condition in (62) then implies

$$(y_k + y_\ell) g_a = h_a \quad . \tag{67}$$

According to the preceding analysis, the decay amplitudes vanish identically if the spin - 1/2 resonances appear also as a parity doublet. If they form a parity triplet (B_a, B_b, B_c), the conditions (63) and (64) can be solved as

$$c_a g_a = \frac{-(x_b + x_c)c}{(x_a - x_b)(x_a - x_c)} \quad , \quad c_a h_a = \frac{-c}{(x_a - x_b)(x_a - x_c)} \quad , \tag{68}$$

and their cyclic permutations with respect to a,b and c. Thus, it follows from (67) and (68) that

$$(y_k + y_\ell)(x_b + x_c)\ c = c \ , \tag{69}$$

and its cyclic permutations with respect to a,b, and c. All these three conditions can be satisfied only if $c=0$.

This implies that $g_a = h_a = 0$ and , subsequently, $f_{ak} = f_{a\ell} = 0$ for all a,b, and c. Thus the decays are also forbidden between a spin -3/2 parity doublet and the spin -1/2 resonances unless the latter form at least a quartet of resonances.

Finally, we consider the case when both spin -1/2 and spin -3/2 resonances appear as parity triplets, (B_a, B_b, B_c) and (B_k, B_ℓ, B_m). We can solve the last three conditions in (62) as

$$f_{ak} c_k = \frac{h_a - (y_\ell + y_m) g_a}{(y_k - y_\ell)(y_k - y_m)} \tag{70}$$

and its cyclic permutations with respect to k,ℓ, and m. The first conditions in (62) are satisfied by the solution (70) as long as g_a and h_a satisfy the first conditions in (63) and (64). The rest of the conditions in (63) and (64) then follow as long as f_{ab} satisfies the conditions (36). The first conditions in (63) and (64) can be written also as

$$\sum_a g'_a c'_a = 0 \quad , \quad \sum_a h'_a c'_a = 0 \ , \tag{71}$$

where

$$g'_a = 2 \sum_b y_b c_b (f_{ba} + \delta_{ba}) \ ,$$

$$h'_a = 2 \sum_b y_b^2 c_b (f_{ba} + \delta_{ba}) + 4 \sum_b y_b c_b (f_{ba} - \delta_{ba}) y_a \tag{72}$$

We now observe that the self-consistency conditions (34) are invariant, because of the conditions (36), under the replacement

$$c'_a \to c'_a + \alpha x_a c_a \ , \tag{73}$$

where α is arbitrary. This means that we can always make one of the primed c_a's vanish. Thus, if we set $c'_a = 0$ and if there is only a triplet of spin -1/2 resonances, the conditions (71) require

$$g'_b h'_c = g'_c h'_b \tag{74}$$

to be satisfied in addition to the conditions (36). If the condition (74) is not satisfied, the conditions (71) imply that all the primed c_a's can be made to vanish simultaneously. This is the case when $c'_a \propto x_a c_a$ and, consequently, c=0 because of (65). For a triplet of spin-1/2 resonances, the conditions (63) and (64) then imply $g_a=h_a=0$ and, thus, $f_{ak} = 0$ because of (70). Therefore, the decays are also forbidden between a triplet of spin-1/2 resonances and a triplet of spin -3/2 resonances, unless the former satisfies an extra condition (74) with the definitions (72), in addition to the self-consistency conditions (36).

We can show [12] that the same conclusion follows also from the most general set of self-consistency conditions for the decays in which the spins of baryon resonances change by one unit. In other words, the decays among baryon resonances in which spin changes by one unit are generally forbidden if the resonances appear at most as parity doublets and parity triplets [12]. This study[12] also suggests that the same conclusion generally holds for the decays in which spin changes by more than one unit. However, this extension certainly requires a further study.

We have shown in the preceding section that the baryon resonances with spin 5/2 and higher appear to form parity doublets. If this is actually the case, the above considerations imply that these high-spin resonances do not decay among themselves. This may explain at least partially the reason why the high-spin resonances can appear without making their widths necessarily too broad. If the high-spin resonances actually appear as parity doublets, the preceding discussions also imply that the baryon resonances with the lower spins form at least quartets so that the high-spin resonances can decay into. This is certainly in agreement with the experiments [11] on nucleon resonances which report five nucleons (including the nucleon) with spin and isospin 1/2. A study [10] of the observed Λ- and Σ-resonances indicates quite convincingly that the Λ- and Σ-resonances with spin 1/2 form quartets which consist of two members with the parity opposite to the others. Most spin -3/2 baryon resonances have been observed [11] to appear as parity triplets. However, the experiments [11] also report the decays of high-spin baryon resonances into Δ(1236) and Σ(1385) both with spin 3/2. This implies that, as long as these high-spin resonances appear as parity doublets, the spin -3/2 resonances must also form quartets. It is certainly worthwhile th search for the new spin -3/2 resonances.

We demonstrate below how the five nucleons mentioned above satisfy the self-consistency conditions (36). These are listed in the Table, together with the notation used below.

Notation	Parity	Mass (MeV)	Γ_{tot} (MeV)	Γ_{el} (MeV)
N_1	+	940	---	---
N_2	+	1466	211	139
N_3	-	1548	116	38
N_4	-	1709	300	236
N_5	+	~ 1751	327	105

Table of nucleon resonances with spin and isospin $\frac{1}{2}$ taken from ref. [11]

First we recall that the observed pion-nucleon coupling constant corresponds to the amplitude given by (42). This is far from what is implied by the solution (41) for a parity doublet, implying that the nucleon cannot possibly form a parity doublet. In fact, the Table shows that N_2 has an even parity and both N_3 and N_4 have an odd parity. Therefore, the smallest set of nucleons to be considered is a triplet that consists of N_1, N_2 and either N_3 or N_4.

We here quote the experimental information [11] that N_3 is primarily an η-nucleon resonance and N_5 is highly inelastic. On the other hand, both N_2 and N_4 are the elastic pion-nucleon resonances in the sense that the pion-nucleon phase-shifts become 90° at both N_2 and N_4. This suggests that we should consider a triplet (N_1,N_2, N_4). In this case, the self-consistency conditions (36) can accommodate two inputs. If we assume, as these inputs, f_{11} given by (42) and f_{12} determined by Γ_{el} (N_2) in the Table, the solution to (36) is uniquely determined as

$$c_1 = 1 \ , \quad c_2 = -1.98 \ , \quad c_4 = 1.98 \ ,$$
$$f_{11} = 1.18 \ , \quad f_{22} = 0.234, \quad f_{44} = -0.079 \qquad (75)$$
$$f_{12} = 0.374 \ , \quad f_{14} = 0.284 \ , \quad f_{24} = -0.956 \ ,$$

which yield the decay rates

$$\Gamma(N_4 \to N_1 + \pi) = 229 \text{ MeV} \ , \quad \Gamma(N_4 \to N_2 + \pi) = 118 \text{ MeV} \qquad (76)$$

The first decay rate in (76) is in excellent agreement with $\Gamma_{el}(N_4)$ in the Table. However, the second decay in (76) is nearly twice the difference $\Gamma_{tot}(N_4) - \Gamma_{el}(N_4)$ implying that f_{24} in (75) is too intense to agree with the experiment. The reason why f_{24} becomes large in magnitude in (75) is simply because N_2 and N_4 appear in the triplet (N_1,N_2,N_4) with the opposite parities and a relatively small mass difference between them. In other words, N_2 and N_4 tend to form a parity doublet, resulting in $|f_{24}|$ of ~ 1 in (75). This means that $|f_{24}|$ cannot be reduced unless there arises another resonance between them so that N_2 and N_4 do not have to behave like a parity doublet. This naturally explains the existence of N_3.

For a quartet of resonances, the conditions (36) can accommodate five inputs. These can include f_{11} given by (42), f_{12}, f_{13}, and f_{14} all determined by the elastic widths of N_2, N_3 and N_4 respectively, in the Table and also $|f_{44}| < 1$, a reasonable requirement. Then, the solution to (36) is given by [2]

$$c_1 = 1 \ , \quad c_2 = -2.02 \ , \quad c_3 = 1.83 \ , \quad c_4 = 1.02 \ ,$$
$$f_{11} = 1.18, \quad f_{12} = 0.374 \ , \ f_{13} = 0.155 \ ,$$
$$f_{14} = 0.288, \quad |f_{44}| < 1 \ , \quad f_{24} \sim -0.4 \ ,$$
$$f_{23} \sim -0.87 \ , \quad f_{22} \sim 0.19 \ , \qquad (77)$$

and f_{33}, f_{34} strongly depending upon f_{44}. We see in the solution (77) that $|f_{24}|$ is reduced by more than a factor of two, thus reducing $\Gamma(N_4 \to N_2 + \pi)$ to less than a quarter of the figure in (76). Thus the solution (77) is totally consistent with the data in the Table. We note that $|f_{23}|$ is quite large in (77) which is due to the fact that N_2 and N_3 tend to form a parity doublet in this quartet. However, we note also in the Table that the mass difference between N_2 and N_3 is smaller than the pion mass and, therefore, this large $|f_{23}|$ does not cause any difficulty. Thus, we see that the quartet of N_1, N_2, N_3 and N_4 is the smallest set of nucleons which is self-consistent in the sense of the conditions (36).

A study [7] is made on the solution of the conditions (34) and (36) together, in the case when the above quartet of nucleons is combined with the observed [11] three nucleon resonances with spin 3/2. It is found that there exists no solution unless N_5 in the Table is added, and the solution with N_5 is very satisfactory. Thus, all the five nucleons in the Table are needed to satisfy the self-consistency conditions.

Further details of the analyses of the observed baryon resonances [11] are referred to the original work [6], [7],[10]. We can state that all the observed resonances are actually needed to satisfy the self-consistency conditions and these conditions are in fact satisfied by these resonances quite well.

VII. SUMMARY

1. The off-mass-shell approach to baryon resonances is a new formulation of bootstrap idea in which the dispersion theory and the current-commutator algebra [1] are incorporated in a systematic and unambiguous manner. We study in this approach the off-mass-shell continuation of the matrix element in which a baryon resonance decays into another baryon resonance with the emission of a single pion, kaon, and η-meson. The off-mass-shell continuation is defined not only with respect to the boson in terms of the divergence of the axial-vector current but also regarding the baryon resonance in terms of the high-spin baryon field which is essentially a product of vector and axial-vector currents and the (spin -1/2) baryon field. These currents are assumed to satisfy the current-commutator algebra [1] . The dispersion relation satisfied by the off-mass-shell decay amplitude with respect to the external mass of the baryon resonance reduces, when it is saturated in terms of the baryon resonances, to the off-mass-shell self-consistency conditions satis-

fied by the physical decay amplitudes among the baryon resonances. The off-mass-shell approach consists of analyzing the baryon resonances in terms of these self-consistency conditions.

2. The off-mass-shell approach, therefore, assumes no assumption other than the conventional bootstrap assumption that the resonances behave more or less like stable particles and, most importantly, can saturate the dispersion relation mentioned above.

3. The most non-trivial point about the off-mass-shell self-consistency conditions is that they do not yield a satisfactory solution unless the baryon resonances form a certain minimum set of resonances. This set of resonances consists of parity doublets which are the pairs of two resonances with the same quantum numbers including the spin but with the opposite parities. This minimum set includes also quartets of resonances which are more or less two parity doublets combined. The off-mass-shell approach suggests that the high-spin baryon resonances (presumably spin 5/2 and higher) appear as parity doublets and the lower-spin baryon resonances (presumably both spin 1/2 and 3/2) form at least parity quartets. This seems to be in general agreement with the experiments.

4. Presumably the most unambiguous way to test the specific predictions of the off-mass-shell approach is to compare further with experiments the parity-doublet mass formula (58). In particular, it is proposed to examine whether the new $\Sigma(7/2^-)$ has the mass given by (56). This mass is quite close to the mass of the observed $\Sigma(5/2^+)$.

5. The detailed studies [6],[7],[10] of the observed baryon resonances in terms of the off-mass-shell self-consistency conditions indicate that the observed reso-

nances are in fact all needed, but not sufficient, to satisfy these conditions. The new resonances which are needed include

$$\Delta(\tfrac{1}{2}^-) \text{ at } \sim 1900 \text{ MeV} \quad [6] ,$$

$$\Lambda(\tfrac{1}{2}^+) \text{ at } \sim 2030 \text{ MeV} \quad [10] ,$$

$$\Lambda(\tfrac{3}{2}^+) \text{ at } \sim 1600 \text{ MeV} \quad [10],$$

$$\Sigma(\tfrac{7}{2}^-) \text{ at } \sim 1920 \text{ MeV} \quad [9] ,$$

where the masses depend on the informations that are not available at present and also the additional new resonances. It is quite worthwhile to search for the above new resonances since the agreement of the self-consistency conditions with the observed resonances is generally quite satisfactory.

6. The detailed studies [6], [7] of the observed nucleon resonances show that all the observed total widths of these resonances are well explained in terms of the decays among themselves with the emission of a single pion.

7. The detailed studies [6],[7],[10] of the observed resonances indicate that the coupling constant with pion appears to be as substantial as the known pion-nucleon coupling constant only with respect to the stable baryons and at most the spin -3/2 baryon resonances which have the least masses among the spin -3/2 resonances with the same quantum numbers. It appears [6],[7],[10] that all the other baryon resonances have almost no strong interactions and, thus, behave more or less like free particles until they decay into other baryon resonances with the emission of a single boson.

8. The type of the axial-vector current that is to be used in the off-mass-shell approach is determined as $\frac{1}{2}$(f+d)-type, based upon the comparison with the experiments of the parity-doublet mass formula [9] and also independently by the analysis [10] of the observed Λ- and Σ-resonances in terms of the off-mass-shell self-consistency conditions. It is of interest that the axial-vector current of f- or d-type appears to be ruled out.

9. The off-mass-shell continuation of the decay matrix element with respect to Λ-resonances is to be defined in this off-mass-shell approach in terms of the field given by (46) which is a specific combination of an octet field and a singlet field [9].

REFERENCES

1. M. GELL-MANN, Phys. Rev. 125, 1067 (1962); Physics 1, 63 (1964).
2. M. SUGAWARA, Phys. Rev. 172, 1423 (1968) discusses the case of spin 1/2.
3. M. SUGAWARA and J. W. MEYER, Phys. Rev. 174, 1709 (1968) discusses the case of spin 3/2.
4. M. SUGAWARA and J. W. MEYER, Off-Mass-Shell Self-Consistency Conditions for Spin -5/2 Nucleon Resonances, Phys. Rev., to be published.
5. M. SUGAWARA, Off-Mass-Shell Self-Consistency Conditions for Arbitrary Spin and Isospin, Phys. Rev., to be published.
6. M. SUGAWARA, Nucleon Resonances of Isospin 3/2, Phys. Rev., to be published.
7. D. J. GEORGE and M. SUGAWARA, Nucleon Resonances of Isospin 1/2, Phys. Rev., to be published.
8. S. L. ADLER, Phys. Rev. 140, B736 (1965).

9. M. SUGAWARA, Complete Set of Off-Mass-Shell Self-Consistency Conditions for Baryon Resonances, to be published.

10. M. SUGAWARA, Λ^* and Σ^* with Spin 1/2 and 3/2, to be published.

11. Review of Particle Properties, Particle Data Group, Lawrence Radiation Laboratory, Aug. 1968 (unpublished).

12. M. SUGAWARA, to be published.

MISCELLANEOUS TOPICS RELATED TO THE ANNIHILATION OF AN ELECTRON-POSITRON PAIR INTO MESONS[†]

By

M. GOURDIN

Laboratoire de Physique Théorique et Hautes Energies,

Orsay, France

The notes are divided into three parts:

In part I, we present the general questions in a model-independent way. The only assumption is that the annihilation electron-positron proceeds via the one-photon exchange approximation.

In part II, we discuss the vector-meson dominance model and try to resolve and to compute the problems initiated in part I in this particular approximation.

In part III, we examine the models for the mixing of the two vector mesons ω and ϕ and we make crude approximations to relate various phenomena.

It must be observed that the repartition into three parts corresponds to a decreasing degree of rigour and to an increasing number of approximations.

---

† Lecture given at VIII. Internationale Universitätswochen für Kernphysik, Schladming, February 24 - March 8, 1969.

PART ONE

The $e^+ e^-$ Annihilation into Hadrons

SECTION A

The One-Photon Exchange Approximation

I. GENERAL STRUCTURE OF THE $e^+ e^-$ ANNIHILATION CROSS SECTION

1. We are interested in the process

$$e^+ + e^- \Rightarrow f$$

where f is an arbitrary final state compatible with the usual conservation laws of electric charge, baryonic charge and leptonic charge (Q=B=L=O).

Using the kinematics as indicated on fig. 1

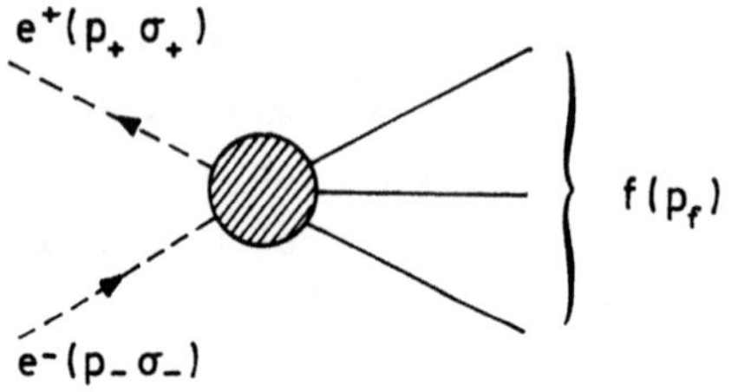

Fig. 1

the differential cross section is given by

$$d\sigma(e^+ e^- \rightarrow f) =$$

$$= \frac{1}{[(p_+p_-)^2-m^4]^{1/2}} \left(\frac{m^2}{4} \sum_{\sigma_+\sigma_-} \right) \sum_{\text{pol.f}} (2\pi)^4 \delta_4(p_f-p_+-p_-) \cdot$$

$$\cdot \; |\langle f|T|e^+e^-\rangle|^2 d\rho_f \qquad (A.1)$$

where m is the electron mass.
The density of final state $d\rho_f$ is written as

$$d\rho_f = \prod_{a\,\varepsilon\,\text{final state}} \frac{N_a}{(2\pi)^3} \frac{d_3 p_a}{E_a} .$$

The normalization factor N_a is $\frac{1}{2}$ for bosons and the mass m_a for fermions; E_a and $\vec{p}_a$ are the energy and momentum of the particle a.

2. In the one-photon exchange approximation, the transition matrix element is factorized into the product of two matrix elements of the electromagnetic current.

$$\langle f|T|e^+e^-\rangle = -\frac{e^2}{s}\langle f|J_\mu^{em}(o)|O\rangle \bar{V}_{\sigma_+}(p_+)\gamma^\mu u_{\sigma-}(p_-) \qquad (A.2)$$

where $u_{\sigma-}(p_-)$ is the Dirac spinor for the electron and $V_{\sigma+}(p_+)$ the Dirac spinor for the positron. The electric charge e is normalized so that

$$\alpha = \frac{e^2}{4\pi} = \frac{1}{137}$$

and the invariant quantity s is defined by

$$s = -(p_+ + p_-)^2$$

e.g. s is the square of the total energy in the centre-of-mass system. It is straightforward to perform the summation over the electron and positron polarizations

$$\frac{m^2}{4} \sum_{\sigma_+,\sigma_-} [\bar{V}_{\sigma_+}(p_+)\gamma^\mu u_{\sigma_-}(p_-)][\bar{V}_{\sigma_+}(p_+)\gamma^\nu u_{\sigma_-}(p_-)]^* =$$

$$= \frac{1}{4}(p_+^\mu p_-^\nu + p_+^\nu p_-^\mu + \frac{s}{2} g^{\mu\nu})$$

and the differential cross section takes the form

$$d\sigma(e^+e^- \to f) = \frac{1}{[(p_+p_-)^2 - m^4]^{1/2}} \frac{e^4}{s^2} - \frac{1}{4}(p_+^\mu p_-^\nu + p_-^\mu p_+^\nu + \frac{s}{2} g^{\mu\nu})\{f\}_{\mu\nu} \qquad (A.3)$$

where the final-state tensor $\{f\}_{\mu\nu}$ is defined by

$$\{f\}_{\mu\nu} = \sum_{\text{pol.f}} (2\pi)^4 \delta_4 (p_f - p_+ - p_-) d\rho_f \langle f | J_\mu^{em}(0) | 0 \rangle \langle 0 | J_\nu^{em}(0) | f \rangle . \qquad (A.4)$$

Let us remark that we can also use the invariant variable s to write the equation (A.4) into

$$d\sigma(e^+e^- \to f) = \frac{8\pi^2\alpha^2}{s^{5/2}(s-4m^2)^{1/2}} (p_+^\mu p_-^\nu + p_-^\mu p_+^\nu + \frac{s}{2} g^{\mu\nu}) \{f\}_{\mu\nu} . \qquad (A.5)$$

3. Let us now work in the e^+e^- C.M. system which is also for storage ring experiments the laboratory system. We define

$$p_+ = (\vec{p}, p_o) , \quad p_- = (-\vec{p}, p_o) , \quad \frac{s}{4} = p_o^2 = \vec{p}^2 + m^2 .$$

The two tensors of equations (A.4) or (A.5) have only space components because of the conservation of the electromagnetic current

$$(p_+^\mu p_-^\nu + p_-^\mu p_+^\nu + \frac{s}{2} g^{\mu\nu}) \Rightarrow \frac{s}{2}(\delta_{mn} - \frac{p_m p_n}{p_o^2})$$

and equation (A.5) becomes

$$d\sigma_{CM}(e^+e^- \Rightarrow f) = \frac{4\pi^2\alpha^2}{s^{3/2}(s-4m^2)^{1/2}} (\delta_{mn} - \frac{p_m p_n}{p_o^2}) \{f\}_{mn} . \qquad (A.6)$$

The electron mass m can be neglected compared to the electron energy in almost all the applications and equation (A.6) reduces to

$$d\sigma_{CM}(e^+e^- \Rightarrow f) = \frac{4\pi^2\alpha^2}{s^2} (\delta_{mn} - \frac{p_m p_n}{\vec{p}^2}) \{f\}_{mn} . \qquad (A.7)$$

The total cross section $\sigma_{tot}(e^+e^- \Rightarrow f)$ is obtained integrating over all the angular variables in (A.4).

$$(\delta_{mn} - \frac{p_m p_n}{\vec{p}^2}) \{f\}_{mn} \Rightarrow \frac{2}{3} \mathrm{Tr}\{f\}$$

and the final expression is simply

$$\sigma_{tot}(e^+e^- \to f) = \frac{8\pi^2\alpha^2}{3}\frac{1}{s^2}\,Tr\{f\} . \qquad (A.8)$$

Going back to the definition (A.4) of the final state tensor $\{f\}_{\mu\nu}$, its trace Tr{f} is given by

$$Tr\{f\} = S_f(2\pi)^4\delta_4(p_f-p_+-p_-)<f|J_\mu^{em}(o)|0><0|J_\nu^{em}(o)|f>g^{\mu\nu} \qquad (A.9)$$

where the symbol S_f means

a) a summation over the polarization of the final-state particles and possibly internal symmetry group summations (isospin...)

b) a phase-space integration

$$S_f \equiv \sum_{pol.f} \int d\rho_f$$

II. THE FINAL STATE $\bar{B}B$ WHERE B IS A SPINLESS MESON

1. We restrict ourselves to a final two-body state $f \equiv \bar{B}B$ where B is a spin zero meson($B = \pi^+, K^+, K^o$) .

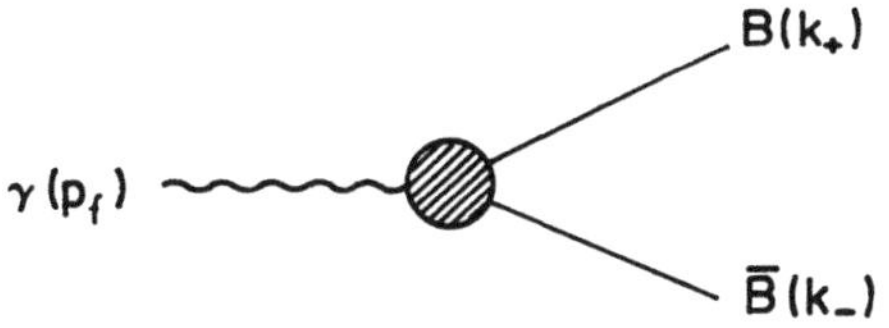

Fig. 2

The tensor $\{\bar{B}B\}_{\mu\nu}$ is simply defined by equation (A.4)

$$\{\bar{B}B\}_{\mu\nu} = \frac{1}{(4\pi)^2}\left\{\frac{d_3k_+}{k_{+o}}\frac{d_3k_-}{k_{-o}}\,\delta_4(p_f-k_+-k_-)\right\}<\bar{B}B|J_\mu^{em}(0)|0> \times$$
$$\times \; <0|J_\nu^{em}(0)|\bar{B}B> .$$

The electromagnetic factor of the B meson is then introduced in the usual way :

$$<\bar{B}B|J_{\mu}^{em}(0)|0> = i(k_+ - k_-)_{\mu} F_B(s)$$

and the tensor $\{\bar{B}B\}_{\mu\nu}$ is written as

$$\{\bar{B}B\}_{\mu\nu} = \frac{1}{(4\pi)^2}\left\{\frac{d_3k_+}{k_{+o}}\frac{d_3k_-}{k_{-o}}\,\delta_4(p_f-k_+-k_-)\right\}(k_+-k_-)_{\mu}(k_+-k_-)_{\nu}|F_B(s)|^2 .$$

2. In the $\bar{B}B$, C.M. system the energy momentum variables are given by

$$k_+ = (\vec{k},k_o) \ , \ k_- = (-\vec{k},k_o) \ , \ \frac{s}{4} = k_o^2 = \vec{k}^2+m_B^2$$

and the phase-space density is simply written

$$\delta_4(p_f-k_+-k_-)\ \frac{d_3k_+}{k_{+o}}\ \frac{d_3k_-}{k_{-o}} \Rightarrow \frac{k}{\sqrt{s}}\, d\Omega_f$$

from which we deduce

$$\{\bar{B}B\}_{mn} = \frac{d\Omega_f}{(2\pi)^2}\ \frac{k}{\sqrt{s}}\ |F_B(s)|^2\ k_m\ k_n \ . \qquad (A.10)$$

Using equation (A.6) the differential cross section for the annihilation $e^+e^- \Rightarrow \bar{B}B$ can take the form

$$\frac{d\sigma(e^+e^-\Rightarrow \bar{B}B)}{d\ \Omega}\ C.M. = \frac{\alpha^2}{8}\ \frac{1}{s}\ \frac{(1-(4m_B^2/s))^{3/2}}{(1-(4m^2/s))^{1/2}}[1-(1-\frac{4m^2}{s})Z^2]\,|F_B(s)|^2 \qquad (A.11)$$

where Z is the cosine of the C.M. angle. Neglecting now the electron mass ($m<<\sqrt{s}$) we obtain

$$\frac{d\sigma(e^+e^-\Rightarrow \bar{B}B)}{d\ \Omega}\ C.M. = \frac{\alpha^2}{8}\ \frac{1}{s}(1-\frac{4m_B^2}{s})^{3/2}(1-Z^2)\,|F_B(s)|^2 . \qquad (A.12)$$

3. The total cross section is computed by integrating the expression (A.12). The result is simply

$$\sigma_{tot}(e^+e^-\Rightarrow \bar{B}B) = \frac{\pi\alpha^2}{3}\ \frac{1}{s}(1-\frac{4m_B^2}{s})^{3/2}|F_B(s)|^2 . \qquad (A.13)$$

An equivalent way to obtain the equality (A.13) is to calculate $Tr\{\bar{B}B\}$ and to use equation (A.8). We have

$$Tr\{\bar{B}B\} = \frac{1}{8\pi} s(1-\frac{4m_B^2}{s})^{\frac{3}{2}} |F_B(s)|^2 . \qquad (A.14)$$

III. THE FINAL STATE $B^o\gamma$ WHERE B^o IS A PSEUDOSCALAR MESON

1. We now study the final state $f \equiv B^o\gamma$ where B^o is a pseudoscalar meson ($B^o = \pi^o, \eta, \eta'$) of mass m_o.

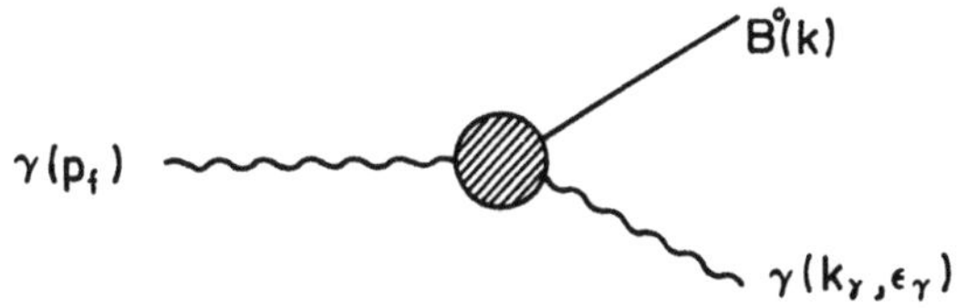

Fig. 3

From Lorentz invariance and parity conservation, the matrix element of the electromagnetic current between the vacuum and the $B^o\gamma$ state has the general structure

$$<B^o\gamma|J_\mu^{em}(0)|0> = e\frac{G(s)}{m_o} \varepsilon_{\mu\rho\sigma\lambda} k^\rho k_\gamma^\sigma \varepsilon_\gamma^\lambda$$

where $\varepsilon_{\mu\rho\sigma\lambda}$ is the completely skew symmetric Ricci tensor.

In the C.M. system we have

$$<B^o\gamma|J_\mu^{em}(0)|0>= e\frac{G(s)}{m_o}\sqrt{s}\ \varepsilon_{mpq}k^p\varepsilon_\gamma^q . \qquad (A.15)$$

After summation over the transverse photon polarization, we easily obtain the space components of the tensor $\{B^o_\gamma\}$

in the C.M. system

$$\{B^o\gamma\}_{mn} = \frac{d\Omega_f}{4\pi} \frac{\alpha}{m_o^2} k^3\sqrt{s}|G(s)|^2[\delta_{mn} - \frac{k_m k_n}{k^2}] \quad (A.16)$$

and the trace of the tensor $\{B^o\gamma\}$

$$Tr\{B^o\gamma\} = 2\alpha \frac{k^3\sqrt{s}}{m_o^2} |G(s)|^2 \quad . \quad (A.17)$$

2. Combining equations (A.7) and (A.16) we obtain the differential cross section in the C.M. system

$$\frac{d\sigma(e^+e^- \Rightarrow B^o\gamma)}{d\Omega} \text{ C.M.} = \frac{\pi\alpha^3}{8} \frac{1}{m_o^2}(1- \frac{m_o^2}{s})^3(1+Z^2)|G(s)|^2 \quad (A.18)$$

where Z is the cosine of the C.M. angle.

The total cross section is simply computed from (A.8) and (A.17)

$$\sigma_{tot}(e^+e^- \Rightarrow B^o\gamma) = \frac{2\pi^2\alpha^3}{3} \frac{1}{m_o^2} (1- \frac{m_o^2}{s})^3|G(s)|^2 . \quad (A.19)$$

3. The form factor G(s) extrapolated at s=o describes the 2γ decay mode of the B^o meson. With the kinematics indicated on fig. 4

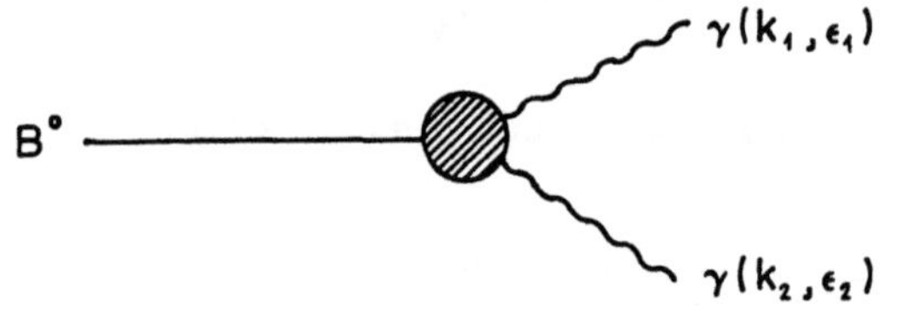

Fig. 4

the transition matrix element for the $B^o \Rightarrow 2\gamma$ decay is written as

$$\langle 2\gamma|T|B^{o}\rangle = \frac{1}{\sqrt{2}}\, e^2\, \frac{G(0)}{m_o}\, \varepsilon_{\mu\nu\rho\sigma}\, k_1^{\mu}\, \varepsilon_1^{\nu}\, k_2^{\rho}\, \varepsilon_2^{\sigma} \; . \quad (A.20)$$

The factor $\frac{1}{\sqrt{2}}$ is due to the Bose statistics satisfied by the 2γ final state.

It is then straightforward to compute the radiative decay width for the B^{o} meson

$$\Gamma(B^{o} \rightarrow 2\gamma) = \frac{\pi\alpha^2}{4}\, m_o |G(0)|^2 \quad . \qquad (A.21)$$

IV. DISCUSSION

A short comment on the one photon exchange approximation. This assumption has been tested for space like photons in two types of reactions:

a) elastic scattering of electrons,

b) inelastic scattering of electrons .

For elastic scattering the ratio of the differential cross section to the Mott cross section is a linear function of $tg^2\frac{\theta}{2}$

$$\frac{d\sigma}{d\sigma_{Mott}} = A(s) + tg^2\,\frac{\theta}{2} B(s)$$

where θ is the laboratory angle of the scattered electron, and for inelastic scattering such a dependence can be generalized to be

$$\frac{d\sigma}{d\sigma_{Mott}} = A(s,w^2) + tg^2\,\frac{\theta}{2} B(s,w^2)$$

where now w^2 is the invariant mass squared of the outgoing hadrons. For the present there is no experimental evidence for two photon exchange, even for large energies.

For timelike photons there exist two types of tests:

a) Specific predictions for the angular distributions (A.12 and A.18), but it is very difficult to measure deviations.

b) Search for pure C=+1 final states which correspond to two photon exchange. (In pure electrodynamics a 2γ final state has an extremely small cross section and it is more convenient to use enhancements due to strong interactions). For instance we can look at a $\pi^o\pi^o$ final state (e.g. via $f^o \to 2\pi^o$) or at the decay of a 1^+ particle, like A_1. This second way b) is more conclusive but it is extremely difficult to see a pure C=+1 final state because of large background. On the basis of the result for space like photons from now on we consider the one photon exchange approximation at least at the energies, where the experiments at Orsay and Novosibirsk are performed to be correct.

SECTION B

The π-Meson Electromagnetic Form Factor

I. DEFINITION

1. We consider the electromagnetic vertex of the charged π meson as represented on fig. 1

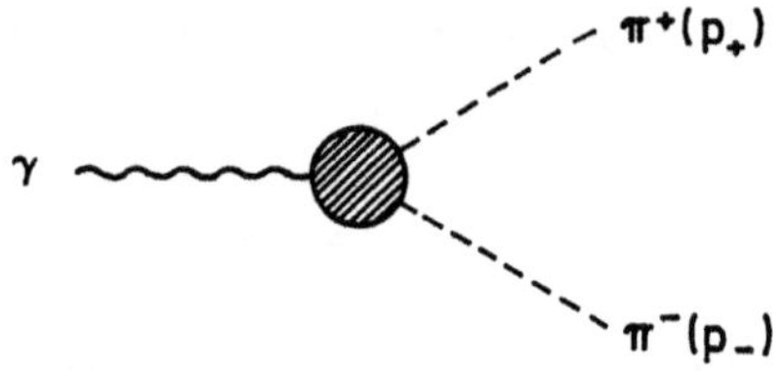

Fig. 1

The structure of the matrix element of the electromagnetic current between the vacuum and a $\pi^+\pi^-$ state is restricted by the Lorentz covariance and the conservation of the electromagnetic current to the general form

$$<\pi^+\pi^-|eJ_\mu^{em}(0)|0> = e\,F_\pi(s)(p_+-p_-)_\mu \qquad (B.1)$$

where

$$s = -(p_+ + p_-)^2 \;.$$

By definition, $F_\pi(s)$ is the π-meson electromagnetic form factor.

2. For a real photon, $s=0$, the vertex function with the three particles on the mass shell reduces to the coupling constant which in the present case is simply the electric charge. The normalization condition of $F_\pi(s)$ is simply

$$F_\pi(0) = 1 \;. \qquad (B.2)$$

3. Let us remark that from TCP invariance, the neutral π meson electromagnetic form factor is identically zero.

II. ANALYTICITY PROPERTIES

We are working in the lowest order with respect to electromagnetic interactions and we consider only the effect of strong interactions on the form factor $F_\pi(s)$.

1. The electromagnetic form factor $F_\pi(s)$ is an analytic function of s in the complex s plane with a cut on the positive real axis from $4m_\pi^2$ to $+\infty$ (Refs. 1,2)

2. On the real negative axis, $s < 0$, the virtual photon is space like. From the hermiticity of the electromagnetic current, the electromagnetic form factor is real in this region.

There exist, at least in principle, several ways

to measure experimentally the π-meson electromagnetic form factor in the space-like region.

a) Elastic Scattering of Charged π Meson on Atomic Electrons

with the lowest order diagram as represented on fig. 2 .

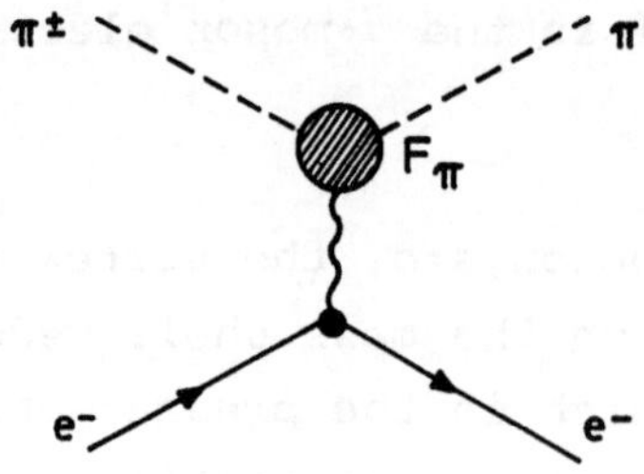

Fig. 2

b) Coulomb Scattering of Charged π Meson on Nucleus

Such an experiment has been performed with a He^4 target by comparing π^+ and π^- Coulomb scattering. The diagram of interest is drawn on fig. 3 .

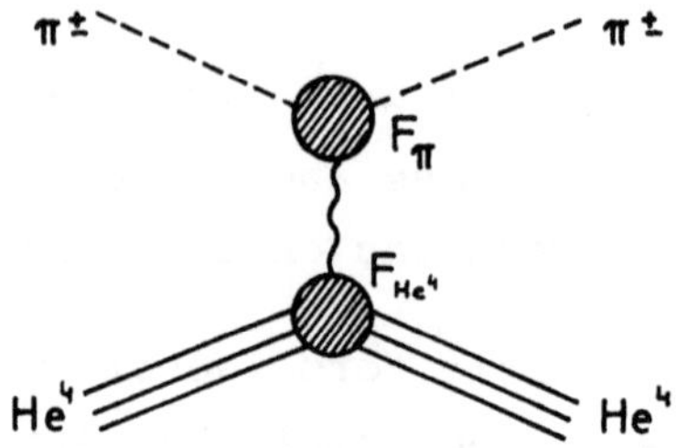

Fig. 3

The He^4 electromagnetic form factor is known from e^--He^4 elastic scattering experiments. The main difficulty in extracting $F_\pi(s)$ is an accurate determination of the nuclear effects.

c) Electroproduction of π^+ on Nucleon

The electroproduction experiment must be performed in a kinematical situation where the so-called photoelectric term, represented on fig. 4, plays an important role.

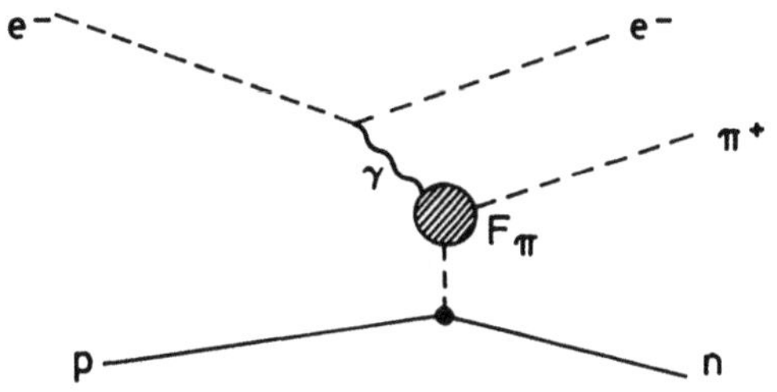

Fig. 4

In such experiments the background is always important and for a large part model dependent.

3. On the <u>real positive axis</u> , s > 0, the virtual photon is time like and the form factor $F_\pi(s)$ becomes complex above the 2π meson threshold $s > 4m_\pi^2$.

The storage ring experiments, as those recently performed in Novosibirsk and Orsay, allow a direct measurement of $|F_\pi(s)|$ by looking at the electron-positron annihilation into a $\pi^+\pi^-$ system. The corresponding lowest order diagram is represented on fig. 5

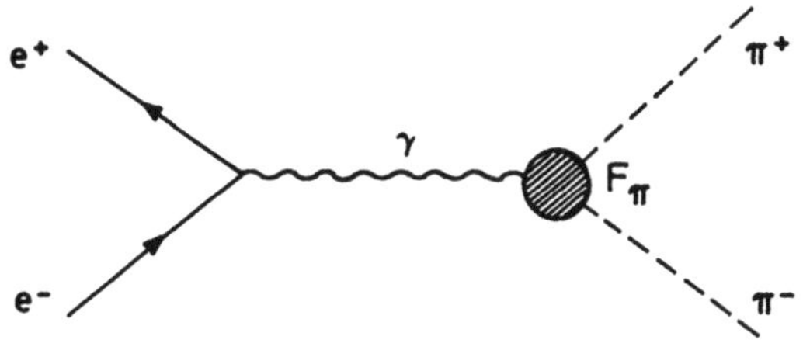

Fig. 5

4. Let us first apply the Cauchy theorem to the function $F_\pi(s)$ around the contour C shown on Fig. 6

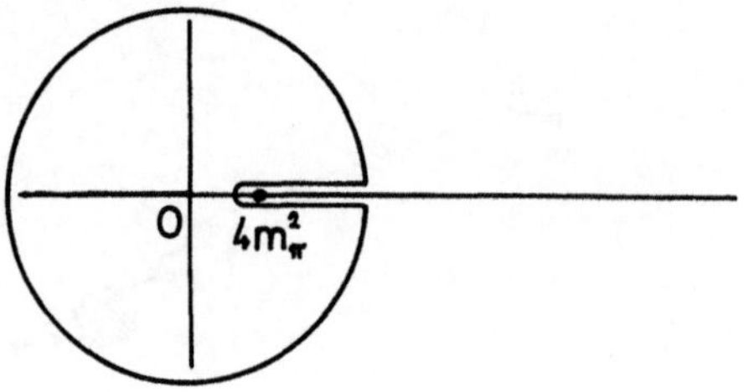

Fig. 6

The discontinuity across the cut is simply ($s > 4m_\pi^2$)

$$\lim_{\varepsilon \to 0^+} [F_\pi(s+i\varepsilon) - F_\pi(s-i\varepsilon)] = 2i \text{ Im } F_\pi(s) \qquad (B.3)$$

and we obtain an unsubtracted dispersion relation

$$F_\pi(s) = \frac{1}{\pi} \int_{4m_\pi^2}^{\infty} \frac{\text{Im } F_\pi(t)}{t-s-i\varepsilon} dt \qquad (B.4)$$

if the spectral function Im $F_\pi(t)$ vanishes sufficiently rapidly at infinity. Assuming the validity of the integral representation (B.4) we can write the normalization (B.2) as a sum rule

$$1 = \frac{1}{\pi} \int_{4m_\pi^2}^{\infty} \frac{\text{Im } F_\pi(t)}{t} dt \quad . \qquad (B.5)$$

5. If the integral (B.4) fails to converge we can perform a subtraction at s=0 and apply the Cauchy theorem to the function $[F_\pi(s) - F_\pi(0)]/s$

$$F_\pi(s) = 1 + \frac{s}{\pi} \int_{4m_\pi^2}^{\infty} \frac{\text{Im } F_\pi(t)}{t(t-s-i\varepsilon)} dt \quad . \qquad (B.6)$$

In order to ensure the convergence of the previous integral, the ratio Im $F_\pi(t)/t$ must vanish for $t\to +\infty$.

III. UNITARITY

1. In order to make useful the dispersion relation (B.4) or (B.6) we compute the spectral function Im $F_\pi(s)$ using the unitarity property of the S matrix. A straightforward calculation gives

$$\text{Im } F_\pi(s) =$$

$$= \theta(s-4m_\pi^2)\sum_n S_n(2\pi)^4\delta_4(p_+ + p_- - p_n)\frac{1}{2}\langle n\left|\frac{(p_+-p_-)^\mu J_\mu^{em}(0)}{(p_+-p_-)^2}\right|0\rangle \times$$

$$\times \langle\pi^+\pi^-|T|n\rangle^* \quad . \qquad \text{(B.7)}$$

The intermediate states $|n\rangle$ are restricted by the energy momentum Dirac distribution and some other conservation laws. For instance all the possible intermediate states must have zero electric charge, zero baryonic charge, and zero leptonic charge.

In the lowest order approximation with respect to electromagnetic interactions the states n are strongly coupled to the $\pi^+\pi^-$ final state and have therefore strangeness S=0 and total isotopic spin I=1 (as a consequence of the generalized Pauli principle). It follows that G-parity is conserved and there can be only even numbers of π mesons.

2. In the region $4m_\pi^2 \leq s \leq 16m_\pi^2$ only a $\pi^+\pi^-$ state can contribute and the so-called "elastic" unitarity relation is diagrammatically represented on fig. 7 .

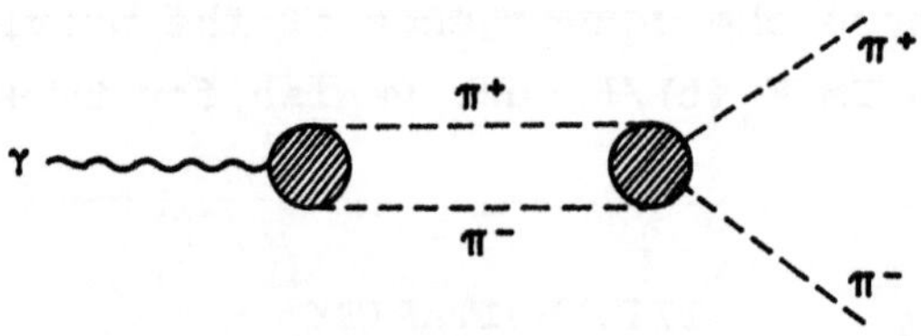

Fig. 7

Formula (B.7) reduces to

$$\mathrm{Im}\, F_\pi(s) =$$

$$= \Theta(s-4m_\pi^2) F_\pi(s) \int \frac{d\Omega_n}{4\pi} Z\left(\frac{k}{8\pi\sqrt{s}}\right) \langle \pi^+\pi^- | T | \pi^+\pi^- \rangle^* \qquad \text{(B.8)}$$

where Z is the cosine of the C.M. scattering angle. The amplitude for elastic $\pi^+\pi^-$ scattering in the total isotopic spin I=1 can be expanded in Legendre polynomials of Z

$$\langle \pi^+\pi^- | T | \pi^+\pi^- \rangle = \frac{8\pi\sqrt{s}}{k} \sum_J (2J+1) f_{1J}(s) P_J(Z) \; . \qquad \text{(B.9)}$$

From the elastic unitarity we have the constraint

$$\mathrm{Im}\, f_{1J}(s) = |f_{1J}(s)|^2$$

or equivalently, using the π-π phase shifts,

$$f_{1J}(s) = \exp\{i\, \delta_{1J}(s)\} \sin\delta_{1J}(s) \; .$$

The angular integration in (B.8) extract the P-wave term in the expansion (B.9) and the final result is simply

$$\mathrm{Im}\, F_\pi(s) = F_\pi(s)\, f_{11}^*(s) \qquad \text{(B.10)}$$

or, using the phase shift representation of $f_{11}(s)$

$$\mathrm{Im}\, F_\pi(s) = F_\pi(s) \exp\{-i\delta_{11}(s)\} \sin\delta_{11}(s) \; . \qquad \text{(B.11)}$$

Equation (B.11) tells us that the phase of the π-meson electromagnetic form factor $F_\pi(s)$ is $\delta_{11}(s)$, the phase of the π-π scattering amplitude in the state I=1, J=1.

IV. THE MUSKHELISHVILI EQUATION

1. Let us combine now the unsubtracted dispersion relation (B.5) with the "elastic" unitarity relation (B.11) used in the complete range of integration neglecting the contributions due to four π mesons and other higher mass intermediate states

$$F_\pi(s) = \frac{1}{\pi} \int_{4m_\pi^2}^{\infty} \frac{F_\pi(t)\exp\{-i\delta_{11}(t)\}\sin\delta_{11}(t)}{t - s - i\varepsilon} dt \ . \qquad (B.12)$$

Equation (B.12) is an homogeneous, singular, integral equation of the Muskhelishvili type [3].

2. A unique solution exists, if the following conditions are satisfied:

$$\int_{4m_\pi^2}^{\infty} \frac{\delta_{11}(s)}{s} ds < \infty \qquad (B.13)$$

$$\delta_{11}(\infty) - \delta_{11}(0) = 0 \qquad (B.14)$$

A subtraction would be needed if (B.13) is not fulfilled. (B.14) follows from the Levinson theorem [6] and the experimental fact that no bound states in the partial wave I=J=1 of the $\pi\pi$ system have been found. Then the solution of (B.12) is uniquely determined using the normalization condition (B.2) :

$$F_\pi(s) = \frac{[\exp\{\rho_{11}(s)-\rho_{11}(0)+i\delta_{11}(s)\}]}{(1-(s/4m_\pi^2))} \qquad (B.15)$$

with

$$\rho_{11}(s) = \frac{1}{\pi} P \int_{4m_\pi^2}^{\infty} \frac{\delta_{11}(t)}{t-s} dt \quad . \tag{B.16}$$

V. MODELS

1. The partial wave amplitude

$$\frac{\sqrt{s}}{k^3} \exp\{i\delta_{11}(s)\} \sin\delta_{11}(s) \tag{B.17}$$

is an analytic function of s in the complex s plane with

a) a right-hand cut on the positive real axis from $4m_\pi^2$ to $+\infty$.

b) a left-hand cut on the negative real axis from $-\infty$ to 0.

Because of the threshold behaviour of the P-wave phase shift $\delta_{11}(s)$ the expression (B.17) is regular at the point $s=4m_\pi^2$.

2. Let us write now the partial wave (B.17) as the ratio of two functions

$$\frac{\sqrt{s}}{k^3} \exp\{i\delta_{11}(s)\} \sin\delta_{11}(s) = \frac{N(s)}{D(s)} \tag{B.18}$$

where N(s) is an analytic function of s except on the left-hand cut; D(s) is an analytic function of s except on the right-hand cut.

The function D(s) satisfies a dispersion relation and we use the Cauchy theorem with the contour C of fig. 6. The discontinuity of D(s) on the right-hand cut is given by the definition (B.18) and we notice that N(s) is real on the right-hand cut

$$\text{Im } D(s) = -\frac{k^3}{\sqrt{s}} N(s) = -\frac{1}{8} \frac{(s-4m_\pi^2)^{3/2}}{s^{1/2}} N(s) . \qquad (B.19)$$

The number of subtractions needed to write the dispersion relation depends obviously on the high-energy behaviour of the left-hand cut contributions represented by the function N(s). For instance, if $\lim_{s\to\infty} N(s) = 0$ one subtraction can be sufficient and the subtraction point can be chosen at s=0,

$$D(s) = D(0) - \frac{s}{8\pi} \int_{4m_\pi^2}^{\infty} \frac{(t-4m_\pi^2)^{3/2}}{t^{3/2}} \frac{N(t)}{t-s-i\varepsilon} dt \qquad (B.20)$$

but if $\lim_{s\to\infty} N(s) = \text{cst}$ two subtractions are necessary and choosing always s=0 as the subtraction point

$$D(s) = D(0) + s\, D'(0) - \frac{s^2}{8\pi} \int_{4m_\pi^2}^{\infty} \frac{(t-4m_\pi^2)^{3/2}}{t^{5/2}} \frac{N(t)}{t-s-i\varepsilon} dt . \qquad (B.21)$$

3. The N/D formalism has been used to construct partial-wave amplitudes like (B.17) from specific assumptions concerning the left-hand cut contributions. Models are constructed in this way making approximations on the form of N(s). Let us briefly describe three possibilities considered in the literature.

a) Frazer and Fulco [7] replace the left-hand cut by a pole on the real negative axis at $s=-s_o$; the function N(s) is approximated by

$$N(s) = s_o/(s+s_o) .$$

It is possible to write for D(s) a dispersion relation in the form (B.20) and to compute explicitly the dispersion integral. The result is for D(s) an expression with two arbitrary parameters D(0) and s_o.

b) Vaughn and Wali [8] approximate the left-hand cut by a double pole and use as numerator function

$$N(s) = (\frac{s_o}{s+s_o})^2$$

In their paper the dispersion relation for D(s) is written in the form (B.21) with two subtractions at s=0 in order to have a more rapid convergence of the dispersion integral. Choosing arbitrarily D'(0) = 1, Vaughn and Wali deduce an expression for D(s) with two adjustable parameters D(0) and s_o.

c) Gounaris and Sakurai [9] make the crudest but simplest assumption of a constant numerator function. The dispersion relation for D(s) is used in the form (B.21) and the result is for D(s) an explicit form with again two arbitrary parameters D(0) and D'(0); N(s)=1.

In all the three models the two free constants are determined by the requirement to have a ρ-meson resonance at the correct mass, m_ρ, and with the correct width Γ_ρ.

4. Let us now study in more detail the Gounaris-Sakurai model. Putting N(t) ≡ 1 in equation (B.21) we compute the dispersion integral and obtain the following expression for D(s)

$$D(s) = D(0) + sD'(0) + [k^2h(s) + \frac{m_\pi^2}{\pi} - \frac{s}{3\pi} - i\frac{k^3}{\sqrt{s}}] \tag{B.22}$$

where the functions k(s) and h(s) are defined by

$$k(s) = (\frac{s}{4} - m_\pi^2)^{1/2}$$
$$h(s) = \frac{2}{\pi}\frac{k}{\sqrt{s}}\log\frac{\sqrt{s}+2k}{2m_\pi} . \tag{B.23}$$

Performing an analytic continuation of the functions k(s) and h(s) below the threshold $0 < s \le 4\, m_\pi^2$

$$k(s) = i(m_\pi^2 - \frac{s}{4})^{1/2}$$

$$h(s) = -\left(\frac{m_\pi^2 - \frac{s}{4}}{s}\right)^{1/2}\left[1 - \frac{2}{\pi}\arctan\left(\frac{s}{m_\pi^2 - \frac{s}{4}}\right)^{1/2}\right]$$

it is straightforward to check that the bracket in equation (B.23) is real and behaves like s^2 for $s \simeq 0$ as expected from equation (B.21).

Such a form of D(s) is called a generalized effective range formula for the phase shift $\delta_{11}(s)$:

$$\frac{k^3}{\sqrt{s}} \cot \delta_{11}(s) = \mathrm{Re}\, D(s) = k^2 h(s) + a + bs \qquad \text{(B.24)}$$

where

$$a = D(0) + \frac{m_\pi^2}{\pi} \quad , \; b = D'(0) - \frac{3}{\pi} \quad . \qquad \text{(B.25)}$$

The existence of an elastic $\pi\pi$ resonance, the ρ meson, in the I=1, J=1 channel determines the constants a and b by two constraints on the phase shift $\delta_{11}(s)$

$$\cot \delta_{11}(m_\rho^2) = 0$$

$$\frac{d}{ds} \cot \delta_{11}(m_\rho^2) = \frac{1}{m_\rho \Gamma_\rho}$$

and the result is

$$\begin{aligned} a &= m_\pi^2\, h(m_\rho^2) + m_\rho^2\left[\frac{k_\rho^3}{m_\rho \Gamma_\rho} + k_\rho^2\, h'(m_\rho^2)\right] \\ b &= -\frac{1}{4}\, h(m_\rho^2) - \left[\frac{k_\rho^3}{m_\rho^2 \Gamma_\rho} + k_\rho^2\, h'(m_\rho^2)\right] \end{aligned} \qquad \text{(B.26)}$$

where

$$k_\rho = k(m_\rho^2) = \left(\frac{m_\rho^2}{4} - m_\pi^2\right)^{\frac{1}{2}} \quad .$$

An effective range formula like equation (B.24) is certainly not valid for large values of $|s|$. In fact the function D(s) as determined by equations (B.22), (B.25) and (B.26) vanishes for a large and negative value

$s_o \simeq -1.2\times10^6\ m_\rho^2$, e.g. outside the range of validity of the model.

As a last remark, the high-energy behaviour of D(s) is controlled by the function h(s) and we have $D(s) \simeq s \times \times \log s$ from which we deduce the high-energy behaviour of the phase shift $\delta_{11}(s) \simeq (\log s)^{-1}$.

5. The electromagnetic form factor $F_\pi(s)$ is an analytic function of s in the complex s plane except on the right-hand cut. In the elastic unitarity approximation the phase of $F_\pi(s)$ is $\delta_{11}(s)$. It follows that the function $\Phi(s)$ given by

$$\Phi(s) = F_\pi(s)\ D(s)$$

is defined everywhere in the complex s plane and analytic except perhaps at the threshold point where it can have poles

$$\Phi(s) = \sum_n \frac{d_n}{(s-4m_\pi^2)^n} \quad . \tag{B.27}$$

In equation (B.27) the sum is limited to a finite number of terms and n is an algebraic integer whose possible values are governed by the asymptotic behaviour of the functions $F_\pi(s)$ and D(s) for large s.

6. In the Gounaris-Sakurai model, the denominator function D(s) is given by the expression (B.22) with the two subtraction constants D(O) and D'(O) related to the ρ-meson resonance parameters m_ρ and Γ_ρ by equations (B.25) and (B.26).

The second assumption made by Gounaris and Sakurai is the simplest choice for the function $\Phi(s)$

$$\Phi(s) = D(O)$$

leading, for the π-meson electromagnetic form factor $F_\pi(s)$

to the asymptotic behaviour $s^{-1} \log^{-1} s$ for large positive s.

$$F_\pi(s) =$$

$$= m_\rho^2(1+d\frac{\Gamma_\rho}{m_\rho}) / \{m_\rho^2 - s + \Gamma_\rho \frac{m_\rho^2}{k_\rho^3}[k^2(h(s)-h(m_\rho^2)) -$$

$$- (s-m_\rho^2)k_\rho^2 h'(m_\rho^2)] - im_\rho\Gamma_\rho(\frac{k}{k_\rho})^3 \frac{m_\rho}{\sqrt{s}}\} \qquad (B.28)$$

where the constant d is given by:

$$d = \frac{3}{\pi}\frac{m_\pi^2}{k_\rho^2}\log\frac{m_\rho+2k_\rho}{2m_\pi} + \frac{m_\rho}{2\pi k_\rho} - \frac{m_\pi^2 m_\rho}{\pi\, k_\rho^3} . \qquad (B.29)$$

Equation (B.28) exhibits the following features.

a) Near the ρ-meson mass $s \simeq m_\rho^2$ the term in bracket in the denominator of equation (B.28) behaves like $(s-m_\rho^2)^2$ and therefore it can be ignored in this region. For practical computations, the denominator of $F_\pi(s)$ reduces to the usual P-wave form

$$m_\rho^2 - s + O[(s-m_\rho^2)^2] - i\, m_\rho\, \Gamma_\rho(\frac{k}{k_\rho})^3 \frac{m_\rho}{\sqrt{s}} .$$

b) The numerator of $F_\pi(s)$ however is not simply m_ρ^2. Because of the existence of a term in bracket in the denominator of $F_\pi(s)$ a correct normalization at s=0 implies the presence of the constant d giving a measure of non-zero width corrections. For instance at $s=m_\rho^2$ the form factor F_π is given by

$$F_\pi(m_\rho^2) = i\,\frac{m_\rho}{\Gamma_\rho}(1+ d\frac{\Gamma_\rho}{m_\rho}) .$$

The numerical value of d for the actual ρ-meson mass is close to d≃0.48 and we then obtain a 14 % effect for $|F_\pi(m_\rho^2)|^2$ which is the measured quantity.

c) As a common property of all the width energy dependent Breit-Wigner expressions the actual maximum of

$|F_\pi(s)|^2$ is not at $s = m_\rho^2$ but is somewhat shifted towards the left.

It is important to keep in mind that the form factor (B.28) is only a model with a limited domain of validity: $|s| \leq 1 GeV^2$. The use of such a form in the high-energy domain can be suspicious and must carefully examined.

VI. EXPERIMENTS

1. The modulus of the π-meson electromagnetic form factor has been measured in the ρ-resonance region by observing the reaction $e^+ + e^- \Rightarrow \pi^+ + \pi^-$.

Experiments of this type have been performed in Novosibirsk [10] and in Orsay [11]. We discuss only the results of the last Orsay experiment.

2. The electron-positron annihilation cross section into a $\pi^+\pi^-$ pair has been measured at five energies around the ρ^0-resonance mass

$\sqrt{s}$ = 645 MeV, 705 MeV, 764 MeV, 825 MeV, 886 MeV

and more than 800 pion pairs have been detected.

3. The data have been first fitted by the simplest Breit-Wigner formula, ignoring the normalization condition at s=0.

$$|F_\pi(s)|^2 = \frac{m_\rho^4}{(m_\rho^2-s)^2+m_\rho^2\Gamma_\rho^2} |F_\pi(0)|^2 .$$

The result of this three-parameter fit is

$$\begin{aligned} m_\rho &= (760 \pm 5.5) \text{ MeV} \\ \Gamma_\rho &= (112 \pm 11.5) \text{ MeV} \\ F_\pi(0) &= 1.1 \pm 0.06 \end{aligned}$$

The χ^2 is 5.7 for two degrees of freedom.

The extrapolation of the simplest Breit-Wigner formula in the low-energy region and at s=0 is very doubtful.

4. The data are better fitted with a P-wave Breit-Wigner formula as for instance the expression (B.28) proposed by Gounaris and Sakurai. Two parameters only are used and the result is

$$m_{\rho^0} = (770 \pm 4)\ \text{MeV}$$

$$\Gamma_{\rho^0} = (111 \pm 6)\ \text{MeV}\ .$$

The χ^2 is now 5.6 for three degrees of freedom.
As expected, the maximum of $|F_\pi(s)|^2$ occurs 10 MeV below the ρ^0 meson mass, fig. 8 .

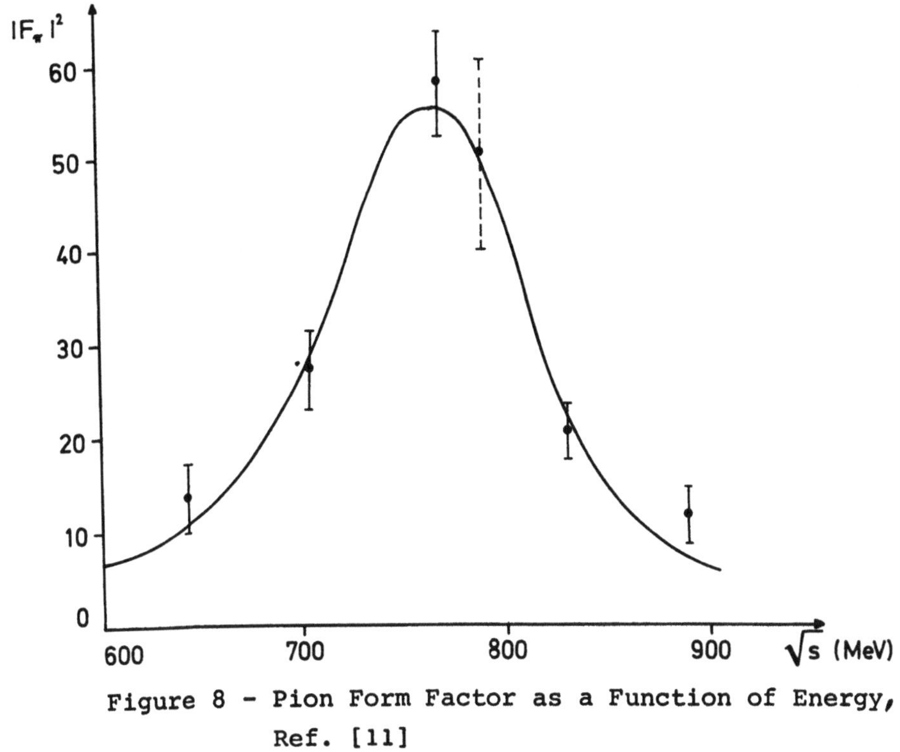

Figure 8 - Pion Form Factor as a Function of Energy, Ref. [11]

SECTION C

Spectral Representations and Sum Rules

I. CONSERVED CURRENTS

1. We first consider a conserved current $J^{\alpha}_{\mu}(x)$ where α is a weight of the adjoint representation of an internal symmetry group like SU(2) or SU(3). It is always possible to choose α so that the current $J^{\alpha}_{\mu}(x)$ is hermitian

$$J^{\alpha}_{\mu}(x)^{*} = J^{\alpha}_{\mu}(x) .$$

By hypothesis the divergence of the current is zero

$$\partial^{\mu} J^{\alpha}_{\mu}(x) = 0$$

where ∂^{μ} is a short notation for $\partial/\partial x_{\mu}$. The space integral of the time component of the current is associated to a conserved quantity

$$Q^{\alpha} = \int J^{\alpha}_{0}(x)\, d_{3}x \quad ;$$

Q^{α} is generally called a charge.

2. A.Källen-Lehmann [12], [13] representation can be written for the vacuum expectation value of the product of two current components. The form of such a representation is determined by the Lorentz covariance and the divergence condition :

$$\langle 0| J^{\alpha}_{\mu}(x) J^{\beta}_{\nu}(o) |0\rangle =$$

$$= \delta^{\alpha\beta} \int_{0}^{\infty} \rho^{\alpha}(m^2) [g_{\mu\nu} - \frac{1}{m^2} \partial_{\mu}\partial_{\nu}] i\Delta^{(+)}(x;m^2)\, dm^2 . \qquad (C.1)$$

The term $\delta^{\alpha\beta}$ expresses the property of the vacuum to belong to the scalar representation of the symmetry group.

The spectral function $\rho^{\alpha}(m^2)$ is real because of the hermitian character of the current.

The invariant distribution $\Delta^{(+)}(x;m^2)$ is defined by its four-dimensional Fourier transform:

$$\Delta^{(\ )}(x;m^2) = \frac{1}{(2\pi)^4}\int e^{ikx}\ \Delta^{(\ )}(k;m^2)\,d_4k \qquad (C.2)$$

$$\Delta^{(\pm)}(k;m^2) = \mp 2i\ \pi\Theta(\pm k_o)\ \delta(k^2+m^2) \qquad (C.3)$$

where Θ is the usual step function and $\delta(k^2+m^2)$ the invariant Dirac distribution on the mass hyperboloid.

3. Analogous Källen-Lehmann representations can be written for the product of the two current components in the reversed order

$$<0|J^{\beta}_{\nu}(0)J^{\alpha}_{\mu}(x)|0> =$$

$$= \delta^{\alpha\beta}\int_0^{\infty} \rho^{\alpha}(m^2)[g_{\mu\nu} - \frac{1}{m^2}\partial_{\mu}\partial_{\nu}]\frac{1}{i}\ \Delta^{(-)}(x;m^2)\ dm^2$$

and for the commutator of the two current components

$$<0|[J^{\alpha}_{\mu}(x),J^{\beta}_{\nu}(o)]|0> =$$

$$= \delta^{\alpha\beta}\int_0^{\infty} \rho^{\alpha}(m^2)[g_{\mu\nu} - \frac{1}{m^2}\ \partial_{\mu}\partial_{\nu}]i\Delta(x;m^2)\ dm^2\ . \qquad (C.4)$$

The causal invariant distribution $\Delta(x;m^2)$ is defined by

$$\Delta \equiv \Delta^{(+)} + \Delta^{(-)}$$

or, using equation (C.3)

$$\Delta(k;m^2) = -2i\pi\varepsilon(k_o)\delta(k^2+m^2) \qquad (C.5)$$

where $\varepsilon(k_o)$ is the discontinuous sign function of k_o.

4. It is now convenient to take the Fourier transform

of both sides of the Källen-Lehmann representation (C.1) in order to study some properties of the spectral function $\rho^{\alpha}(m^2)$

$$\int e^{-ikx}\langle 0|J^{\alpha}_{\mu}(x)J^{\alpha}_{\nu}(0)|0\rangle d_4x = 2\pi\Theta(k_o)\left[g_{\mu\nu}-\frac{k_{\mu}k_{\nu}}{k^2}\right]\rho^{\alpha}(-k^2) \ . \tag{C.6}$$

We introduce a complete set of intermediate states $|n\rangle$

$$\langle 0|J^{\alpha}_{\mu}(x)J^{\alpha}_{\nu}(0)|0\rangle = \sum_n S_n\langle 0|J^{\alpha}_{\mu}(x)|n\rangle\langle n|J^{\alpha}_{\nu}(0)|0\rangle \ .$$

From space-time translation invariance

$$\langle 0|J^{\alpha}_{\mu}(x)|n\rangle = \exp\{ip_n x\}\langle 0|J^{\alpha}_{\mu}(0)|n\rangle \ .$$

The x integration in the left-hand side of equation (C.6) is then easily performed and we obtain

$$2\pi\Theta(k_o)\left[g_{\mu\nu}-\frac{k_{\mu}k_{\nu}}{k^2}\right]\rho^{\alpha}(-k^2) =$$

$$= \sum_n S_n(2\pi)^4\delta_4(p_n-k)\langle 0|J^{\alpha}_{\mu}(0)|n\rangle\langle n|J^{\alpha}_{\nu}(0)|0\rangle \ .$$

Let us recall that the symbol S_n defined in Section A implies a summation over the polarizations of the intermediate state n and a phase-space integration. Contracting with the tensor $g^{\mu\nu}$ we finally obtain

$$\rho^{\alpha}(s) = \tag{C.7}$$

$$= \frac{1}{6\pi}\Theta(k_o)\sum_n S_n(2\pi)^4\delta_4(p_n-k)\langle 0|J^{\alpha}_{\mu}(0)|n\rangle\langle n|J^{\beta}_{\nu}(0)|0\rangle g^{\mu\nu}$$

where $s+k^2 = 0$.

It is now straightforward to prove that $\rho^{\alpha}(s)$ is a definite positive quantity using for instance the property, for a conserved current, to have only space-components in the centre - of - mass system

$$\rho^{\alpha}(s) > 0 \ . \tag{C.8}$$

II. NON-CONSERVED CURRENTS

1. The current $J^{\alpha}_{\mu}(x)$ is always hermitian but its four divergence is different from zero. Therefore, two spectral functions are needed to write Källen-Lehmann representations like (C.1) and (C.4)

$$<0|J^{\alpha}_{\mu}(x)J^{\beta}_{\nu}(0)|0> = \tag{C.9}$$

$$= \delta^{\alpha\beta}\int_{0}^{\infty}\{\rho^{\alpha}_{(1)}(m^2)[g_{\mu\nu}-\frac{1}{m^2}\partial_{\mu}\partial_{\nu}]-\rho^{\alpha}_{(o)}(m^2)\partial_{\mu}\partial_{\nu}\}i\Delta^{(+)}(x;m^2)dm^2$$

$$<0|[J^{\alpha}_{\mu}(x),J^{\beta}_{\nu}(0)]|0> = \tag{C.10}$$

$$= \delta^{\alpha\beta}\int_{0}^{\infty}\{\rho^{\alpha}_{(1)}(m^2)[g_{\mu\nu}-\frac{1}{m^2}\partial_{\mu}\partial_{\nu}]-\rho^{\alpha}_{(o)}(m^2)\partial_{\mu}\partial_{\nu}\}i\Delta(x;m^2)dm^2 .$$

When the current is conserved the spectral function $\rho^{\alpha}_{(o)}$ vanishes identically.

Using the same méthod as in the previous section it can be shown that

$$\rho^{\alpha}_{(1)}(s) \geq 0$$

$$\rho^{\alpha}_{(0)}(s) \geq 0 \quad . \tag{C.11}$$

Of course, if one of the spectral functions is identically zero, the second one is definite positive.

III. EQUAL TIME COMMUTATORS

1. From equations (C.2) and (C.5), the distribution $\Delta(x;m^2)$ satisfies the integral representation

$$\Delta(\vec{r},t;m^2) = -\frac{1}{(2\pi)^3}\int e^{i\vec{k}\vec{r}}\frac{\sin\omega_k t}{\omega_k}d_3k \tag{C.12}$$

where $x=(\vec{r},t)$ and $\omega_k = \sqrt{\vec{k}^2+m^2}$.

Equation (C.12) is used to prove the following properties of Δ and its derivatives at time t=0:

$$
\begin{aligned}
&\Delta(\vec{r},0;m^2) = 0\\
&\partial_j\Delta(\vec{r},0;m^2) = 0\\
&\partial_o\Delta(\vec{r},0;m^2) = -\frac{\partial}{\partial t}\Delta(\vec{r},0;m^2) = \delta_3(\vec{r})\\
&\partial_j\partial_k\Delta(\vec{r},0;m^2) = 0\\
&\partial_o\partial_o\Delta(\vec{r},0;m^2) = 0 \ . \qquad \text{(C.13)}
\end{aligned}
$$

2. Let us go back to Källen-Lehmann representation of the vacuum expectation value of the commutator of two components of a conserved current

$$<0|[J^\alpha_\mu(x),J^\beta_\nu(0)]|0> = \delta^{\alpha\beta} \int_0 \rho^\alpha(m^2)[g_{\mu\nu} - \frac{1}{m^2}\partial_\mu\partial_\nu]i\Delta(x;m^2)\,dm^2. \qquad \text{(C.4)}$$

As a consequence of the relations (C.13) we easily obtain the vacuum expectation value of equal time commutators [14]

$$<0|[J^\alpha_k(\vec{r},0),J^\beta_\ell(0)]|0> = 0 = <0|[J^\alpha_o(\vec{r},0),J^\beta_o(0)]|0> \qquad \text{(C.14)}$$

$$<0|[J^\alpha_o(\vec{r},0),J^\beta_\ell(0)]|0> = \frac{1}{i}\partial_\ell\delta_3(\vec{r})\delta^{\alpha\beta}\int_0^\infty \frac{\rho^\alpha(m^2)}{m^2}\,dm^2 \ . \qquad \text{(C.15)}$$

3. If now J^α is a non-conserved current, using the same method we easily check that equation (C.14) remains true and equation (C.15) is simply replaced by [15]

$$<0|[J^\alpha_o(\vec{r},0),J^\beta_\ell(0)]|0> = \frac{1}{i}\partial_\ell\delta_3(\vec{r})\delta^{\alpha\beta}\int_0 [\frac{\rho^\alpha_{(1)}(m^2)}{m^2} + \rho^\alpha_{(o)}(m^2)]\,dm^2 . \qquad \text{(C.16)}$$

4. The equal time commutator of the time component with a space component of a current density has the mini-

mal structure

$$[J_0^\alpha(\vec{r},0), J_\ell^\beta(0)] = C_\gamma^{\alpha\beta} J_\ell^\gamma(0)\,\delta(\vec{r}) + S^{\alpha\beta}\,\frac{1}{i}\,\partial_\ell\,\delta(\vec{r}) \qquad (C.17)$$

where $C_\gamma^{\alpha\beta}$ are the skew-symmetric structure constants of the group.

The Schwinger term [16] is defined as $S^{\alpha\beta}$. The vacuum expectation value of the Schwinger term - or the Schwinger term itself if it is a C-number - is given by

$$<0|S^{\alpha\beta}|0> = \delta^{\alpha\beta} C^\alpha$$

where the quantity C^α is given by equation (C.15),

$$C^\alpha = \int_0^\infty \frac{\rho^\alpha(m^2)}{m^2}\,dm^2 \qquad (C.18)$$

for a conserved current and by equation (C.16)

$$C^\alpha = \int_0^\infty \left[\frac{\rho^\alpha_{(1)}(m^2)}{m^2} + \rho^\alpha_{(o)}(m^2)\right]dm^2 \qquad (C.19)$$

for a non-conserved current.

In both cases, the positivity properties of the spectral functions (equations (C.8) and (C.11)) force away C^α to be positive and therefore nonzero.

IV. ELECTROMAGNETIC CURRENT

1. The electromagnetic current is a conserved current. The spectral function $\rho^{em}(s)$ is defined by eq. (C.7)

$$\rho^{em}(s) = \frac{1}{6\pi}\sum_n S_n (2\pi)^4 \delta_4(p_n - k) <0|J_\mu^{em}(0)|n><n|J_\nu^{em}(0)|0> g^{\mu\nu}.$$

Comparing with equation (A.9) we obtain

$$\rho^{em}(s) = \sum_n \rho_n^{em}(s)$$

$$\rho_n^{em}(s) = \frac{1}{6\pi}\,\mathrm{Tr}\{n\} \quad .$$

The spectral function $\rho_n^{em}(s)$ is then related to the total cross section for e^+e^- annihilation into a final state n by using formula (A.8) in the one photon exchange approximation

$$\rho_n^{em}(s) = \frac{s^2}{16\pi^3\alpha^2}\sigma_{tot}(e^+e^- \to n) . \quad (C.20)$$

2. The electromagnetic current can be decomposed into an isovector and an isoscalar component, disregarding other possibilities

$$J^{em} = J^{I_3} + \frac{1}{2} J^Y . \quad (C.21)$$

J^{I_3} is the third component of the isotopic spin current, J^Y is the hypercharge current.

The equality (C.21) implies obviously the Gell-Mann-Nishijima relation for the charges $Q=I_3+\frac{1}{2}Y$.

In the framework of SU(3) symmetry, it is convenient to use the weights associated to particles of the adjoint representation. The U spin scalar electromagnetic current is written as

$$J^{em} = J^3 + \frac{1}{\sqrt{3}} J^8 .$$

J^3 corresponds to a ρ^0 meson (isovector), J^8 to a ϕ^8 meson (isoscalar). The corresponding relations for the spectral functions are

$$\rho^3(s) = \rho^{I_3}(s) = \frac{s^2}{16\pi^3\alpha^2}\sigma_{tot}(e^+e^- \to I=1) \quad (C.22)$$

$$\rho^8(s) = \frac{3}{4}\rho^Y(s) = \frac{3s^2}{16\pi^3\alpha^2}\sigma_{tot}(e^+e^- \to I=0) \quad (C.23)$$

$$\rho^{em}(s) =$$

$$= \rho^3(s) + \frac{1}{3}\rho^8(s) = \frac{s^2}{16\pi^3\alpha^2}\sigma_{tot}(e^+e^- \to \text{hadrons}) . \quad (C.24)$$

3. The Schwinger terms C^3 and C^8 or more precisely the vacuum expectation value of the Schwinger terms are immediately expressed as integrals over the total cross sections for e^+e^- annihilation into hadrons:

$$C^3 = \frac{1}{16\pi^3\alpha^2} \int_{4m_\pi^2}^{\infty} s.\sigma_{tot}(e^+e^- \Rightarrow I=1)\,ds \qquad (C.25)$$

$$C^8 = \frac{3}{16\pi^3\alpha^2} \int_{9m_\pi^2}^{\infty} s.\sigma_{tot}(e^+e^- \Rightarrow I=0)\,ds \;. \qquad (C.26)$$

V. SUM RULES

1. The so-called first Weinberg sum rule [17], [18] assumes an exact SU(3) symmetry and the value of the Schwinger terms is independent of the weight α

$$C^\alpha = C \quad .$$

As a consequence for the actual problem

$$C^3 = C^8 \;. \qquad (C.27)$$

2. The SU(3) symmetry is obviously broken and using a specific model, Sugawara [19] tries to evaluate the corrections to the Weinberg sum rule due to the breaking of the symmetry

$$C^\alpha = C\,\eta^\alpha.$$

The parameter η^α which is unity in an exact SU(3) symmetry is related by Sugawara to the mass spectrum of the vector meson octuplet

$$\eta^\alpha = (m^\alpha/m)^2 \;.$$

In particular, equation (C.31) has to be replaced by

$$C^3 = ((m^3)^2/(m^8)^2)\,C^8$$

and using now a Gell-Mann-Okubo formula for the masses

of the vector meson octuplet

$$4m_{K^*}^2 = m_\rho^2 + 3m_8^2$$

we obtain the Sugawara sum rule in the form

$$c^3 = ((3m_\rho^2/(4m_{K^*}^2 - m_\rho^2)\ c^8 \quad . \tag{C.28}$$

SECTION D

Vacuum Polarization

1. MODIFICATION TO THE PHOTON PROPAGATOR

1. Let us denote by p_μ the energy momentum four vector of a virtual photon ($p^2+s=0$). The photon propagator $D_{\mu\nu}(p)$ has the general structure

$$i\ D_{\mu\nu}(p) = \frac{g_{\mu\nu}}{s} - (g_{\mu\nu} + \frac{p_\mu p_\nu}{s})\ \frac{\pi(s)}{s} \tag{D.1}$$

corresponding to the formal sum of diagrams represented on fig. 1

Fig. 1

The function $\pi(s)$ possesses analytic properties and it is possible to write a Källen-Lehmann [12] representation for $\pi(s)$

$$\frac{\pi(s)}{s} = \frac{1}{\pi} \int_0^\infty \frac{\mathrm{Im}\pi(t)}{t(t-s-i\varepsilon)}\ dt \quad . \tag{D.2}$$

2. The spectral function $\mathrm{Im}\pi(t)$ can be computed in terms of observable matrix elements of the electromagnetic current between the vacuum and a complete set of inter-

mediate states f.

It is then convenient to split $\pi(s)$ into its various contributions

$$\pi(s) = \sum_n \pi^{(n)}(s)$$

The spectral function associated to $\pi^{(n)}(s)$ is simply given by

$$\mathrm{Im}\,\pi^{(n)}(t) =$$

$$= \frac{e^2}{6t} S_n (2\pi)^4 \delta_4 (p_n - p) \langle n | J^{em}_\mu(0) | 0 \rangle \langle 0 | J^{em}_\nu(0) | n \rangle g^{\mu\nu} . \qquad (D.3)$$

Comparing such an expression with equation (A.4) we easily obtain a more condensed form of the spectral function

$$\mathrm{Im}\,\pi^{(n)}(t) = \frac{e^2}{6t} \mathrm{Tr}\{n\} . \qquad (D.4)$$

3. Let us now consider the hadronic contributions to the lowest order with respect to electromagnetic interactions. The expression (A.8) of the total cross section for the annihilation process

$$e^+ + e^- \rightarrow n$$

can be used to transform equation (D.4) into [20]

$$\mathrm{Im}\,\pi^{(n)}(t) = \frac{t}{4\pi\alpha} \sigma_{tot}(e^+e^- \rightarrow n) \qquad (D.5)$$

and the corresponding modification of the photon propagator due to the intermediate state n takes the form

$$\pi^{(n)}(s) = \frac{s}{4\pi^2\alpha} \int_0^\infty \frac{\sigma_{tot}(e^+e^- \rightarrow n)}{t-s-i\varepsilon} dt . \qquad (D.6)$$

For instance the spectral function associated to the intermediate state $\bar{B}B$, where B is a spinless meson, is simply given by the Bouchiat-Michel formula [21]

$$\mathrm{Im}\pi^{(\bar{B}B)}(t) = \frac{\alpha}{12}\left(1-\frac{4m_B^2}{t}\right)^{\frac{3}{2}}|F_B(t)|^2\Theta(t-4m_B^2) \quad . \qquad (D.7)$$

4. The hadronic contributions can be split into two parts depending on their isospin character

$$\pi^{(I=0)}(s) = \frac{s}{4\pi^2\alpha}\int_{9m_\pi^2}^{\infty}\frac{\sigma_{tot}(e^+e^-\Rightarrow I=0)}{t-s-i\varepsilon}dt$$

$$\pi^{(I=1)}(s) = \frac{s}{4\pi^2\alpha}\int_{4m_\pi^2}^{\infty}\frac{\sigma_{tot}(e^+e^-\Rightarrow I=1)}{t-s-i\varepsilon}dt$$

$$\pi(s) = \pi^{(I=0)}(s) + \pi^{(I=1)}(s) \quad .$$

5. The modification of the annihilation cross section

$$e^+ + e^- \Rightarrow f$$

due to vacuum polarization effects is simply the multiplication of the usual uncorrected cross section by the factor

$$|1 - \pi(s)|^2 \quad .$$

II. HADRONIC CONTRIBUTIONS TO THE MUON ANOMALOUS MAGNETIC MOMENT

1. The general method to obtain the hadronic contributions to the muon anomalous magnetic moment, $a_\mu=\frac{1}{2}(g_\mu-2)$, due to vacuum polarization corrections is well known [22]. The resulting expression, corresponding to the class of Feynman diagrams shown on fig. 2

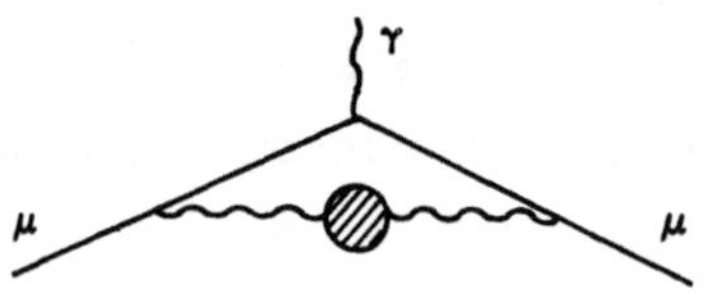

Fig. 2

has the following structure

$$a_\mu(\text{hadrons}) = \frac{1}{\pi}\int_0^\infty \frac{dt}{t}\, \mathrm{Im}\, \pi^{(\text{hadrons})}(t) K_\mu^{(2)}(t) \qquad (D.8)$$

where $\mathrm{Im}\,\pi^{(\text{hadrons})}(t)$ is the hadronic spectral function defined in the previous section and $K_\mu^{(2)}(t)$ is the second-order vertex function given by

$$K_\mu^{(2)}(t) = \frac{\alpha}{\pi}\int_0^1 dx \,\frac{x^2\,(1-x)}{x^2 + \frac{t}{m_\mu^2}(1-x)} \qquad (D.9)$$

where m_μ is the muon mass.

2. The explicit form of $K_\mu^{(2)}(t)$ is known [23], [24]. Obviously

$$K_\mu^{(2)}(0) = \frac{\alpha}{2\pi}$$

and for $t \geq 4m_\mu^2$ a convenient parametrization is the following

$$K_\mu^{(2)}(t) = \frac{\alpha}{\pi}\{\frac{y^2}{2}(2-y^2)+(1+y^2)(1+y^2)\,\frac{\log(1+y)-y+y^2/2}{y^2} +$$

$$+ \frac{1+y}{1-y}\,y^2 \log y\}$$

where

$$y = \frac{1-(1-(4m_\mu^2/t))^{1/2}}{1+(1-(4m_\mu^2/t))^{1/2}} \quad .$$

For large t, $K_\mu^{(2)}(t)$ goes to zero as 1/t and it appears that the high-energy contributions to a_μ(hadrons) are depressed by the factor $K_\mu^{(2)}(t)$ and the integral (D.8) is dominated by the low values of t, in particular by those values of t where electron-positron annihilation cross sections have been recently measured.

3. We shall, of course, work in the one-photon exchange approximation for the e^+e^- annihilation into hadrons. The final state has then total isotopic spin I=0 and I=1.

It is convenient to split a_μ(hadrons) correspondingly

$$a_\mu(\text{hadrons}) = a_\mu(I=0) + a_\mu(I=1) \quad .$$

Combining now equations (D.5) and (D.8) we obtain

$$a_\mu(I=0) = \frac{1}{4\pi^2\alpha} \int_{9m_\pi^2}^{\infty} \sigma_{tot}(e^+e^- \to I=0)\, K_\mu^{(2)}(t)\,dt$$

$$a_\mu(I=1) = \frac{1}{4\pi^2\alpha} \int_{4m_\pi^2}^{\infty} \sigma_{tot}(e^+e^- \to I=1)\, K_\mu^{(2)}(t)\,dt \quad .$$

SECTION E

High-Energy Behaviour for Annihilation Cross Section

1. We have written in Sections C and D integrals involving the total cross section for electron-positron annihilation into hadrons.

a) The Schwinger terms are proportional to

$$\int s\, \sigma_{tot}(s)\,ds \quad . \tag{E.1}$$

b) The vacuum polarization contributions involve integrals of the type

$$\int \frac{1}{s}\, \sigma_{tot}(s)\,ds \quad . \tag{E.2}$$

We must now examine the important problem of the convergence of the integrals (E.1)and (E.2). Of course, if integral (E.1) converges so does integral (E.2).

2. The integral (E.1) converges if for large s, $s^2\sigma_{tot}(s)$ tends to zero sufficiently rapidly. For instance

$$s^2\sigma_{tot}(s) \text{ behaves like } s^{-\varepsilon} \qquad \varepsilon > 0$$

$$s^2\sigma_{tot}(s) \text{ behaves like } (\log s)^{-\alpha} \qquad \alpha > 1 \quad .$$

Sufficient conditions of convergence for integral (E.2) are that $\sigma_{tot}(s)$ behaves like $s^{-\varepsilon}$ ($\varepsilon>0$) or $(\log s)^{-\alpha}$ ($\alpha>1$) at infinity.

3. The quark model offers a possibility to evaluate the Schwinger terms and to study the high-energy behaviour of the total cross section $\sigma_{tot}(e^+e^- \Rightarrow \text{hadrons})$.

It has been shown by Gribov, Ioffe and Pomeranchuk [25] that the Schwinger terms like C^3 or C^8 are infinite and that the expected high-energy behaviour of the total cross section is

$$\sigma_{tot}(s) \simeq \frac{\text{const.}}{s} \quad .$$

Therefore (E.1) diverges linearly and (E.2) converges.

4. The algebra of field model is a more convergent theory and the Schwinger terms are finite. For instance the value of C^3 is simply $C^3 = m_\rho^2/f_\rho^2$ where f_ρ is defined by

$$J_\mu^{I=1} = - \frac{m_\rho^2}{f_\rho} V_\mu^\rho \quad .$$

5. Let us discuss now the total cross section in one particular channel for instance $\sigma_{tot}(e^+e^- \Rightarrow \pi^+\pi^-)$. From the results of Section A, in the one-photon exchange approximation

$$\sigma_{tot}(e^+e^- \Rightarrow \pi^+\pi^-) \simeq \frac{\text{const.}}{s} |F_\pi(s)|^2 \quad .$$

Nothing is known about the high-energy behaviour of the π-meson electromagnetic form factor and we can try to make speculations. Let us look as a guide at the nucleon electromagnetic form factors. From elastic electron-proton scattering experiments ($s < 0$) we know that the nucleon electromagnetic form factors decrease rapidly for large $|s|$ at least like $|s|^{-1}$, perhaps like $|s|^{-2}$ or more rapidly.

Assuming an analogous behaviour, in the time-like

region for the π-meson electromagnetic form factor we can expect the quantity $s^3\sigma_{tot}(s)$ to be bounded by some constant or to tend to zero when s goes to infinity. Therefore the $\pi^+\pi^-$ contributions to integrals like (E.1) will be finite.

As a last remark, in the Gounaris-Sakurai model, such a condition is fulfilled

$$\sigma_{tot}(e^+e^- \rightarrow \pi^+\pi^-) \approx s^{-3}\log^{-2}s$$

but at infinite energy the model is perhaps meaningless.

References - PART I

1. G. F. CHEW, R. KARPLUS, S. G. GASIOROWICZ and F. ZACHARIASEN, Phys. Rev. 110, 265 (1958).
2. P. FEDERBUSH, M. L. GOLDBERGER and S. B. TREIMAN, Phys. Rev. 112, 642 (1958).
3. N. I. MUSKHELISHVILI, Trud.Tbil.Mat.Inst.10,1 (1941).
4. R. OMNÈS, Nuovo Cimento 8, 316 (1958),
5. M. GOURDIN and A. MARTIN, Nuovo Cimento 8, 699 (1958).
6. G. F. CHEW, S Matrix Theory, Benjamin, New York,1961.
7. W. R. FRAZER and J. FULCO, Phys. Rev. Letters 2, 365 (1959); Phys. Rev. 117, 1603, 1609 (1960).
8. M. T. VAUGHN, K. C. WALI, Phys. Rev. Letters 21, 938 (1968).
9. G. GOUNARIS and J. SAKURAI, Phys. Rev. Letters 21, 244 (1968).
10. V. AUSLANDER et al., Phys. Letters 25B, 433 (1967).
11. J. AUGUSTIN et al., Phys. Rev. Letters 20, 129 (1968); Phys. Letters 28B, 508 (1969).
12. G. KÄLLÉN, Helv. Phys. Acta 25. 417 (1952).
13. H. LEHMANN, Nuovo Cimento 11, 342 (1954).
14. K. JOHNSON, Nuclear Physics 25, 431 (1961).

15. S. OKUBO, Nuovo Cimento 44A, 1015 (1966).

16. J. SCHWINGER, Phys. Rev. Letters 3, 296 (1959).

17. S. WEINBERG, Phys. Rev. Letters 18, 507 (1967).

18. T. DAS, V. S. MATHUR and S. OKUBO, Phys. Rev. Letters 18, 507 (1967); Phys. Rev. Letters 19, 470 (1967).

19. H. SUGAWARA, Phys. Rev. Letters 21, 772 (1968).

20. R. GATTO, Nuovo Cimento 28, 658 (1963).

21. C. BOUCHIAT and L. MICHEL, Journ. Phys. Rad. 22, 121 (1961).

22. Loyal DURAND III, Phys. Rev. 128,441 (1962).

23. S. J. BRODSKY and E. de RAFAEL, Phys. Rev. 168, 1620 (1968).

24. B. E. LAUTRUP and E. de RAFAEL, Phys. Rev. 174, 1835 (1968).

25. V. N. GRIBOV, B. L. IOFFE and I. Ya. POMERANCHUK, Phys. Letters 24B, 554 (1967).

PART TWO

Vector-Meson Dominance Model

SECTION F

Decay of the Vector Mesons into a Lepton-Antilepton Pair

I. THE POLE MODEL

1. We are interested in the decay of a vector meson V ($V=\rho,\omega,\phi$) into a lepton-antilepton pair $\ell^+\ell^-$ ($\ell=e,\mu$). We use a pole model as drawn on fig. 1

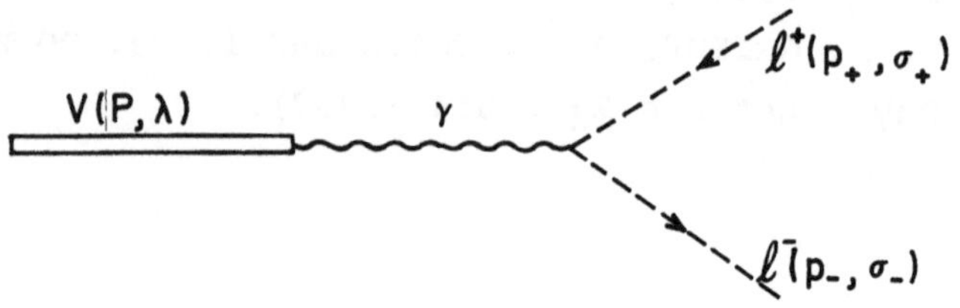

Fig. 1

2. The transition matrix element is factorized into the product of two matrix elements of the electromagnetic current

$$\langle \ell^+\ell^-|T|V\rangle = \frac{e^2}{m_V^2}\,\bar{u}_{\sigma_-}(p_-)\gamma^\mu v_{\sigma_+}(p_+)\langle 0|J_\mu^{em}(0)|V\rangle \qquad (F.1)$$

where the conservation law $m_V^2 = -(p_+ + p_-)^2$ has been used.

The vector meson-photon transition is represented by a coupling constant m_V^2/f_V (Refs. [1],[2],[3])

$$\langle 0|J_\mu^{em}(0)|V\rangle = \frac{m_V^2}{f_V}\,e_\mu(P,\lambda) \qquad (F.2)$$

where $e_\mu(P,\lambda)$ is the polarization vector for the vector meson of energy momentum P and helicity λ. Let us recall the subsidiary condition

$$P^{\mu} e_{\mu}(P,\lambda) = 0 \quad .$$

3. In the algebra of field model [4] the electromagnetic current is identified with the vector-meson fields and the couplings $1/f_V$ are the coefficients of proportionality in such an identification

$$J_{\mu}^{em} = - \sum_{V} \frac{m_V^2}{f_V} V_{\mu} \quad . \qquad (F.3)$$

4. A straightforward calculation gives the partial width for the decay of a vector meson V into a lepton pair $\ell^+\ell^-$

$$\Gamma(V \Rightarrow \ell^+\ell^-) = \frac{4\pi\alpha^2}{3} \frac{m_V}{f_V^2}(1+2\frac{m_\ell^2}{m_V^2})(1-\frac{4m_\ell^2}{m_V^2})^{\frac{1}{2}} \quad .$$

The lepton mass m_ℓ being small with respect to the vectormeson mass m_V, up to terms $O[m_\ell^4/m_V^4]$ we simply have

$$\Gamma(V \Rightarrow \ell^+\ell^-) = \frac{4\pi\alpha^2}{3} \frac{m_V}{f_V^2} \quad . \qquad (F.4)$$

II. VECTOR-MESON DOMINANCE

1. We consider the e^+e^- annihilation into a final state f

$$e^+ + e^- \Rightarrow f$$

with a threshold $s_f : s > s_f$.

If a vector meson $V(\rho,\omega,\phi)$ is physically realizable as an intermediate unstable state ($m_V^2 > s_f$) and if V can decay strongly in the state f, the reaction $e^+e^- \Rightarrow f$ is dominated, in the neighbourhood of $s \simeq m_V^2$ by the V meson contribution described in the chain

$$e^+ + e^- \Rightarrow f \quad .$$

2. At the V-meson mass the total cross section is factorized as

$$\sigma_{tot}(e^+e^- \Rightarrow V \Rightarrow f)_{s=m_V^2} = \frac{12\pi}{m_V^2} \frac{\Gamma(V \Rightarrow e^+e^-)}{\Gamma_V} \frac{\Gamma(V \Rightarrow f)}{\Gamma_V} . \quad (F.5)$$

Formula (F.5) is used to extract, from experiment, the partial decay widths $\Gamma(V \Rightarrow e^+e^-)$.

3. The ω meson and the ϕ meson can be strongly coupled to the same hadronic state, for instance the $\pi^+\pi^-\pi^o$ state. But the interference between the ω- and ϕ-meson contributions is always extremely small and, in fact, completely negligible.

More generally, the V-meson contribution is important only in an energy range approximately defined by

$$|\sqrt{s} - m_V| \lesssim \Gamma_V \quad .$$

It is clear that the two domains

$$|\sqrt{s} - m_\omega| \lesssim \Gamma_\omega \qquad \text{and} \qquad |\sqrt{s} - m_\phi| \lesssim \Gamma_\phi$$

cannot overlap.

4. The situation is different for the ρ meson and ω meson where m_ω belongs to the range $|\sqrt{s} - m_\rho| \lesssim \Gamma_\rho$. If the final state is a pure hadronic state, like 2π , 3π , the total isotopic spin and therefore the G parity of the final state are well defined. The ρ meson and the ω meson being opposite values of the G parity, and ρ-ω interference will occur only via electromagnetic interactions, and formula (F.5) holds. If, however, the final state contains in addition to hadrons, photons or leptons, for instance a $\pi^o\gamma$, a $\eta\gamma$ state, the G parity of the final state is no more defined, the ρ-meson and ω-meson contributions occur on the same footing and the ρ-ω interference can be important and measurable.

III. THE $\rho^o \Rightarrow e^+e^-$ DECAY

1. The final state f is a $\pi^+\pi^-$ pair. At the total energy equal to the ρ-meson mass, $s=m_\rho^2$, formula (F.5) is simply [5]

$$\frac{\Gamma(\rho \Rightarrow e^+e^-)}{\Gamma_\rho} = \frac{m_\rho^2}{12\pi} \sigma_{tot}(e^+e^- \Rightarrow \pi^+\pi^-)_{s=m_\rho^2} \quad . \tag{F.6}$$

2. In the Orsay experiment [6] the measured cross section at $s=m_\rho^2$ is

$$\sigma_{tot}(e^+e^- \Rightarrow \pi^+\pi^-)_{s=m_\rho^2} = (1.64 \pm 0.20)10^{-30} \text{ cm}^2 . \tag{F.7}$$

The best fit of the π-meson electromagnetic form factor with a Gounaris-Sakurai models(B.40) has given

$$m_\rho = (770 \pm 4) \text{ MeV}$$

$$\Gamma_\rho = (111 \pm 6) \text{ MeV}$$

from which we deduce, using equation (F.6)

$$\frac{\Gamma(\rho \Rightarrow e^+e^-)}{\Gamma_\rho} = (6.63 \pm 0.85)10^{-5} \tag{F.8}$$

and

$$\Gamma(\rho \Rightarrow e^+e^-) = (7.36 \pm 0.7) \text{ keV} \quad . \tag{F.9}$$

3. The result (F.9) can be equivalently formulated in terms of the ρ-γ coupling constant f_ρ: using equation (F.4), we find

$$f_\rho^2/4\pi = 1.86 \pm 0.18 \quad . \tag{F.10}$$

IV. THE $\omega \Rightarrow e^+e^-$ DECAY

1. The final state f is $\pi^+\pi^-\pi^o$. At the total energy

equal to the ω-meson mass $s=m_\omega^2$ formula (F.5) is

$$\frac{\Gamma(\omega \to e^+e^-)}{\Gamma_\omega} = \frac{\Gamma_\omega}{\Gamma(\omega \to 3\pi)} \frac{m_\omega^2}{12\pi} \sigma_{tot}(e^+e^- \to \pi^+\pi^-\pi^0)_{s=m_\omega^2}$$

2. In the Orsay experiment [7] the ω mass is fixed to be m_ω = 783 MeV and the ω total width is found to be

$$\Gamma_\omega = (16.2 \pm 3.2) \text{ MeV}$$

e.g. somewhat larger than the world data average value of [8]

$$\Gamma_\omega = (12.2 \pm 1.3) \text{ MeV} .$$

Including the possibility of non-resonating background, we choose the width as given by the average of the world data and the measured cross section at $s = m_\omega^2$ is

$$\sigma_{tot}(e^+e^- \to 3\pi)_{s=m_\omega^2} = (1.65 \pm 0.31)\ 10^{-30} \text{cm}^2 . \tag{F.12}$$

3. The branching ratio $\Gamma(\omega \to 3\pi)/\Gamma_\omega$ is given by the world average [8]

$$\frac{\Gamma(\omega \to 3\pi)}{\Gamma_\omega} = 0.907 \pm 0.008$$

and using equation (F.6) we obtain

$$\frac{\Gamma(\omega \to e^+e^-)}{\Gamma_\omega} = (7.6 \pm 1.4) 10^{-5} \tag{F.13}$$

$$\Gamma(\omega \to e^+e^-) = (0.94 \pm 0.18) \text{ keV} . \tag{F.14}$$

4. The result (F.14) can be expressed in terms of the ω-γ coupling constant f_ω: using equation (F.4), we find

$$f_\omega^2/4\pi = 14.8 \pm 2.8 . \tag{F.15}$$

V. THE $\phi \to e^+e^-$ DECAY

1. The final state f can be K^+K^-, $K^0\bar{K}^0$ or $\pi^+\pi^-\pi^0$. At the total energy equal to the ϕ-meson mass $s=m_\phi^2$ formula (F.5) is

$$\frac{\Gamma(\phi\to e^+e^-)}{\Gamma_\phi} = \frac{\Gamma_\phi}{\Gamma(\phi\to K^+K^-)} \frac{m_\phi^2}{12\pi} \sigma_{tot}(e^+e^-\to K^+K^-)_{s=m_\phi^2} \quad (F.16)$$

$$\frac{\Gamma(\phi\to e^+e^-)}{\Gamma_\phi} = \frac{\Gamma_\phi}{\Gamma(\phi\to K^0\bar{K}^0)} \frac{m_\phi^2}{12\pi} \sigma_{tot}(e^+e^-\to K^0\bar{K}^0)_{s=m_\phi^2} \quad (F.17)$$

$$\frac{\Gamma(\phi\to e^+e^-)}{\Gamma_\phi} = \frac{\Gamma_\phi}{\Gamma(\phi\to\pi^+\pi^-\pi^0)} \frac{m_\phi^2}{12\pi} \sigma_{tot}(e^+e^-\to \pi^+\pi^-\pi^0)_{s=m_\phi^2} \ . \quad (F.18)$$

2. In the Orsay experiment [9] only the $K^0\bar{K}^0$ (K_SK_L because of PC) and the $\pi^+\pi^-\pi^0$ decay modes of the ϕ meson were detected and identified by looking at a $\pi^+\pi^-$ pair in various kinematical situations.

The ϕ-meson mass has been taken at its world data average value m_ϕ = 1019.3 MeV and the ϕ-meson width measured by this experiment

$$\Gamma_\phi = (4.2 \pm 0.9) \text{ MeV} \ . \quad (F.19)$$

The measured cross section at $s=m_\phi^2$ for a K^0 $\bar{K}^0$ final state is

$$\sigma_{tot}(e^+e^- \to K^0\bar{K}^0)_{s=m_\phi^2} = (1.17 \pm 0.19)\ 10^{-30}\ \text{cm}^2. \quad (F.20)$$

3. The comparison of the total cross sections for $e^+e^- \to K^0\bar{K}^0$ and for $e^+e^- \to \pi^+\pi^-\pi^0$ at $s=m_\phi^2$ will give, in the vector meson dominance model, the ratio of the partial decay widths for $\phi\to\pi^+\pi^-\pi^0$ and $\phi\to K^0\bar{K}^0$.

The value obtained in the Orsay experiment

$$\frac{\Gamma(\phi \Rightarrow \pi^+\pi^-\pi^0)}{\Gamma(\phi \Rightarrow K^0\bar{K}^0)} = 0.667 \pm 0.157 \qquad (F.21)$$

is in disagreement with the value 0.354 deduced from other experiments [8].

On the other hand, assuming isotopic spin invariance and taking into account the phase-space corrections and the electromagnetic corrections, we find the theoretical prediction for the ratio of the partial decay widths for $\phi \Rightarrow K^+K^-$ and $\phi \Rightarrow K^0\bar{K}^0$ is [10]

$$\frac{\Gamma(\phi \Rightarrow K^+K^-)}{\Gamma(\phi \Rightarrow K^0\bar{K}^0)} = 1.60 \qquad (F.22)$$

again in disagreement with the value 1.21 deduced from actual experiments [8]. Assuming that the other decay modes ($\pi^0\gamma$, $\eta\gamma$, etc.) are small we deduce, from (F.21) and (F.22), the three branching ratios

$$\frac{\Gamma(\phi \Rightarrow K^0\bar{K}^0)}{\Gamma_\phi} = 0.306 \pm 0.015$$

$$\frac{\Gamma(\phi \Rightarrow \pi^+\pi^-\pi^0)}{\Gamma_\phi} = 0.204 \pm 0.040$$

$$\frac{\Gamma(\phi \Rightarrow K^+K^-)}{\Gamma_\phi} = 0.490 \pm 0.025 \quad . \qquad (F.23)$$

4. Using the results (F.19), (F.20) and (F.23) we obtain from equation (F.17)

$$\frac{\Gamma(\phi \Rightarrow e^+e^-)}{\Gamma_\phi} = (3.98 \pm 0.62)10^{-4} \qquad (F.24)$$

$$\Gamma(\phi \Rightarrow e^+e^-) = (1.67 \pm 0.25) \text{ keV} \quad . \qquad (F.25)$$

5. The partial width (F.25) can be expressed in terms of the ϕ-γ coupling constant f_ϕ from equation (F.4) and we find

$$f_\phi^2/4\pi \simeq 11.0 \pm 1.6 \quad . \qquad (F.26)$$

SECTION G

Sum Rules and Schwinger Terms

I. GENERALITIES

1. We have studied in Section C an integral representation of the Schwinger terms C^3 and C^8 in terms of the total crosssections for e^+e^- annihilation into hadrons (equations (C.29) and (C.30))

$$C^8 = \frac{3}{16\pi^3\alpha^2} \int_{9m_\pi^2}^{\infty} s\, \sigma_{tot}(e^+e^- \Rightarrow I=0)\, ds \qquad (G.1)$$

$$C^3 = \frac{1}{16\pi^3\alpha^2} \int_{4m_\pi^2}^{\infty} s\, \sigma_{tot}(e^+e^- \Rightarrow I=1)\, ds \; . \qquad (G.2)$$

Let us make the assumption that the integrals (G.1) and (G.2) converge, e.g. that the total cross sections for e^+ e^- annihilation into hadrons tend to zero at infinity more rapidly than s^{-2} (see Section E).

2. The first Weinberg sum rule [11] is simply the equality between the two Schwinger terms C^3 and C^8 (equation C.31)

$$C^3 = C^8 = 0 \quad . \qquad (G.3)$$

The Sugawara sum rule [12] takes into account, in a model-dependent way, the breaking of SU(3) symmetry

$$C^3 - \frac{3m_\rho^2}{4m_{K^*}^2 - m_\rho^2} C^8 = 0 \quad . \qquad (G.4)$$

3. In order to estimate the Schwinger terms C^8 and C^3 we make the following assumptions :

For C^8 we use a vector-meson dominance model taking

into account the ω-meson and ϕ-meson contributions in narrow-width approximation.

II. ISOSCALAR TERM

1. The ω meson and ϕ meson are two narrow-width-resonances and we will use a representation of the total cross section $\sigma(e^+e^- \Rightarrow I=0)$ where the energy dependence is concentrated in the denominator of two Breit-Wigner formulae

$$s\sigma(e^+e^- \Rightarrow I=0) \simeq$$

$$\simeq \sum_{V=\omega,\phi} \sigma(e^+e^- \Rightarrow V \Rightarrow I=0)_{s=m_V^2} \frac{m_V^4\Gamma_V^2}{(m_V^2-s)^2+m_V^2\Gamma_V^2} \quad . \qquad (G.5)$$

Let us recall that the peak cross section $\sigma(e^+e^- \Rightarrow V \Rightarrow I=0)_{s=m_V^2}$ gives a measurement of the partial decay width $V \Rightarrow e^+e^-$ as shown in section F

$$\sigma(e^+e^- \Rightarrow V \Rightarrow I=0) = \frac{12\pi}{m_V^2} \frac{\Gamma(V \Rightarrow e^+e^-)}{\Gamma_V} \qquad V=\omega,\phi \quad . \qquad (G.6)$$

Combining equations (G.5) and (G.6) we obtain

$$s\ \sigma(e^+e^- \Rightarrow I=0) \simeq 12\pi \sum_{V=\omega,\phi} \Gamma(V \Rightarrow e^+e^-) \frac{m_V^2\Gamma_V}{(m_V^2-s)^2+m_V^2\Gamma_V^2} \quad . \qquad (G.7)$$

We then obtain the following estimate of the isoscalar Schwinger term

$$C^8 = \frac{9}{4\pi\alpha^2}[m_\omega\Gamma(\omega \Rightarrow e^+e^-)+m_\phi\Gamma(\phi \Rightarrow e^+e^-)] \quad . \qquad (G.8)$$

It can be useful to introduce the vector meson-photon coupling constants f_V as in Section F

$$\Gamma(V \Rightarrow e^+e^-) = \frac{4\pi\alpha^2}{3} \frac{m_V}{f_V^2}$$

and equation (G.8) is equivalently written as

$$C^8 = 3\left[\frac{m_\omega^2}{f_\omega^2} + \frac{m_\phi^2}{f_\phi^2}\right] \quad . \qquad \text{(G.9)}$$

3. Using now the Orsay results [7], [9] discussed in Section F (equations F.15 and F.26) we obtain the numerical value

$$C^8 = (3.24 \pm 0.55)10^{-2}\text{GeV}^2 \quad . \qquad \text{(G.10)}$$

III. ISOVECTOR TERM

1. As a first estimate we calculate the isovector Schwinger term C^3 taking into account only the ρ-meson contribution in a narrow-width approximation. Expressions analogous to (G.8) and (G.9) are immediately written

$$C^3 \simeq \frac{3}{4\pi\alpha^2} m_\rho \Gamma(\rho \Rightarrow e^+e^-) \qquad \text{(G.11)}$$

$$C^3 \simeq \frac{m_\rho^2}{f_\rho^2} \qquad \text{(G.12)}$$

Using the Orsay data [6] as given by equation (F.9) we deduce

$$C^3 \simeq (2.53 \pm 0.25)10^{-2}\text{GeV}^2 \quad . \qquad \text{(G.13)}$$

2. A narrow-width approximation for the ρ-meson resonance is not correct and we must perform a more reliable calculation of C^3.

We now assume that the total cross section $\sigma_{tot}(e^+e^+ \Rightarrow I=1)$ is dominated by the $\pi^+\pi^-$ contributions.

$$C^3 \simeq \frac{1}{48\pi^2}\int_{4m_\pi^2}^{\infty} (1-\frac{4m_\pi^2}{s})^{\frac{3}{2}} |F_\pi(s)|^2 ds \quad . \qquad \text{(G.14)}$$

The model of the π-meson electromagnetic form factor proposed by Gounaris and Sakurai [5] (equation B.28) has been used in the entire range of integration whereas it has been experimentally tested only in the ρ^{o}-meson region.

We must keep in mind that such an extrapolation can be doubtful especially in a calculation where the high-energy region plays an important role.

With the parameters

$$m_{\rho} = (770 \pm 4)\ \text{MeV} \qquad \Gamma_{\rho} = (111 \pm 6)\ \text{MeV}$$

the result is simply

$$C^{3} = (2.35 \pm 0.17)10^{-2}\ \text{GeV}^{2}\ . \qquad (G.15)$$

IV. SUM RULES

1. Assuming the expressions (G.8) and (G.11) to well approximate the integrals (G.1) and (G.2), Das, Mathur and Okubo [13] write the first Weinberg sum rule (G.3) in the form

$$\frac{1}{3}\, m_{\rho}\ \Gamma(\rho \rightarrow e^{+}e^{-}) = m_{\omega}\ \Gamma(\omega \rightarrow e^{+}e^{-}) + m_{\phi}\ \Gamma(\phi \rightarrow e^{+}e^{-})\ . \qquad (G.16)$$

The comparison with experiment of the sum rules (G.16) is made using for C^{3} and C^{8} the numerical estimates (G.10) and (G.13). Equation (G.3) is then written

$$(-0.71 \pm 0.80) = 0\ . \qquad (G.17)$$

It is a priori better to check the first Weinberg sum rule using for C^{3} the more elaborate calculation (G.15). Equation (G.3) now gives

$$(-0.89 \pm 0.72) = 0\ . \qquad (G.18)$$

The left-hand sides of equations (G.17) and (G.18) are not

incompatible with zero.

2. For the Sugawara sum rule (G.4) we obtain

$$(+0.31 \pm 0.63) = 0 \qquad (G.19)$$

in a narrow-width approximation for C^3 and

$$(+0.13 \pm 0.55) = 0 \qquad (G.20)$$

in a more sophisticated calculation (G.15). Again the left-hand sides of equations (G.19) and (G.20) are not incompatible with zero.

SECTION H

Modifications to the Photon Propagator due to Mesons

1. The modification of the e^+e^- annihilation cross section due to vacuum polarization is the multiplication by the function

$$I(s) = |1 - \pi(s)|^2 \qquad (H.1)$$

of the cross section calculated in the one photon exchange approximation with the bare photon propagator $g_{\mu\nu}/s$.

2. We have discussed in Section D the structure of the function $\pi(s)$ and the contribution of hadrons to $\pi(s)$ is related to the total cross section for e^+e^- annihilation into hadrons by

$$\pi^{(\text{hadrons})}(s) = \frac{s}{4\pi\alpha^2} \int_0^\infty \frac{\sigma_{tot}(e^+e^- \Rightarrow \text{hadrons})}{t-s-i\varepsilon} dt \;.$$

3. The details of the calculation can be found in a recent preprint by E. Cremmer and the author :" Vacuum

Polarization Effects in e^+e^- Collision " (Orsay preprint 69/17).

The function I(s) can be expressed in the following form

$$I(s) = \frac{1}{|1+\pi^{\phi}(s)|^2} - 2\mathrm{Re}\,\pi^{I=1}(s) - 2\mathrm{Re}\,\pi^{\omega}(s)\,. \quad (H.2)$$

In the peak approximation the ω contribution is

$$I_{\omega}(s) = \frac{6s}{\alpha}\,\frac{\Gamma(\omega\to e^+e^-)}{m_{\omega}}\,\frac{s-m_{\omega}^2}{(m_{\omega}^2-s)^2+m_{\omega}^2\Gamma_{\omega}^2}\,.$$

The peak (narrow width) approximation is not valid for the ρ contribution. We assume that the hadronic states with I=1 are dominated by the P wave $\pi^+\pi^-$ system. Then we end up with the integral

$$\pi^{I=1}(s) = \frac{\alpha}{12\pi}\, s \int_{4m_{\pi}^2}^{\infty} \left(1-\frac{4m_{\pi}^2}{t}\right)^{\frac{3}{2}} |F_{\pi}(t)|^2 \, \frac{dt}{t(t-s-i\varepsilon)}$$

which can be numerically evaluated using for $F_{\pi}(t)$ the Gounaris-Sakurai model.

The ϕ contribution is quite large due to the relatively great $\Gamma(\phi\to e^+e^-)/\Gamma_{\phi}$ so that we sum up all the diagrams corresponding to the iteration of the second order contributions.

4. The results are shown on fig. 1,3 and 2 and compared with the peak approximation generally used. On fig. 4 we have plotted the total cross section $\sigma_{tot}(e^+e^-\Rightarrow \pi^+\pi^-)$ using the parameter determined by the Orsay experiment. We have two curves with and without the vacuum polarization factor I(s). Fig. 4 exhibits, close to the ω-meson mass, an interference type effect due to a zero of $\mathrm{Re}\,\pi(s)$ and $\sqrt{s} \simeq 780$ MeV; in this region both the ρ and ω mesons contribute to $\pi(s)$ and cancel at this point.

Let us note that we take into account, via the vacuum

polarization, the electromagnetic decay of the ω meson into a $\pi^+\pi^-$ system. Fig. 5 shows the total cross section $\sigma_{tot}(e^+e^- \Rightarrow K^o\bar{K}^o)$ with and without the vacuum polarization factor I(s) the effect of which is essentially to shift the cross section towards the right of some MeV.

Let us finally recall that to extract from experiment the various resonance parameters like mass and width we must compare the experimental data with the theoretical cross sections where are included the vacuum polarization factor I(s) and the radiative corrections also functions of the same parameters.

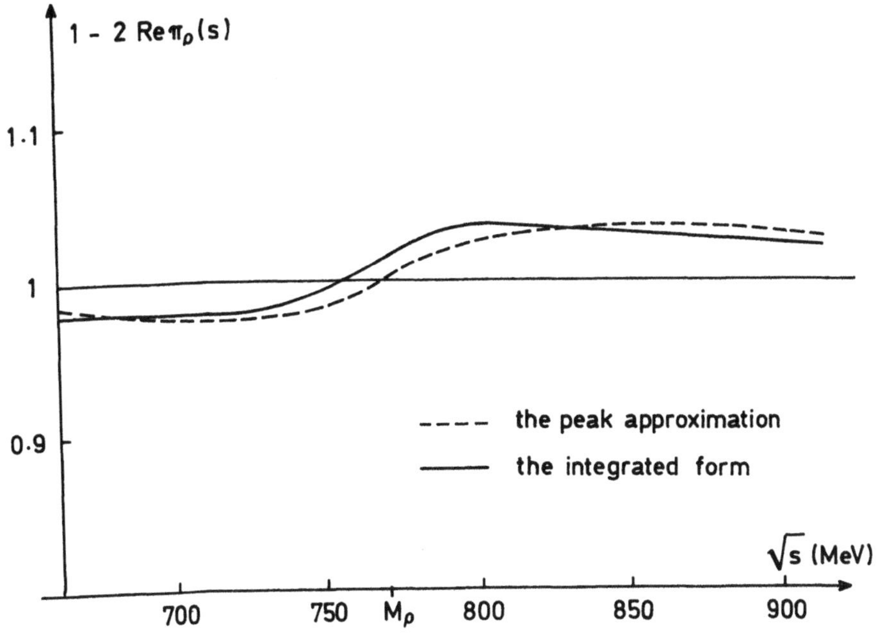

Figure 1 - The ρ-meson contribution.

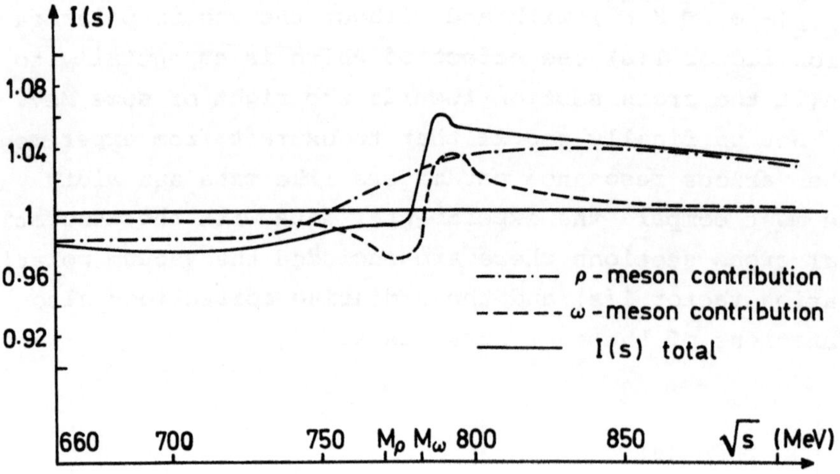

Figure 2 - The ρ-ω region.

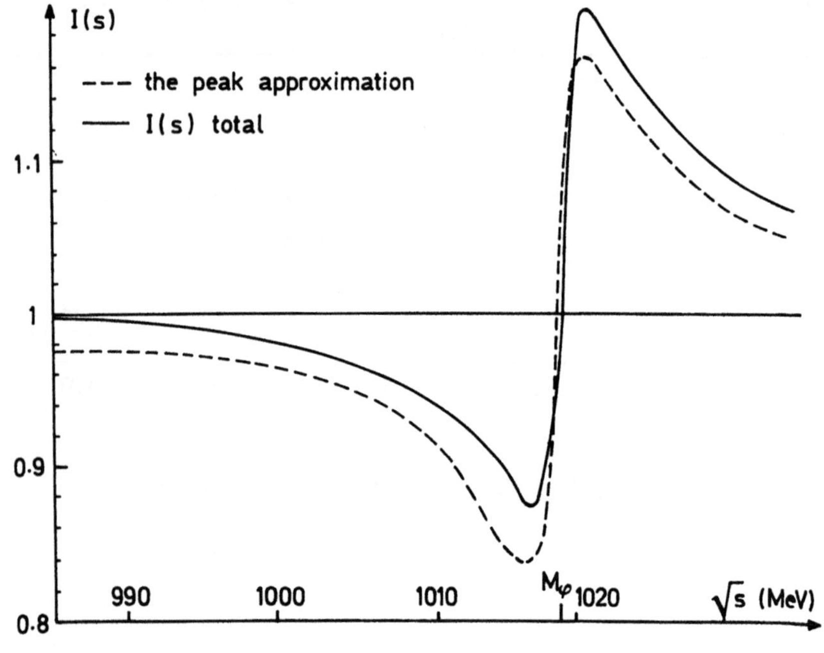

Figure 3 - The φ-meson contribution.

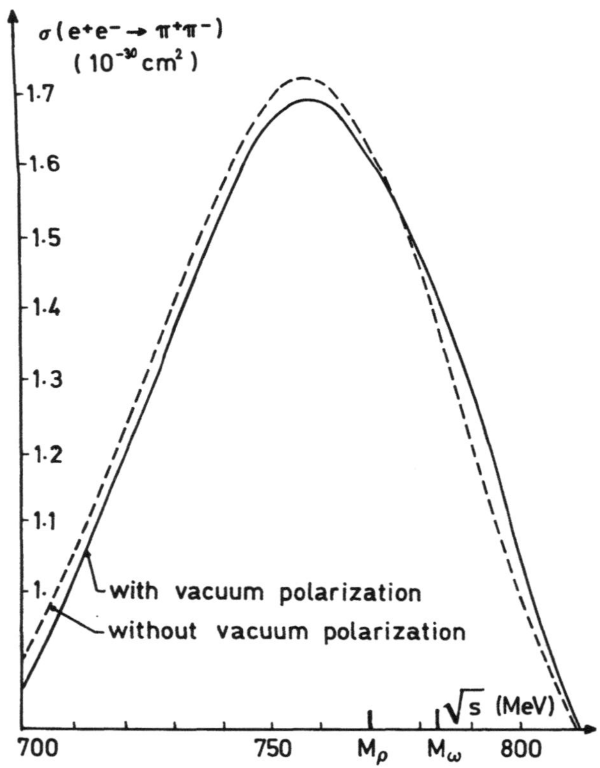

Figure 4 - The total cross section $\sigma_{tot}(e^+e^- \to \pi^+\pi^-)$.

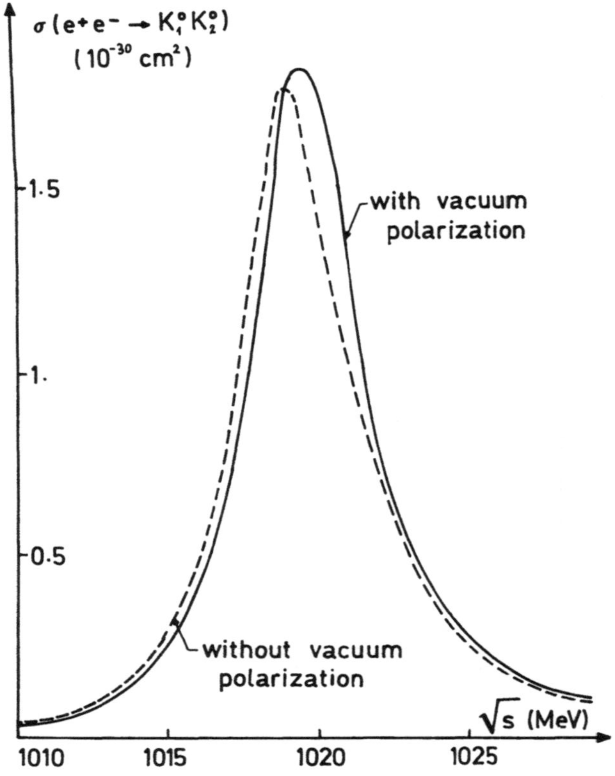

Figure 5 - The total cross section $\sigma_{tot}(e^+e^- \to K^o\bar{K}^o)$.

SECTION I

Mesonic Contributions to the Meson Anomalous Magnetic Moment

1. We report here only the result of a calculation by E. de Rafael and the author ("Hadronic Contributions to the Muon g Factor, CERN preprint TH 986). The following contribution has been obtained by using the Orsay data

$$a_\mu(\text{hadron}) \simeq (6.5 \pm 0.5)\ 10^{-8}\ . \qquad (I.1)$$

2. The result in equation (I.1) when added together with the second [14] and fourth [15] order calculations from quantum electrodynamics, leads to the prediction

$$\frac{1}{2}(g_\mu - 2)_{th} = (116560.7 \pm 0.8)\ 10^{-8}$$

to be compared with the experimental result

$$\frac{1}{2}(g_\mu - 2)_{exp} = (116616 \pm 31)\ 10^{-8}\ .$$

SECTION J

On the $e^+e^- \Rightarrow \pi^o\gamma$ Reaction

This section closely follows a paper by the author and E. Cremmer: "A Calculation for the $e^+e^- \Rightarrow \pi^o\gamma$ Reaction", to be published in Nuclear Physics B .

I. GENERALITIES

1. We now study the e^+e^- annihilation into a $\pi^o\gamma$ system. Such a reaction enters the general framework of the processes $e^+e^- \Rightarrow B^o\gamma$ for which general considera-

tions have been developed in Section A.III.

2. The electromagnetic form factor $G_{\pi^0}(s)$ is normalized so as to describe the $\pi^0 \rightarrow 2\gamma$ decay for s=o and it can be divided into two parts depending on the isotopic-spin character of the virtual photon

$$G(s) = G^{I=1}(s) + G^{I=0}(s) . \qquad (J.1)$$

On the other hand, in the $\pi^0 \rightarrow 2\gamma$ decay, one photon is isovector and the other one isoscalar. It follows that

$$G^{I=1}(0) = G^{I=0}(0) = \frac{1}{2} G(0) . \qquad (J.2)$$

Using the experimental result (from the tables of the Particle Data Group)

$$\Gamma(\pi^0 \rightarrow 2\gamma) = (7.4 \pm 1.5) \text{ eV}$$

we obtain

$$|G(0)| = (3.62 \pm 0.36) 10^{-2} . \qquad (J.3)$$

The model predicts

$$\frac{\Gamma(\rho^0 \rightarrow \pi^0\gamma)}{\Gamma_\rho} \simeq (5.5 \pm 1.1) 10^{-4} ,$$

which has to be compared with the results of our analysis of ρ^- photoproduction data which give [16]

$$\frac{\Gamma(\rho \rightarrow \pi\gamma)}{\Gamma_\rho} < 4 . 10^{-3} .$$

II. NUMERICAL COMPUTATIONS

1. We have computed the total cross section $\sigma(e^+e^- \rightarrow \pi^0\gamma)$ using the π-meson electromagnetic form factor in the form proposed by Gounaris and Sakurai and which fits very well the Orsay results with the parameters

$$m_\rho = (770 \pm 4)\ \text{MeV} \qquad \Gamma_\rho = (111 \pm 6)\ \text{MeV}\ .$$

The total cross section can be divided into three terms

$$\sigma_{tot} = \sigma_{tot}^{I=0} + \sigma_{tot}^{I=1} + \sigma_{interference}\ .$$

2. Let us first consider the 600 - 900 MeV region for the total energy $\sqrt{s}$. The dominant contribution is due to the ω meson and a total width

$$\Gamma_\omega = (11.9 \pm 1.5)\ \text{MeV}$$

has been used in the Breit-Wigner expression. The following analytic expression has been used:

$$\sigma_{tot}^{I=0} = \frac{2\pi^2\alpha^3}{3}\frac{1}{m_0^2}\left(1-\frac{m_0^2}{s}\right)^3\left|\frac{f_{\omega\pi^0\gamma}}{f_\omega}\right|^2\frac{m_\omega^4}{(m_\omega^2-s)^2+m_\omega^2\Gamma_\omega^2}$$

$$\sigma_{tot}^{I=1} = \frac{2\pi^2\alpha^3}{3}\frac{1}{m_0^2}\left(1-\frac{m_0^2}{s}\right)^3\frac{|G(0)|^2}{4}|F_\pi(s)|^2$$

$$\sigma_{interference} =$$

$$= \frac{2\pi^2\alpha^3}{3}\frac{1}{m_0^2}\left(1-\frac{m_0^2}{s}\right)^3 G(0)\ \frac{f_{\omega\pi^0\gamma}}{f_\omega}|F_\pi(s)|^2\frac{m_\omega^4}{(m_\omega^2-s)^2+m_\omega^2\Gamma_\omega^2}\ \times$$

$$\times\ \frac{\left[\left(1-\frac{s}{m_\rho^2}\right)\left(1-\frac{s}{m_\omega^2}\right)+\frac{\Gamma_\rho}{m_\rho}\frac{\Gamma_\omega}{m_\omega}\left(\frac{k}{k_\rho}\right)^3\frac{m_\rho}{\sqrt{s}}\right]}{1+d\,\frac{\Gamma_\rho}{m_\rho}}\ .$$

The constant d has been previously defined and has the numerical value $d \approx .48$.

3. The isoscalar part of the total cross section at $s=m_\omega^2$ is found to be

$$\sigma_{tot}^{I=0}(e^+e^- \Rightarrow \pi^0\gamma)_{s=m_\omega^2} \simeq (1.85 \pm 0.50) 10^{-31}\ \mathrm{cm}^2 .$$

The presence of the ρ meson can be detected only via the interference term. Because of the inequality $\Gamma_\rho >> \Gamma_\omega$ the interference term $\sigma_{interference}$ is always positive and acts as an additive contribution which is maximal at the ω-mass where the ratio $\sigma_{interf.}/\sigma_{total}$ is about 11%.

The value of the total cross section at the ω mass is expected, by this model, to be

$$\sigma_{tot}(e^+e^- \Rightarrow \pi^0\gamma)_{s=m_\omega^2} \simeq (2.09 \pm 0.60) 10^{-31}\ \mathrm{cm}^2 .$$

The value of the ratio $\sigma_{interf.}/\sigma_{tot}$ increases outside the ω peak. For instance

at 700 MeV: $\sigma_{interf.}/\sigma_{tot} \simeq 43\ \%$

at 850 MeV: $\sigma_{interf.}/\sigma_{tot} \simeq 37\ \%$

Unfortunately, at these energies the total cross sections are small and it will be very hard to obtain precise data. It is clear that the actual experimental situation is not accurate enough to allow a precise determination of the $\rho^0 \Rightarrow \pi^0\gamma$ partial width when we look at the interference term in the neighbourhood of the ω peak . Nevertheless, the existence of the interference term can be exhibited by the fact that the experimental cross section cannot be fitted by a simple Breit-Wigner-formula due to the ω term only. In this respect, precise measurements below and above the peak will be of the highest interest.

The curves of fig. 1 represent the total cross sections with and without the interference term.

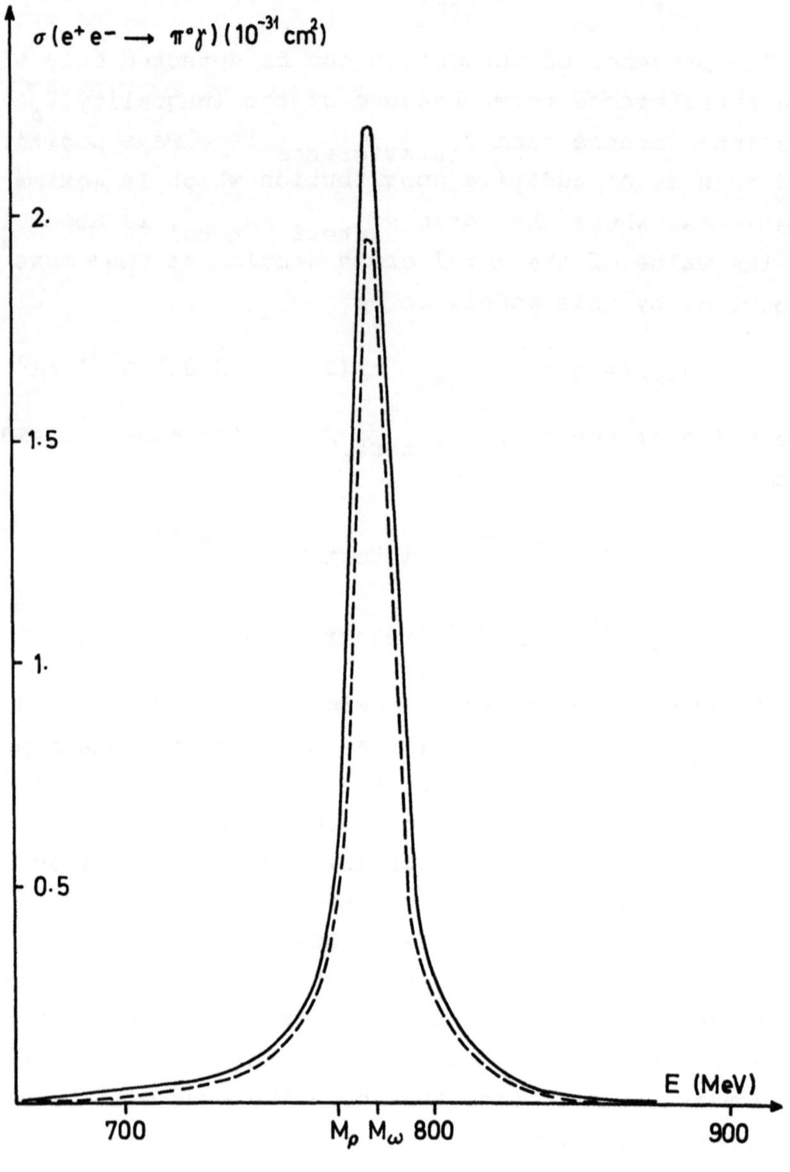

Figure 1

References - PART II

1. M. GELL-MANN and F. ZACHARIASEN, Phys. Rev. 124, 953 (1961).
2. Y. NAMBU and J. J. SAKURAI, Phys. Rev. Letters, 8, 79 (1962).
3. M. GELL-MANN, D. SHARP and W. A. WAGNER, Phys.Rev. Letters 8, 261 (1962).
4. N. KROLL, T. D. LEE and B. ZUMINO, Phys. Rev. 157, 1376 (1967).
5. G. GOUNARIS and J. J. SAKURAI, Phys. Rev. Letters 21, 244 (1968).
6. J. E. AUGUSTIN et al., Phys. Letters 28B, 508 (1969).
7. J. E. AUGUSTIN et al., Phys. Letters 28B, 513 (1969).
8. Particle Data Group.
9. J. E. AUGUSTIN et al., Phys. Letters 28B, 517 (1969).
10. E. CREMMER and M. GOURDIN, Nuclear Physics (to be published).
11. S. WEINBERG, Phys. Rev. Letters 18, 507 (1967).
12. H. SUGAWARA, Phys. Rev. Letters 21, 772 (1968).
13. T. DAS, V. MATHUR, and S. OKUBO, Phys. Rev. Letters 19, 470 (1967).
    R. J. OAKES and J. J. SAKURAI, Phys. Rev. Letters 19, 1266 (1967).
14. J. SCHWINGER, Phys. Rev. 75, 1912 (1949).
15. A. PETERMANN, Helv. Phys. Acta 30, 407 (1957); Phys. Rev. 105, 1931 (1957);
    C. M. SOMMERFIELD, Phys. Rev. 107, 328 (1957); Ann. Phys. (N.Y.) 5, 26 (1958);
    H. SUURA and E. H. WICHMANN, Phys. Rev. 105, 1930 (1957);
    H. M. Y. TERENT'EV, Zh.Eksper.Teor.Fiz. 43, 619 (1962); Soviet Phys. JETP 16, 444 (1963).
16. R. ERBE et al., Nuovo Cimento 48A, 262 (1967).

PART III

SECTION K

The ω-ϕ Mixing

I. MASS FORMULA

1. Let us consider the nine vector mesons ρ, K^*, ω, ϕ with the experimental masses

$$m_\rho = 770 \text{ MeV} \qquad m_{K^*} = 891.5 \text{ MeV}$$

$$m_\omega = 783.5 \text{ MeV} \qquad m_\phi = 1019.5 \text{ MeV}$$

In the SU(3) symmetry, we need an octuplet $\underline{8}$ and a singulet $\underline{1}$ to classify these mesons. The attribution is unambiguous for the ρ and K^* mesons

$$\rho \text{ and } K^* \in \underline{8}$$

But the two mesons ω and ϕ have the same quantum numbers

$$J = 1 \quad P = -1 \quad I = 0 \quad C = -1 \quad G = -1$$

and we cannot a priori decide which belongs to the octuplet and which to the singulet .

2. But the SU(3) symmetry is broken as exhibited by the previous mass spectrum and it is usually assumed that the breaking can be described by a term which transforms like the I=0, Y=0 component of an octuplet. One consequence is the famous Gell-Mann [1] - Okubo [2] mass formula which is written for the vector meson octuplet as

$$4m_{K^*}^2 = m_\rho^2 + 3m_8^2 \tag{K.1}$$

where m_8 is the mass of the I=0, Y=0 particle, ϕ_8 accompanying the ρ and K^* mesons in the octuplet. We also define as ω_1 the vector meson singulet of mass m_1.

From the relation (K.1) we obtain

$$m_8 = 928.5 \text{ MeV} \tag{K.2}$$

and the inequality

$$m_\omega < m_8 < m_\phi \quad .$$

3. Neither the ω meson nor the ϕ meson satisfy the relation (K.2) and we can imagine two explanations

a) The Gell-Mann-Okubo mass formula does not hold for vector mesons but in this case why is it so successful for the baryon octuplet ($J = \frac{1^+}{2}$) and the baryon decuplet ($J = \frac{3^+}{2}$).

b) The identification of the physical particles ω and ϕ with the pure SU(3) weights, ω_1 and ϕ_8 is not possible. Various authors [3] have then suggested that there exists a configuration mixing and the ω and ϕ mesons are only a superposition of the states ω_1 and ϕ_8 .

Of course in an exact SU(3) symmetry all the masses are equal in the octuplet and there is no mixing.

We now consider the two-dimensional vector-space of isoscalar vector mesons. A first orthogonal basis is written using the SU(3) weight

$$V_o = \begin{vmatrix} \phi_8 \\ \omega_1 \end{vmatrix} \tag{K.3}$$

The squared mass matrix M_o^2 is real and symmetrical because of time-reversal invariance

$$M_o^2 = \begin{vmatrix} m_8^2 & \delta m^2 \\ \delta m^2 & m_1^2 \end{vmatrix} \tag{K.4}$$

The existence of non-diagonal elements is due to the breaking of SU(3) symmetry and therefore to the possibility to have transition between the singulet ω_1 and the octuplet ϕ_8.

The physical particles ϕ and ω are associated to the eigenvectors of the squared mass matrix.

$$V = \begin{vmatrix} \phi \\ \omega \end{vmatrix} \tag{K.5}$$

In the physical basis, the squared mass matrix is diagonal with, as eigenvalues, the squared physical masses

$$M^2 = \begin{vmatrix} m_\phi^2 & 0 \\ 0 & m_\omega^2 \end{vmatrix} \tag{K.6}$$

We describe the relation between V_o and V by a linear transformation R

$$V = R\, V_o \tag{K.7}$$

The induced transformation on the squared mass matrix is simply

$$M^2 = R\, M_o^2\, R^{-1} \tag{K.8}$$

The masses being real quantities the only possible transformation R is an orthogonal one and equation (K.7) is written

$$\begin{aligned} \phi &= \cos\theta\ \phi_8 + \sin\theta\ \omega_1 \\ \omega &= -\sin\theta\ \phi_8 + \cos\theta\ \omega_1 \end{aligned} \tag{K.9}$$

or in the inverse form

$$\begin{aligned} \phi_8 &= \cos\theta\ \phi - \sin\theta\ \omega \\ \omega_1 &= \sin\theta\ \phi + \cos\theta\ \omega \end{aligned} \tag{K.10}$$

The angle θ is called the mixing angle.

From relation (K.8) we can deduce three equalities

$$m_8^2 = \cos^2\theta\ m_\phi^2 + \sin^2\theta\ m_\omega^2 \tag{K.11}$$

$$m_1^2 = \sin^2\theta\ m_\phi^2 + \cos^2\theta\ m_\omega^2 \tag{K.12}$$

$$\delta m^2 = \sin\theta \cos\theta \, (m_\phi^2 - m_\omega^2) \qquad \text{(K.13)}$$

We easily check the conservation of trace and determinant

$$\mathrm{Tr}\, M^2 = \mathrm{Tr}\, M_o^2 \qquad\qquad \det M^2 = \det M_o^2$$

The trace relation is used to compute the mass m_1

$$m_8^2 + m_1^2 = m_\phi^2 + m_\omega^2$$

It follows

$$m_1 = 889.5 \text{ MeV} \qquad \text{(K.14)}$$

Combining now the Gell-Mann-Okubo formula (K.1) with equation (K.11) we obtain the mixing angle θ in terms of physical masses

$$4m_{K^*}^2 - m_\rho^2 = 3[\cos^2\theta m_\phi^2 + \sin^2\theta m_\omega^2] \qquad \text{(K.15)}$$

Choosing now θ in the first quadrant, we find

$$\theta = 40.2^\circ \qquad \text{(K.16)}$$

Let us emphasize that the inequality $m_\omega < m_8 < m_\phi$ is crucial to obtain the real angle θ.

4. The squared mass m_V^2 is a solution of the equation Re $F(s) = 0$ where $F(s)$ is the coefficient of $g_{\mu\nu}$ in the inverse vector-meson propagator. On this basis it has been suggested [4] that a formula of the Gell-Mann-Okubo type can be used for the inverse square masses. Equation (K.15) becomes

$$4m_{K^*}^{-2} - m_\rho^{-2} = 3[\cos^2\theta_{\text{inv.}} m_\phi^{-2} + \sin^2\theta_{\text{inv.}} m_\omega^{-2}] \qquad \text{(K.17)}$$

and the mixing angle $\theta_{\text{inv.}}$ is numerically given by

$$\theta_{\text{inv.}} = 28.6^\circ \qquad \text{(K.18)}$$

5. In the quark model, the three vector mesons of hypercharge zero are generally defined as [5]

$$\rho^{o} = \frac{1}{\sqrt{2}}(q_1\bar{q}_1 - q_2\bar{q}_2)$$

$$\omega = \frac{1}{\sqrt{2}}(q_1\bar{q}_1 + q_2\bar{q}_2)$$

$$\phi = - q_3\bar{q}_3$$

where $q_1q_2q_3$ are the three quarks and $\bar{q}_1\bar{q}_2\bar{q}_3$ the three antiquarks as shown in fig. 1

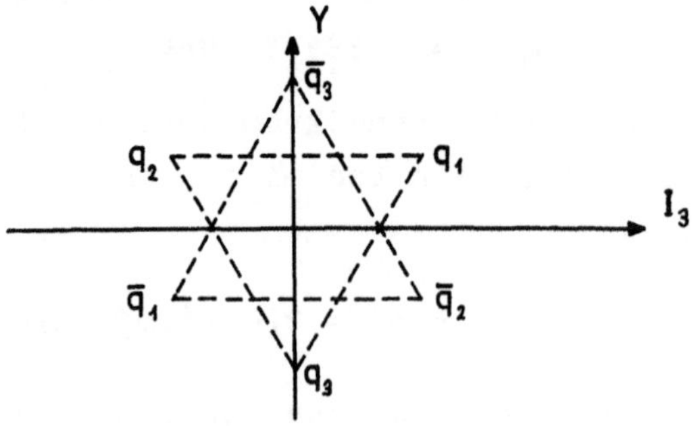

Fig. 1

The corresponding mixing angle is then defined by
$\cos\theta_q = \sqrt{2/3}$, $\sin\theta_q = \sqrt{1/3}$

$$\theta_q = 35.2^{o} \qquad (K.19)$$

II. MIXING MODELS

1. In the SU(3) theory the electromagnetic current is assumed to transform like a U spin scalar of an adjoint representation. As an immediate consequence the photon cannot be coupled to the ω_1 meson.

The baryonic current is a SU(3) invariant and therefore it is not coupled to the ϕ_8 meson.

These properties are the basis of the application of

the ω-ϕ mixing theory to transition matrix elements. Because of the large mass differences between the physical vector mesons a difficulty remains in the application of the mixing theory. In fact, it is not possible to formulate model-independent prescriptions and only experiments will allow a choice between various possibilities.

2. Model I

A first possibility is to write relations for the T-matrix elements even if they have dimensions. We then have

$$\begin{aligned} \langle 0|J^{em}|\phi\rangle &= \cos\theta \langle 0|J^{em}|\phi_8\rangle \\ \langle 0|J^{em}|\omega\rangle &= -\sin\theta \langle 0|J^{em}|\phi_8\rangle \\ \langle 0|J^{em}|\rho\rangle &= \sqrt{3}\ \langle 0|J^{em}|\phi_8\rangle \end{aligned} \qquad (K.20)$$

$$\begin{aligned} \langle 0|J^{B}|\phi\rangle &= \sin\theta \langle 0|J^{B}|\omega_1\rangle \\ \langle 0|J^{B}|\omega\rangle &= \cos\theta \langle 0|J^{B}|\omega_1\rangle \end{aligned} \qquad (K.21)$$

The vector index of the current densities and the polarization vector of the vector mesons have been forgotten in equations (K.20) and (K.21) as unessential.

The various coupling constants are defined as in Section F (equation F.2)

$$\langle 0|J^{em}|V\rangle = m_V^2/f_V \qquad V=\rho,\omega$$

$$\langle 0|J^{B}|V\rangle = m_V^2/g_V \qquad V=\omega,\phi$$

As auxiliary parameters we can introduce two angles θ_Y and θ_B by

$$\frac{f_\phi}{f_\omega} = -\tan\theta_Y \qquad \frac{g_\omega}{g_\phi} = \tan\theta_B$$

From equations (K.20) we obtain

$$\frac{m_\phi^2}{f_\phi} = \frac{1}{\sqrt{3}} \cos\theta \frac{m_\rho^2}{f_\rho}$$

$$\frac{m_\omega^2}{f_\omega} = -\frac{1}{\sqrt{3}} \sin\theta \frac{m_\rho^2}{f_\rho} \tag{K.22}$$

and in particular

$$\tan \theta_Y = \left(\frac{m_\phi}{m_\omega}\right)^2 \tan\theta \tag{K.23}$$

From equations (K.21) it follows

$$\tan \theta_B = (m_\omega/m_\phi)^2 \tan\theta \tag{K.24}$$

Let us recall the relation between the coupling constant f_V and the partial decay width of a vector meson into a lepton-antilepton pair (equation F.4)

$$\Gamma(V \Rightarrow e^+e^-) = \frac{4\pi\alpha^2}{3} \frac{m_V}{f_V^2} \tag{K.25}$$

Eliminating the θ angle between the two equations (K.22) and using the relation (K.25) we obtain a sum rule

$$m_\rho^3\Gamma(\rho \Rightarrow e^+e^-) - 3[m_\omega^3\Gamma(\omega \Rightarrow e^+e^-) + m_\phi^3\Gamma(\phi \Rightarrow e^+e^-)] = 0 \tag{K.26}$$

in disagreement with experiment.

The sum rule (K.26) is the second Weinberg sum rule [6] in the vector-meson pole approximation

$$\int_0^\infty [\rho^3(s) - \rho^8(s)]ds = 0$$

What happens now for the first Weinberg sum rule [6]? It is not satisfied and we obtain, from equations (K.22), the relation

$$C^3 = \left(m_\rho^2\left[\frac{\cos^2\theta}{m_\phi^2} + \frac{\sin^2\theta}{m_\omega^2}\right]\right)^{-1} C^8 \tag{K.27}$$

The first factor on the right-hand side of equation (K.22) can be interpreted as a correction to the first Weinberg sum rule because of the breaking of SU(3) symmetry.

Whereas the sum rule (K.26) is independent of the mixing angle θ, this is not the case for the sum rule (K.27). With $\theta = 40.2^{\circ}$ the correcting factor is 1.36 and again in disagreement with experiment.

3. Model II

Let us now adopt an opposite point of view and write the various relations on the dimensionless part of the current matrix elements, e.g. on the inverse coupling constants f_V^{-1}.

$$\frac{1}{f_\phi} = \frac{1}{\sqrt{3}} \cos\theta \frac{1}{f_\rho}$$
$$\frac{1}{f_\omega} = -\frac{1}{\sqrt{3}} \sin\theta \frac{1}{f_\rho} \qquad \text{(K.28)}$$

and the three angles θ, θ_Y and θ_B are equal. Such a model is often called a mass-mixing model.

Eliminating the θ angle between the two equations (K.28) and using the relation (K.25) we obtain a sum rule

$$\frac{\Gamma(\rho\to e^+e^-)}{m_\rho} - 3\left[\frac{\Gamma(\omega\to e^+e^-)}{m_\omega} + \frac{\Gamma(\phi\to e^+e^-)}{m_\phi}\right] = 0 \qquad \text{(K.29)}$$

in good agreement with experiment. A numerical estimate of (K.29) gives, if one uses the Orsay data,

$$(1.15 \pm 1.80) = 0 \quad .$$

The sum rule (K.29) can be derived from a spectral function sum rule

$$\int_0^\infty \frac{\rho^3(s) - \rho^8(s)}{s^2}\, ds = 0$$

which expresses the SU(3) symmetry of hadronic isovector and isoscalar contributions to the charge renormalization.

It is straightforward, from equation (K.28), to write the relation the relation between the two Schwinger terms C^3 and C^8. Using, in addition, the relation (K.15) for the θ mixing angle, we obtain the Sugawara sum rule [7]

$$C^3 - \frac{3m_\rho^2}{4m_{K^*}^2 - m_\rho^2} C^8 \qquad \text{(K.30)}$$

in good agreement with experiment.

4. Model III

We use an intermediate point of view between Model I and Model II and we write the mixing relations in the form

$$\frac{m_\phi}{f_\phi} = \frac{1}{\sqrt{3}} \cos\theta \frac{m_\rho}{f_\rho}$$
$$\frac{m_\omega}{f_\omega} = -\frac{1}{\sqrt{3}} \sin\theta \frac{m_\rho}{f_\rho} \qquad \text{(K.31)}$$

The angle θ_Y is here related to θ by

$$\tan\theta_Y = \frac{m_\phi}{m_\omega} \tan\theta$$

and for the angle θ_B we have the expression

$$\tan\theta_B = \frac{m_\omega}{m_\phi} \tan\theta \quad .$$

This model is the usual version of the current mixing model. The elimination of the mixing angle θ leads to the Das-Mathur-Okubo sum rule [8]

$$m_\rho \Gamma(\rho \Rightarrow e^+e^-) - 3[m_\omega \Gamma(\omega \Rightarrow e^+e^-) + m_\phi \Gamma(\phi \Rightarrow e^+e^-)] = 0 \qquad \text{(K.32)}$$

discussed in Section G and which agrees reasonably well

with experiment.

Two versions of the Model III have been discussed in the literature

a) Model III_a by Das, Mathur and Okubo [8]. They use the mixing angle (K.16) $\theta = 40.2^o$.

b) Model III_b by Oakes and Sakurai [9]. They use the mixing angle (K.18) $\theta_{inv.} = 28.6^o$.

5. The first comparison of the predictions of these models with the Orsay experiments is made by computing the partial decay widths $\Gamma(\omega \Rightarrow e^+e^-)$ and $\Gamma(\phi \Rightarrow e^+e^-)$ with, as an input, the experimental value of $\Gamma(\rho \Rightarrow e^+e^-)$

$$\Gamma(\rho \Rightarrow e^+e^-)_{exp} = (7.36 \pm 0.7) \text{ keV}$$

Reaction	Model I	Model II	Model III_a	Model III_b	Experiment
$\Gamma(\omega \Rightarrow e^+e^-)$ keV	.97±.10	1.04±.10	1.00±.10	0.55±.06	.94±.18
$\Gamma(\phi \Rightarrow e^+e^-)$ keV	.62±.06	1.89±.19	1.07±.11	1.43±.14	1.64±.24

Table I

Another way to compare the theoretical predictions with experiment is to look at the angle θ_Y previously defined by

$$\frac{f_\phi}{f_\omega} = - \tan \theta_Y$$

Using equation (K.25) we obtain

$$\tan^2 \theta_Y = \frac{m_\phi}{m_\omega} \frac{\Gamma(\omega \Rightarrow e^+e^-)}{\Gamma(\phi \Rightarrow e^+e^-)}$$

The results are given in Table II

	Model I	Model II	Model III_a	Model III_b	Experiment
$\tan^2\theta_Y$	2.03	0.71	1.21	0.50	0.74±0.16

Table II

If the experimental data are correct we can obtain the following conclusions

a) Model I is excluded by experiment

b) Model III_b and all other models with a mixing angle $\theta_{inv.}$ seem to be excluded.

c) Model II agrees with the experimental data within the errors and is favourable especially as far as the angle θ_Y is concerned.

d) Model III_a cannot be excluded.

SECTION L

Mesonic Form Factors

I. THE π-MESON ELECTROMAGNETIC FORM FACTOR

1. In the vector-meson dominance model the π-meson electromagnetic form factor is represented by the diagram of fig. 1

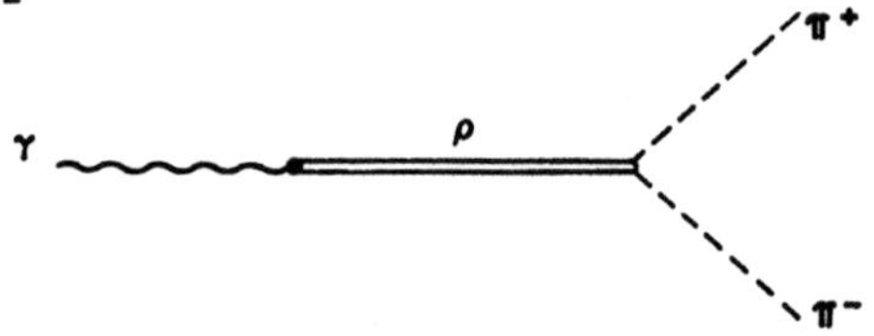

Fig. 1

A straightforward calculation gives

$$F_\pi(s) \approx \frac{m_\rho^2}{f_\rho} \frac{1}{m_\rho^2 - s - i m_\rho \Gamma_\rho(s)} f_{\rho\pi\pi} \qquad (L.1)$$

where $\Gamma_\rho(s)$ is an energy-dependent width

$$\Gamma_\rho(m_\rho^2) = \Gamma_\rho$$

$$\Gamma_\rho(s) \text{ is real for } s > 4m_\pi^2$$

$$i\Gamma_\rho(s) \text{ is real for } s < 4m_\pi^2 \quad .$$

2. The normalization condition at s=0 is $F_\pi(0) = 1$. In the narrow-width approximation we deduce the famous relation [10], [11]

$$\frac{f_{\rho\pi\pi}}{f_\rho} = 1 \quad . \tag{L.2}$$

3. We can try to do something more elaborate comparing the expression (L.1) with a π-meson electromagnetic form factor à la Sakurai. In this model, the value of F_π at the π-meson peak is (see Section B)

$$F_\pi(m_\rho^2) = i\,\frac{m_\rho}{\Gamma_\rho}\,(1+ d\,\frac{\Gamma_\rho}{m_\rho}) \tag{L.3}$$

From equation (L.1) we have

$$F_\pi(m_\rho^2) = i\,\frac{m_\rho}{\Gamma_\rho}\,\frac{f_{\rho\pi\pi}}{f_\rho} \tag{L.4}$$

Identifying the results (L.3) and (L.4) we obtain a non-zero width correction to equation (L.2)

$$\frac{f_{\rho\pi\pi}}{f_\rho} = 1 + d\,\frac{\Gamma_\rho}{m_\rho} \tag{L.5}$$

Using the experimental values of the Orsay experiments

$m_\rho = (770 \pm 4)$ MeV $\quad \Gamma_\rho = (111 \pm 6)$ MeV $\quad d = 0.48$

we predict

$$\frac{f_{\rho\pi\pi}}{f_\rho} = 1.069 \pm 0.004$$

or for the squared coupling constants

$$\frac{f^2_{\rho\pi\pi}}{f^2_{\rho}} = 1.143 \pm 0.008 \tag{L.6}$$

4. The coupling constant $f_{\rho\pi\pi}$ is related to the $\rho\Rightarrow\pi\pi$ width by the formula

$$\Gamma(\rho\Rightarrow \pi\pi) = \frac{m_\rho}{12}\left(1-\frac{4m_\pi^2}{m_\rho^2}\right)^{\frac{3}{2}} \frac{f^2_{\rho\pi\pi}}{4\pi}$$

Using the experimental value $\Gamma(\rho\Rightarrow \pi\pi) = (111 \pm 6)$ MeV we obtain

$$\frac{f^2_{\rho\pi\pi}}{4\pi} = 2.13 \pm 0.13 \tag{L.7}$$

The coupling constant f_ρ is related to the $\rho\Rightarrow e^+e^-$ partial width by the formula

$$\Gamma(\rho\Rightarrow e^+e^-) = \frac{\alpha^2}{3} m_\rho \frac{4\pi}{f^2_\rho}$$

From the Orsay value $\Gamma(\rho\Rightarrow e^+e^-) = (7.36 \pm 0.7)$ keV we deduce

$$\frac{f^2_{\rho}}{4\pi} = 1.86 \pm 0.18 \tag{L.8}$$

Combining now the results (L.7) and (L.8) the ratio of the squared coupling constants is found to be

$$\frac{f^2_{\rho\pi\pi}}{f^2_{\rho}} = 1.145 \pm 0.12 \tag{L.9}$$

The result is in excellent agreement with the prediction (L.6) and compatible with the value 1 as given by the naive model (L.2)

II. THE K-MESON ELECTROMAGNETIC FORM FACTORS

1. We have two electromagnetic form factors $F_{K^+}(s)$ and

$F_{K^0}(s)$ normalized to the electric charge of the K mesons

$$F_{K^+}(0) = 1 \qquad F_{K^0}(0) = 0$$

In the isotopic spin formalism

$$\begin{aligned} F_{K^+}(s) &= F_K^{I=0}(s) + F_K^{I=1}(s) \\ F_{K^0}(s) &= F_K^{I=0}(s) - F_K^{I=1}(s) \end{aligned} \tag{L.10}$$

with the normalization condition

$$F_K^{I=0}(0) = \frac{1}{2} \qquad F_K^{I=1}(0) = \frac{1}{2} \tag{L.11}$$

2. We use the vector-meson dominance model to construct the isoscalar and isovector K-meson form factors (fig.2 and fig. 3)

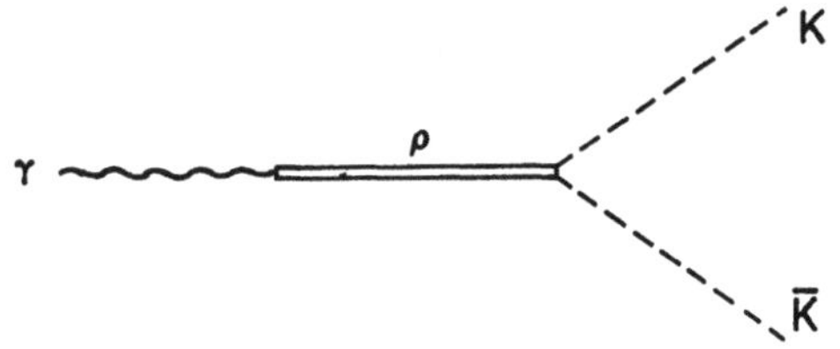

Fig. 2

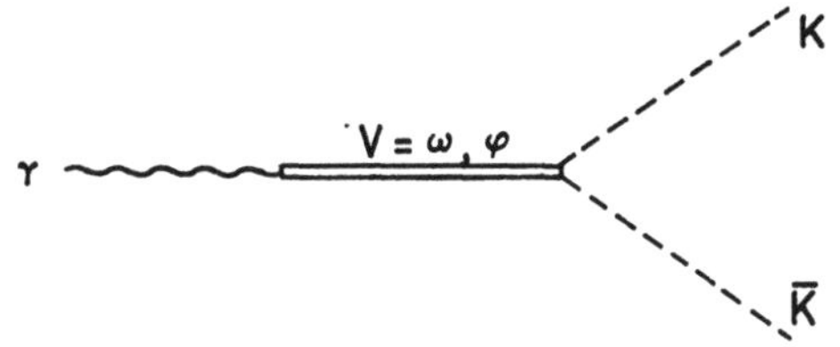

Fig. 3

The model gives

$$F_K^{I=1}(s) = \frac{m_\rho^2}{f_\rho} \frac{1}{m_\rho^2 - s - i m_\rho \Gamma_\rho(s)} f_{\rho K\bar{K}} \tag{L.12}$$

$$F_K^{I=0}(s) = \frac{m_\omega^2}{f_\omega} \frac{1}{m_\omega^2 - s - i\, m_\omega \Gamma_\omega(s)} f_{\omega K\bar{K}} + \frac{m_\phi^2}{f_\phi} \frac{1}{m_\phi^2 - s - i\, m_\phi \Gamma_\phi(s)} \times$$

$$\times\ f_{\phi K\bar{K}} \qquad \text{(L.12)}$$

Comparing with equation (L.1) we obtain

$$F_K^{I=1}(s) \simeq \frac{f_{\rho K\bar{K}}}{f_{\rho\pi\pi}} F_\pi(s) \qquad \text{(L.14)}$$

3. The normalization conditions (L.11) give, in the narrow-width approximation, the following relations [12]

$$\frac{f_{\rho K\bar{K}}}{f_\rho} = \frac{1}{2} \quad . \qquad \text{(L.15)}$$

$$\frac{f_{\omega K\bar{K}}}{f_\omega} + \frac{f_{\phi K\bar{K}}}{f_\phi} = \frac{1}{2} \qquad \text{(L.16)}$$

Let us remark that a correct normalization of equation (L.14) at s=0 implies the SU(3) relation

$$f_{\rho K\bar{K}} = \frac{1}{2} f_{\rho\pi\pi}$$

which combined with the equality (L.2) gives the condition (L.15)

III. THE K-MESON BARYONIC FORM FACTORS

1. The baryonic current is an isoscalar and we have only one K-meson baryonic form factor

$$G_{K^+}(s) = G_{K^0}(s) = G_K(s) \quad .$$

The K-meson baryonic charge is zero and we deduce the normalization condition at s=0

$$G_K(0) = 0 \quad . \qquad \text{(L.17)}$$

2. Using again a vector dominance model, we obtain the following form for $G_K(s)$

$$G_K(s) = \frac{m_\omega^2}{g_\omega} \frac{1}{m_\omega^2 - s - i\, m_\omega \Gamma_\omega(s)} f_{\omega K\bar{K}} +$$

$$+ \frac{m_\phi^2}{g_\phi} \frac{1}{m_\phi^2 - s - i\, m_\phi \Gamma_\phi(s)} f_{\phi K\bar{K}} \qquad \text{(L.18)}$$

where g_ω and g_ϕ, defined in Section K, describe the coupling of the baryonic current with the isoscalar vector mesons ω and ϕ.

3. The normalization condition at s=0 (L.18) gives the relation

$$\frac{f_{\omega K\bar{K}}}{g_\omega} + \frac{f_{\phi K\bar{K}}}{g_\phi} = 0 \qquad \text{(L.19)}$$

4. Equations (L.16) and (L.19) are used to calculate the strong coupling constant $f_{\phi K\bar{K}}$ in terms of f_ω, f_ϕ, g_ω, and g_ϕ

$$f_{\phi K\bar{K}} = \frac{1}{2} f_\phi \Big/ \left(1 - \frac{f_\phi}{f_\omega} \frac{g_\omega}{g_\phi}\right) \qquad \text{(L.20)}$$

Considering successively the three models for the ω-ϕ mixing discussed in Section K we obtain

$$\text{Model I} \quad f_{\phi K\bar{K}} = \frac{\sqrt{3}}{2} \cos\theta \, f_\rho (m_\phi/m_\rho)^2 \qquad \text{(L.21)}$$

$$\text{Model II} \quad f_{\phi K\bar{K}} = \frac{\sqrt{3}}{2} \cos\theta \, f_\rho \qquad \text{(L.22)}$$

$$\text{Model III} \quad f_{\phi K\bar{K}} = \frac{\sqrt{3}}{2} \cos\theta \, f_\rho (m_\phi/m_\rho) \qquad \text{(L.23)}$$

The comparison with experiment will be made in the next section.

SECTION M

Vector-Meson Decay in Pseudoscalar Mesons

I. GENERALITIES

1. We restrict ourselves to the decay of the vector mesons ρ, K^* and ϕ into a $\bar{B}B$ system where B is a pseudoscalar meson.

The coupling constant $f_{V\bar{B}B}$ is related to the decay width $\Gamma(V \rightarrow \bar{B}B)$ by the relation [3]

$$\Gamma(V \rightarrow \bar{B}B) = \frac{m_V}{12} \left(1-4m_B^2/m_V^2\right)^{\frac{3}{2}} \frac{f^2_{V\bar{B}B}}{4\pi} \qquad \text{(M.1)}$$

2. The phase space factors are calculated from the masses. Taking the $\rho \rightarrow 2\pi$ decay as a reference we obtain the following table

Vector Meson	Final State	Phase Space Ratio
ρ	$\pi\pi$	1
K^+	$K\pi$	0.385
ϕ	K^+K^-	0.0247
	$K^o\bar{K}^o$	0.0158

Table 1

3. The SU(3) symmetry predictions are well known. The vector mesons belong to an octuplet and a singulet. The pseudoscalar mesons belong to an octuplet.

We have then to construct invariant SU(3) $V\bar{B}B$ couplings. From the generalized Pauli principle the $\bar{B}B$ system must be skew symmetric in the exchange of the SU(3) indices. It follows that

a) we have only one reduced amplitude

b) the decay $\omega_1 \Rightarrow \bar{B}B$ is forbidden.

The Clebsch-Gordan coefficients are given as follows

$$f^2_{K^*K\pi} = f^2_{\phi_8 K^+K^-} = f^2_{\phi_8 K^0\bar{K}^0} = \frac{3}{4} f^2_{\rho\pi\pi} \qquad (M.2)$$

II. THE K^* PROBLEM

1. The prediction of the exact unitary symmetry is

$$\frac{\Gamma(K^* \Rightarrow K\pi)}{\Gamma(\rho \Rightarrow \pi\pi)} = 0.385 \times \frac{3}{4} = 0.289 \qquad (M.3)$$

Using the Orsay result

$$\Gamma(\rho \Rightarrow \pi\pi) = (111 \pm 6)\ \text{MeV} \qquad (M.4)$$

we find the SU(3) prediction to be

$$\Gamma(K^* \Rightarrow K\pi) = (32 \pm 1.9)\ \text{MeV} \qquad (M.5)$$

2. The world average values for the ρ and K^* widths as quoted by the particle data group [13] are

$$\Gamma(\rho^0 \Rightarrow \pi^+\pi^-) = (140 \pm 5)\ \text{MeV} \qquad (M.6)$$

$$\Gamma(K^* \Rightarrow K\pi) = (49 \pm 1)\ \text{MeV} \qquad (M.7)$$

The value (M.7) of the K^* width disagrees with the SU(3) calculation (M.5). If one uses the world average value (M.6) instead of the Orsay result (M.4) one finds

$$\Gamma(K^* \Rightarrow K\pi) = (38 \pm 1.5)\ \text{MeV} \qquad (M.8)$$

also in disagreement with experiment.

3. The determination of the ρ-meson width is a delicate problem and the large difference between the numbers (M.4) and (M.6) illustrates this well-known fact.

Nevertheless it seems that colliding-beam experiments constitute one of the cleanest ways to measure the ρ-meson parameters. It must be observed that the Novosibirsk and Orsay experiments have given results for the ρ-meson width below the value (M.5).

A possible question is now: how can we believe the K^* width (M.7) ? We will assume in what follows that this is the case and we will use the results (M.4) and (M.7) as the correct ones.

4. The disagreement between equations (M.5) and (M.7) shows a breaking of the SU(3) symmetry in the vector-meson decay and we now try to take into account SU(3) non-invariant amplitudes in a phenomenological way.

We assume, as usual, the symmetry breaking amplitude to transform like the I=0, Y=0 weight of an octuplet. For the $V\bar{B}B$ transitions we have three possibilities. If, in addition, we assume a universal vector-meson coupling as proposed by Sakurai [14] two of the three amplitudes vanish and we finally obtain a one-parameter solution for the coupling constants [15]

$$f^2_{\rho\pi\pi} = f_o^2(1+\alpha)^2 \tag{M.9}$$

$$f^2_{K^*K\pi} = \frac{3}{4} f_o^2\left(1-\frac{\alpha}{2}\right)^2 \tag{M.10}$$

$$f^2_{\phi_8 K^+K^-} = f^2_{\phi_8 K^o\bar{K}^o} = \frac{3}{4} f_o^2(1-\alpha)^2 \tag{M.11}$$

From the experimental ρ meson and K^* meson widths we deduce

$$\alpha = -0.138 \pm 0.017 \tag{M.12}$$

which is a reasonable order of magnitude for a first-order violation of SU(3) symmetry.

III. THE ϕ PROBLEM

1. The new question is: how to apply the ω-ϕ mixing? The most naive way is certainly to write

$$f_{\phi K\bar{K}} = \cos\theta \; f_{\phi_8 K\bar{K}} \qquad (M.13)$$

The prediction of the exact unitary symmetry is

$$\frac{\Gamma(\phi \to K\bar{K})}{\Gamma(\rho \to \pi\pi)} = (0.0405)\left(\frac{3}{4}\cos^2\theta\right) = 0.0177 \; .$$

Using the value (M.4) for the ρ-meson width, we obtain

$$\Gamma(\phi \to K\bar{K}) = (1.96 \pm 0.12) \text{ MeV} \qquad (M.14)$$

to compare with the Orsay value

$$\Gamma(\phi \to K\bar{K}) = (3.39 \pm 0.73) \text{ MeV} \qquad (M.15)$$

Again, the exact SU(3) calculation, with the ω-ϕ mixing (M.13), is in disagreement with experiment.

The prediction of the first order broken symmetry is obtained using equation (M.11) and the value of the breaking parameter α as deduced from the K^* width (equation (M.12)). The result is

$$\Gamma(\phi \to K\bar{K}) = (3.41 \pm 0.14) \text{ MeV} \qquad (M.16)$$

in very good agreement with experiment.

2. A more sophisticated way to apply the ω-ϕ mixing is to relate the coupling constant $f_{\phi K\bar{K}}$ to f_ρ assuming a pole model approximation for the electromagnetic and baryonic K-meson form factors. Such a connexion, as shown in Section L, is model dependent and for the three types

of the ω-ϕ mixing models described in Section L, the results are the following

Model I $\quad f_{\phi K\bar{K}} = \frac{\sqrt{3}}{2} \cos\theta \; f_\rho (m_\phi/m_\rho)^2$

Model II $\quad f_{\phi K\bar{K}} = \frac{\sqrt{3}}{2} \cos\theta \; f_\rho$

Model III $\quad f_{\phi K\bar{K}} = \frac{\sqrt{3}}{2} \cos\theta \; f_\rho (m_\phi/m_\rho)$

The corresponding predictions are

Model I $\quad \Gamma(\phi \Rightarrow K\bar{K}) = (5.26 \pm 0.32)$ MeV

Model II $\quad \Gamma(\phi \Rightarrow K\bar{K}) = (1.73 \pm 0.10)$ MeV

Model III_a $\quad \Gamma(\phi \Rightarrow K\bar{K}) = (3.00 \pm 0.18)$ MeV [7]

Model III_b $\quad \Gamma(\phi \Rightarrow K\bar{K}) = (3.96 \pm 0.22)$ MeV [9]

3. In Prof. Jacob's lectures the relations

$$f^2_{K^* K\pi} = \frac{3}{4} f_{\rho\pi\pi} \, f_{\rho K\bar{K}}$$

and

$$\frac{1}{2} f^2_{\phi K\bar{K}} = f^2_{\rho K\bar{K}}$$

have been derived from the Veneziano model. Combining these relations with

$$f^2_{\rho\pi\pi}/4\pi = 2.13 \pm 0.13$$

and

$$f^2_{\phi K\bar{K}}/4\pi = 1.60 \pm 0.37$$

(obtained from $\Gamma(\rho \Rightarrow 2\pi) = (111 \pm 6$ MeV) and $\Gamma(\rho \Rightarrow K\bar{K}) =$ $= (3.4 \pm 0.8$ MeV) we get

$$\Gamma(K^* \Rightarrow K\pi) \approx 40 \text{ MeV !}$$

References - PART III

1. M. GELL-MANN, Phys. Rev. 125, 1067 (1962).
2. S. OKUBO, Prog. Theor. Phys. 27, 949 (1962); 28, 24 (1962).
3. J. J. SAKURAI, Phys. Rev. Letters 9, 472 (1962).
4. S. COLEMAN and H. J. SCHNITZER, Phys. Rev. 134, B863 (1964).
5. S. OKUBO, Phys. Letters 5, 165 (1963).
6. S. WEINBERG, Phys. Rev. Letters 18, 507 (1967).
7. H. SUGAWARA, Phys. Rev. Letters 21, 772 (1968).
8. T. DAS, V. S. MATHUR and S. OKUBO, Phys. Rev. Letters 18, 761 (1967).
9. R. J. OAKES and J. J. SAKURAI, Phys. Rev. Letters 19, 1266 (1967).
10. M. GELL-MANN and F. ZACHARIASEN, Phys. Rev. 124, 953 (1961).
11. M. GELL-MANN, D. SHARP and W. G. WAGNER, Phys. Rev. Letters 8, 261 (1962).
12. T. DAS, V. S. MATHUR and S. OKUBO, Phys. Rev. Letters 19, 470 (1967).
    See also current algebra results
    V. S. MATHUR, L. K. PANDIT and R. E. MARSHAK, Phys. Rev. Letters 16, 947 (1966). W. W. WADA, Phys. Rev. Letters 16, 956 (1960); M. P. KHANNA and A. VAIDYA, Nuovo Cim. 49A, 341 (1967).P.DIVAKARAN et al.,Phys.Rev.Lett.19,535 (1967)
13. Particle Data Group (Jan. 1969).
14. J. J. SAKURAI, Ann. of Phys. 11, 1 (1960).
15. M. GOURDIN, Unitary Symmetries (North-Holland Publ. Company 1967).

EXPERIMENTAL TESTS OF CPT, CP, C AND T INVARIANCES[†][*]

By

G. CONFORTO

The Enrico Fermi Institute and Department of Physics,
The University of Chicago, Chicago,
Illinois, 60637, USA

It is not desirable to cultivate a respect for the law, so much as for the right.
- Thoreau, Civil Disobedience

1. INTRODUCTION

To say that a law of nature is "invariant" under some operation corresponds to the impossibility of obtaining some kind of information. Until about ten years ago, three symmetries were thought to hold in nature:

i) invariance under P, which would make it impossible to objectively define the concept of right or left-handness;

ii) invariance under C, which would make it impossible to objectively define the concept of particle or antiparticle;

iii) invariance under T, which would make it impossible to objectively define the concept of positive or

[†] Lecture given at the VIII. Internationale Universitätswochen für Kernphysik,Schladming, Feb. 24 - March 8, 1969.
[*]Work supported in part by NSF Grant GU-2175.

negative time.

It is very easy to construct models that violate P, C, or T separately: it is enough to make a certain real constant complex. It can be shown, however, that it is impossible, within the framework of quantum field theory (which is more or less the only framework we have to treat relativistic particles) to construct Hamiltonians which violate the product CPT [1].

The history of overthrown of parity is known. It started with the so-called τ-θ puzzle that led Lee and Yang to suggest that parity might be violated in all weak interactions [2]. The first experiment which conclusively showed that parity was violated was performed by C. S. Wu and collaborators [3]. Within a very short time it became evident that parity violating effects were widespread and generally large in all branches of weak interaction physics. Subsequent experiments led to the two-component neutrino theory, which implies a breakdown of charge conjugation invariance.

These facts could,however, be included in the theoretical apparatus in a rather natural way, by replacing the violated P and C invariances with one combined symmetry, corresponding to the conservation of the product CP. This new symmetry would make it impossible for an observer to decide whether he is looking at particles in the real world or at anti-particles in a mirror.

If CP holds, then

$$CP|K^{o}\rangle = - |\bar{K}^{o}\rangle$$

$$CP|\bar{K}^{o}\rangle = - |K^{o}\rangle$$

and if one defines

$$|K_1\rangle = \frac{|K^{o}\rangle - |\bar{K}^{o}\rangle}{\sqrt{2}}$$

$$|K_2\rangle = \frac{|K^{o}\rangle - |\bar{K}^{o}\rangle}{\sqrt{2}}$$

one has $$CP|K_1\rangle = +|K_1\rangle$$

$$CP|K_2\rangle = -|K_2\rangle \quad .$$

The K_1 (CP = +1) decays mainly into $\pi^+\pi^-$ and $\pi^o\pi^o$, both of which have CP = +1 since $P=(-1)^\ell$ and $C=(-1)^\ell$ for the $\pi^+\pi^-$ system (ℓ is the angular momentum of the π's) and P=+1 and C=+1 for the $\pi^o\pi^o$ system.

The K_2 (CP=-1) decays mainly into 3π's and leptonic modes and it is clear that its decay into $\pi^+\pi^-$ or $\pi^o\pi^o$ is strictly forbidden by CP conservation.

The decay

$$K_2 \rightarrow \pi^+ + \pi^-$$

however, was observed by Christenson et al. [4], to occur with a branching ratio of $\sim$ 0.2 % .

Can one then conclude that CP is violated? Not quite. Some peculiar ad hoc models which have been proposed, as yet not in disagreement with experiments, actually offer explanations of the experimental observations which save CP invariance. Although CP non-invariance is, therefore, not an inevitable consequence, it is generally accepted as the most reasonable interpretation of the present data [5].

Under CPT invariance, if CP does not hold, T invariance must also be broken.

On the basis of the history of parity violation, which showed that in physics everything that is not forbidden is allowed, the program, some three or four years ago, was then very clear. On the one hand, one must look for further evidence bearing on CP outside the system of neutral K^o and tests of time reversal invariance were very much in order. On the other hand, one had to completely delineate the CP violation observed in K_2 decays.

In the following sections a rather arbitrary and personal selection of some of the published results of this program will be briefly presented.

2. CPT

As mentioned in Section 1, a violation of CPT would be a very serious blow at our most deeply held ideas. Therefore, experimental tests of this invariance are of great importance, particularly now that violations of P, C, and CP have been established.

The most important consequences of CPT invariance that can be tested experimentally are that for a particle and its anti-particle:

i) masses are equal;
ii) lifetimes are equal;
iii) magnetic and electric moments are opposite;
iv) interactions with gravitational field are equal;
v) to the extent that final state interactions are negligible, spectra and angular distributions of decay products are equal and polarizations are opposite [6].

Equality of partial decay rates for a particle and its antiparticle follows from CPT only if final state interactions can be neglected [6], [7], [8]: it is,however, a direct consequence of CP invariance.

No violation of CPT invariance has been observed as yet. The accuracies within which the above predictions of CPT have been tested are typically the following [9]:

i) masses: $\sim 10^{-17}$ for K^o-$\bar{K}^o$ [10], $\sim 10^{-5}$ for $e^\pm$, $\sim 10^{-4}$ for $\mu^\pm$, $\sim 10^{-3}$ for $K^\pm$, $\sim 10^{-2}$ for $\pi^\pm$;
ii) lifetimes: 10^{-3} for $K^\pm$, $\mu^\pm$, $\pi^\pm$;
iii) magnetic moments: 10^{-6} for $\mu^\pm$, 10^{-4}-10^{-5} for $e^\pm$;

iv) spectra and polarizations: 10^{-1}-10^{-2} for $\mu^{\pm}$ and $\pi^{\pm}$.

A unique possibility of testing CPT invariance is offered by the K^o-$\bar{K}^o$ system, where the so-called Bell-Steinberger unitarity relation [10] could be compared with experiments [11]. The present experimental informations, however, are not yet complete and safe enough to allow this test.

3. CP

a) Non-K^o systems

All tests performed in this field have given negative results. The most accurate one seems to be the comparison of the decay rates $\Gamma(K^+\to\pi^+\pi^+\pi^-)$ and $\Gamma(K^-\to\pi^-\pi^-\pi^+)$, see Section 2, which turn out to be equal within about one tenth of one percent [9].

b) K^o system

i) Non-leptonic decays

Since the original discovery of the decay $K_2\to\pi^+\pi^-$, much progress has been made in this field. More information on the decay $K_2\to\pi^+\pi^-$ has been obtained and the CP violating decay $K_2\to\pi^o\pi^o$ has also been established. The decays in question are generally described in terms of the following notations:

One defines

$$|K_L\rangle = p|K^o\rangle - q|\bar{K}^o\rangle$$
$$|K_S\rangle = p|K^o\rangle + q|\bar{K}^o\rangle$$

where

$$|p|^2 + |q|^2 = 1$$

and

$$\varepsilon = \frac{p-q}{p+q} .$$

With the definition

$$\varepsilon' = \frac{i}{\sqrt{2}} \frac{\text{Im } A_2}{A_o} \exp\{i(\delta_2-\delta_o)\}$$

where A_2 and A_o are the amplitude for K^o to go to two pions in the I=2 and I=0 states, respectively, A_o is taken to be real and δ_2 and δ_o are the $\pi\pi$ scattering phase shifts in the I=2 and I=0 states, respectively, one has

$$\eta_{+-} = \frac{\text{Amplitude } (K_L \to \pi^+\pi^-)}{\text{Amplitude } (K_S \to \pi^+\pi^-)} = \varepsilon + \varepsilon'$$

$$\eta_{oo} = \frac{\text{Amplitude } (K_L \to \pi^o\pi^o)}{\text{Amplitude } (K_S \to \pi^o\pi^o)} = \varepsilon - 2\varepsilon'$$

η_{+-} and η_{oo} are complex quantities, and the experimental program consists, therefore, in the determination of four real numbers: the two moduli and the two phases. The experimental situation can perhaps be summarized as follows (for details see ref. 12) :

- accurate and consistent measurements of $|\eta_{+-}|$ have been obtained. The currently accepted value of $|\eta_{+-}'|$ is $(1.90 \pm 0.05) \times 10^{-3}$.
- measurements of ϕ_{+-} and $|\eta_{oo}|$ have been performed, but the experimental situation is still far from being clear. The results of some experiments seem to keep changing in time and many discrepancies are present in the data. The existence of the decay $K_L \to \pi^o\pi^o$ is, however, beyond doubt.
- no measurement of ϕ_{oo} has been obtained, as yet. Some experiments to measure this quantity are in progress.

Reliable data on the quantities η_{+-} and η_{oo} would allow an unambiguous test of T invariance [10],[11],[13]. As it will be clarified later, there is so far no direct proof that T invariance is violated and the widely accepted opinion that this must happen is based on firm belief in CPT invariance.

ii) Leptonic decays

The only other positive evidence for CP violation obtained so far comes from the study of the decays

$$K_L \to \pi^{\mp} + \ell^{\pm} + \nu$$

where ℓ stands for either the μ meson or the electron. The observations were of an unequal number of positive and negative leptons resulting from these decay processes, which is a clear lack of charge symmetry. More specifically, a Berkeley-SLAC collaboration [14] found

$$\frac{N(\mu^+)}{N(\mu^-)} - 1 = (8.1 \pm 2.7) \times 10^{-3}$$

and a group from Columbia University [15] reported the result

$$\frac{N(e^+)}{N(e^-)} - 1 = (4.48 \pm 0.72) \times 10^{-3} \quad .$$

These two results are more or less consistent and their weighted average gives

$$\frac{N(\ell^+)}{N(\ell^-)} - 1 = (4.7 \pm 0.7) \times 10^{-3} \quad .$$

Under the assumption of the validity of CPT invariance and of the $\Delta S = \Delta Q$ rule, which allows the decays $K^o \to \pi^-\ell^+\nu$ and $\bar{K}^o \to \pi^+\ell^-\nu$ but forbids the decays $K^o \to \pi^+\ell^-\nu$ and $\bar{K}^o \to \pi^-\ell^+\nu$, it is easy to show that

$$\frac{N(\ell^+)}{N(\ell^-)} = 1 + 4 \operatorname{Re} \varepsilon \ .$$

From experiments one has then

$$\operatorname{Re} \varepsilon = (1.2 \pm 0.2) \times 10^{-3} \ .$$

This determination rests, as mentioned, on the assumption of complete validity of the $\Delta S = \Delta Q$ selection rule. One must, therefore, inquire further about the validity of this rule and the effect that a violation of this rule would have on the above result.

With the definition

$$x = \frac{\text{Amplitude } (\bar{K}^0 \to \pi^- \ell^+ \nu)}{\text{Amplitude } (K^0 \to \pi^- \ell^+ \nu)} ,$$

one can show that

$$\frac{N(\ell^+)}{N(\ell^-)} = 1 + 4 \text{ Re } \varepsilon \left(\frac{1 - |x|^2}{|1-x|^2} \right)$$

The effect of a $\Delta S = - \Delta Q$ amplitude is to dilute the asymmetry.

The question of whether $\Delta S = - \Delta Q$ amplitudes are present is a critical issue not only in connection with the question above, but also with regard to another possible violation of CP.

CP invariance requires, in fact, Im x = 0 [15]. Experimentally, this question is investigated by looking at the time distribution of ℓ^+ and ℓ^- from K^0 decays. For an initial K^0 state, the decay rates (the ± refers to the sign of the lepton) are given by

$$R(K^0)^{\pm} = \{ |1+x|^2 \exp\{-\Gamma_2 t\} + |1-x|^2 \exp\{-\Gamma_1 t\} + $$
$$+ [\pm 2(1-|x|^2)\cos\delta t -$$
$$- 4 \text{ Im } x \sin\delta t] \exp\{- \frac{\Gamma_1 + \Gamma_2}{2} t\} \}$$

where Γ_1 and Γ_2 are the inverse of the lifetimes of K_S and K_L, respectively and δ is the K_L-K_S mass difference. By fitting this formula to experimental distributions, values of Re x and Im x can be obtained. Several experiments have been performed so far, (for details, see ref. 12) and their combined results yield

$$\text{Re } x = 0.14 \pm 0.05$$

$$\text{Im } x = 0.12 \pm 0.05 \ .$$

So, even if one accepts these experiments at their face value, there is no compelling indication of CP violation.

4. C

C invariance is known to be broken in weak interactions. In the past years it has been proposed that the effect observed in K_L decay could be due to a breakdown of C invariance in electromagnetic interactions involving hadrons [17], [18]. Indeed, it has been pointed out that C violation in electromagnetic interactions of hadrons could be maximal and still have escaped observation so far.

The most popular test ground for these ideas are the η decays

$$\eta \rightarrow \pi^+ + \pi^- + \pi^o \quad \text{and}$$

$$\eta \rightarrow \pi^+ + \pi^- + \gamma \quad .$$

The experimental results are generally presented in terms of the quantity

$$A = \frac{N^+ - N^-}{N^+ + N^-} \quad ,$$

where N^+ is the number of events with the energy of the π^+ greater than the energy of the π^- and N^- the number of events with the energy of the π^- greater than the energy of the π^+ .

A non-vanishing value of A is clearly a direct proof of C violation. Experimentally, the situation can perhaps be presented as follows:

i) $\underline{\eta \rightarrow \pi^+ + \pi^- + \pi^o}$. The original excitement generated by the results

$$A = (5.8 \pm 3.4)\ \% \quad [19]$$

$$A = (7.2 \pm 2.8)\ \% \quad [20]$$

$$A = (8.7 \pm 5.3)\ \% \quad [21]$$

was cooled off by the more recent results

$$A = (0.3 \pm 1.0)\ \% \quad [22]$$

$$A = (-6.1 \pm 4.0)\ \% \quad [23]$$

The field has recently been reopened by an experiment performed by a Columbia-Brookhaven group, which reported an asymmetry

$$A = (1.5 \pm 0.5)\ \% \quad [24]\ .$$

ii) $\underline{\eta \to \pi^+ + \pi^- + \gamma}$. Four measurements of A have been obtained so far:

$$A = (-2.0 \pm 17.0)\ \% \quad [25]$$

$$A = (\ 1.5 \pm 2.5)\ \% \quad [26]$$

$$A = (-4.0 \pm 8.0)\ \% \quad [27]$$

$$A = (\ 2.4 \pm 1.4)\ \% \quad [28]$$

all of them more or less compatible with A = 0. The result of ref. 24 is suggestive of C non-conservation. Further experimental data are, as usual, badly needed.

5. T

Direct tests of time reversal invariance can be obtained in experiments sensitive to terms of the type

$$\underline{\sigma} \cdot \underline{p}_1 \times \underline{p}_2$$

where $\underline{\sigma}$ is the polarization of anyone of the fermions involved in the reaction and $\underline{p}_1$ and $\underline{p}_2$ are any two independent momentum vectors. Under time reversal, both linear and angular momentum change sign. To the extent that final state interactions can be neglected, time reversal invariance, therefore, requires the expectation value of $\underline{\sigma} \cdot \underline{p}_1 \times \underline{p}_2$ to be zero .

Most of the experiments mentioned below are tests of this prediction.

a) Electromagnetic Interactions

i) $\Sigma^0 \rightarrow \Lambda^0 + e^+ + e^-$

The appearance of a correlation of the form $\underline{\sigma} \cdot \underline{p}_1 \times \underline{p}_2$ in this decay can be tested by measuring the polaritation of the Λ^0 in the direction perpendicular to the decay plane in the Σ^0 center of mass system. The presence of such a polarization would manifest itself in an asymmetry in the angular distribution of the protons from Λ^0 decay in the Λ^0 center of mass system. Experimentally this asymmetry is found to be

$$A = 0.014 \pm 0.021$$

quite consistently with no violation of time reversal invariance [29].

ii) Electron Scattering on Polarized Protons

A time reversal invariance violating asymmetry of the type $\underline{\sigma} \cdot \underline{p}_1 \times \underline{p}_2$ might be detectable in inelastic scattering of electrons of initial momentum $\underline{p}_1$ and final momentum $\underline{p}_2$ onto protons of spin orientation $\underline{\sigma}$ [30]. It can be shown that such a term cannot be present in elastic scattering, and can only appear in inelastic scattering. The interpretation of experimental results rests on the hypothesis that the process can be described by one photon exchange and the choice of the specific excited state produced in the inelastic collision is also critical for the conclusion. Within an accuracy of the order of 5% of the maximum possible effect, the experimental result obtained at the Cambridge Electron Accelerator by a group from Harvard University [31] shows no evidence for a T violation.

iii) Polarization of the Recoil Deuteron in e-d Elastic Scattering

Contrary to the situation in the case of elastic scattering from particles of spin 1/2, elastic electron scattering from particles of spin 1 or greater can retain T violating terms [32]. In the case of the electron-deuteron scattering, the presence of a term of the type $\underline{\sigma}\cdot\underline{p}_1\times\underline{p}_2$ could be investigated by measuring the polarization of the recoil deuteron in the direction perpendicular to the scattering plane. An experiment performed at SLAC gave for this polarization the result 0.070±0.083, [33] in agreement with time reversal invariance.

iv) $n + p \to d + \gamma$ and $\gamma + d \to n + p$

The comparison of the differential cross sections for the two reactions permits a test of detailed balance and hence of time reversal invariance. Differential cross sections for the reaction $\gamma + d \to n + p$ are well known. Preliminary results from a Princeton group measuring differential cross sections for the reaction $n + p \to d + \gamma$ for neutron kinetic energies between 200 and 600 MeV seem to indicate a large violation of time reversal invariance [34]. A similar experiment is also in progress at Berkeley. Final conclusions on this question should probably await more definitive experimental results.

b) Weak Interactions

i) $n \to p + e^- + \nu$

Time reversal invariance in this decay is tested by measuring the correlation $\underline{\sigma}_n \cdot \underline{p}_e \times \underline{p}_\nu$ involving the spin of neutron $\underline{\sigma}_n$, the electron momentum $\underline{p}_e$ and the neutrino momentum $\underline{p}_\nu$. In the V,A theory, neutron beta decay is described in terms of the vector coupling constant G_V and the axial vector coupling constant G_A. Measurements of the correlation $\underline{\sigma}_n \cdot \underline{p}_e \times \underline{p}_\nu$ allow experimental determinations of the phase δ of G_A relative G_V. Old experiments gave $\delta = (176.1 \pm 6.4)^\circ$ [9]. A recent experiment performed in Russia gave the result $\delta = (178.7 \pm 1.3)^\circ$ [35]. Final state interactions are negligible and the results are entirely consistent with T invariance which requires δ to be 0 or 180 degrees.

ii) $K \to \pi + \mu + \nu$

Parity violation in the decay of the μ meson provides a handle on its polarization and with this information it is possible to study the correlation $\underline{\sigma}_\mu \cdot \underline{p}_\pi \times \underline{p}_\mu$. Measurements of this type have been carried out for both charged and neutral K mesons.

Assuming the validity of the $\Delta S = \Delta Q$ selection rule, the decay $K \to \pi + \mu + \nu$ is always described in terms of the two form factors f_+, associated with the sum of the π and K fourmomenta and f_-, associated with the difference. From measurements of the correlation $\underline{\sigma} \cdot \underline{p}_\pi \times \underline{p}_\mu$ experimental values for the quantity Im ξ, where ξ is the ratio f_-/f_+, can be obtained. With complementary determinations of Re ξ, one can express the experimental results in terms of the angle $\phi = \arg(\xi)$. The most definitive experiment, done using neutral K mesons, yields the result $\phi = (180.5 \pm 2.2)^\circ$ [36]. Final state interactions have been estimated to be small and the results are again consistent with the prediction $\phi = 0^{o}$ or $\phi = 180^{o}$ based on time reversal invariance.

iii) $\Lambda^{o} \rightarrow \pi^{-} + p$

In this two-body decay, in which only one independent momentum vector is available, it is necessary to use a different combination of vectors to investigate T invariance. The correlation studied in experiments is $\underline{\sigma}_p \cdot \underline{\sigma}_\Lambda \times \underline{p}_p$, where $\underline{\sigma}_p$ and $\underline{\sigma}_\Lambda$ are the polarizations of the decay proton and of the Λ hyperon, respectively, and $\underline{p}_p$ is the momentum of the proton. The decay in question is completely described by the s and p wave amplitudes and the correlation $\underline{\sigma}_p \cdot \underline{\sigma}_\Lambda \times \underline{p}_p$ is sensitive to the angle Δ the phase of the s amplitude relative to the p amplitude. Final state interactions in this case are important but can be reliably calculated. Assuming the validity of the $|\Delta I| = 1/2$ rule, on the basis of the known π-p scattering phase shifts, time reversal invariance predicts for the angle Δ the value $\Delta=(6.5\pm1.5)^{o}$ [37]. Two experiments to measure Δ have been recently performed and the combined result gives $\Delta = (7.5\pm3.9)^{o}$, indicating again no violation of time reversal invariance.

c) Electric Dipole Moment of the Neutron

On of the most spectacular evidences of a breakdown of time reversal invariance would be the experimental observation of a non-vanishing electric dipole moment of the neutron.

Such a moment can arise only if both parity and time reversal invariances are violated [38], [39].

A number of theoretical estimates of the neutron electric dipole moment have been made, based on the magnitude of the CP violation in K^{o} decay. Depending upon the model, they range from $\sim 10^{-19}$ to $\sim 10^{-28}$ e cm [40].

In particular, if the T violation is due to electromagnetic interactions, one can predict an electric dipole moment of about 10^{-20} e cm. If the violation is a $\sim 10^{-3}$ effect in weak interactions, a dipole moment of the value of 10^{-23} e cm is anticipated. In the case of a CP violating superweak interaction [41] then the expected number would be about 10^{-28} e cm.

Experimentally, the value on the electric dipole moment of the neutron is presently known to be $(2\pm2)\ 10^{-23}$ e cm [12]. To the extent that the various theoretical estimates of the electric dipole moment of the neutron are reliable, the experimental limit available rules out some of the proposed models for CP violation [42].

6. CONCLUSIONS

From what it has been discussed above it is evident that, unfortunately, the history of parity violation is not repeating itself. There is still no direct and convincing evidence for T violation anywhere and the breakdown of CP invariance is still confined to neutral K meson phenomena. We still have not found the key to the ultimate origin of the CP violation.

ACKNOWLEDGMENTS

I am deeply indebted to Professor P. Urban and to Dr. H. Kühnelt for my very pleasant stay in Schladming during which, in spite of the wonderful snow and weather conditions, these notes were almost entirely written.

REFERENCES

1. See, for instance, R. F. STREATER and A. S. WIGHTMAN, PCT, Spin, Statistics and All That, Benjamin, New York, 1964.
2. T. D. LEE and N. YANG, Phys. Rev. 104, 254 (1956).
3. C. S. WU, E. AMBLER, R. W. HAYWARD, D. D. HOPPES and R. P. HUDSON, Phys. Rev. 105, 1413 (1957).
4. J. H. CHRISTENSON, J. W. CRONIN, V. L. FITCH and R. TURLAY, Phys. Rev. Letters 13, 138 (1964).
5. H. J. LIPKIN, Phys. Rev. Letters 22, 213 (1969).
6. T. D. LEE, R. OEHME and C. N. YANG, Phys. Rev. 106, 340 (1957); G. LÜDERS and B. ZUMINO, ibid. 385 (1957).
7. S. OKUBO, Phys. Rev. 109, 984 (1958); CHOW KUANG-CHAO, Nucl. Phys. 9, 652 (1958)-(1959).
8. Y. UEDA and S. OKUBO, Phys. Rev. 139, B1591 (1965); ibid. E 141, 1582 (1967).
9. For a source of references on the experimental data see: N. BARASH-SCHMIDT, A. BARBARO-GALTIERI, G. CONFORTO, L. R. PRICE, M. ROOS, A. H. ROSENFELD, P. SÖDING and C. G. WOHL, Reviews of Mod. Physics, January 1969.
10. J. S. BELL and J. STEINBERGER, Proceedings of the Oxford Conference on Elementary Particles.
11. G. CHARPAK and M. GOURDIN, The K^0-$\bar{K}^0$ System, CERN report 67-18.
12. J. W. CRONIN, Proceedings of the XIV International Conference on High Energy Physics, Vienna 1968, p. 279.
13. R. C. CASELLA, Phys. Rev. Letters 21, 1128 (1968).
14. D. DORFAN, J. ENSTROM, D. RAYMOND, M. SCHWARTZ, S. WOJCICKI, P. H. MILLER, M. PACIOTTI, Phys. Rev. Lett. 19, 987 (1967).
15. S. BENNETT, D. NYGREN, H. SAAL, J. STEINBERGER, J. SUNDERLAND, Phys. Rev. Letters 19, 993 (1967).

16. R. G. SACHS and S. B. TREIMAN, Phys. Rev. Letters 8, 137 (1962).
17. J. BERNSTEIN, G. FEINBERG and T. D. LEE, Phys. Rev. 139B, 1650 (1965).
18. S. BARSHAY, Phys. Letters 17, 78 (1965).
19. C. BALTAY, P. FRANZINI, G. LÜTJENS, M. SCHWARTZ, J. C. SEVERIENS, D. ZANELLO; F. S. CRAWFORD, Jr., S. M. FLATTE, E. C. FOWLER, G. GOLDHABER, S. GOLDHABER (deceased), R. A. GROSSMAN, L. J. LOYD, L. R. PRICE, B. SHEN, F. T. SOLMITZ, M. L. STEVENSON, F. LOEFLER, R. McILWAIN, M. MEER, M. FOSTER, M. GOOD, R. HARTUNG, R. MATSEN, M. PETERS, D. REEDER, H. W. J. FOELSCHE, F. JAMES, J. JOHNSON and H. KRAYBILL, Phys. Rev. 149, 1044 (1966).
20. C. BALTAY, P. FRANZINI, J. KIM, L. KIRSCH, D. ZANELLO, J. L. FRANZINI, R. LOVELESS, J. McFADYEN, H. YARGER, Phys. Rev. Letters 16, 1224 (1966).
21. E. C. FOWLER, Bull. Am. Phys. Soc. 11, 380 (1966).
22. A. M. CNOPS, G. FINOCCHIARO, J. C. LASSALLE, P. MITTNER, P. ZANELLA, J. P. DUFEY, B. GOBBI, M. A. POUCHON, and A. MULLER, Phys. Letters 22, 546 (1966).
23. A. LARRIBE, A. LEVEQUE, A. MULLER, E. PAULI, D. REVEL, T. TALLINI, P. J. LITCHFIELD, L. K. RANGAN, A. M. SEGAR, J. R. SMITH, P. J. FINNEY, C. M. FISHER, and E. PICKUP, Phys. Letters 23, 600 (1966).
24. M. GORMLEY, E. HYMAN, W. LEE, T. NASH, J. PEOPLES, C. SCHULTZ, S. STEIN, Phys. Rev. Letters 21, 402 (1968).
25. F. CRAWFORD and L. R. PRICE, Phys. Rev. Letters 16, 333 (1966).
26. R. BOWEN, A. M. CNOPS, G. FINOCCHIARO, P. MITTNER, J. P. DUFEY, B. GOBBI, N. A. POUCHON, A. MULLER, Phys. Letters 24B, 206 (1967).
27. P. H. LITCHFIELD, L. K. RANGAN, A. M. SEGAR, J. R. SMITH, A. LARRIBE, A. LEVEQUE, A. MULLER, E. PAULI,

D. REVEL, B. TALLINI, Phys. Letters 24B, 486 (1967).

28. M. GORMLEY, E. HYMAN, W. LEE, T. NASH, J. PEOPLES, C. SCHULTZ, S. STEIN, Phys. Rev. Letters 21, 399 (1968).

29. M. BAGGETT, R. G. GLASSER and B. KEHOE, Paper submitted to the XIV International Conference on High Energy Physics, Vienna 1968.

30. N. CHRIST and T. D. LEE, Phys. Rev. 139B, 1650 (1965).

31. J. R. CHEN, J. SANDERSON, J. A. APPEL, G. GLADDING, M. GOITEIN, K. HANSON, D. C. IMRIE, T. KIRK, R. MADARAS, R. V. POUND, L. PRICE, R. WILSON and C. ZAJDE, Phys. Rev. Letters 21, 1279 (1968); Paper submitted to the XIVth International Conference on High Energy Physics, Vienna 1968.

32. I. KOBZAREV, L. B. OKUN' and M. V. TERENT'EV, JETP Letters 2, 289 (1965).

33. R. PREPOST, R. M. SIMONDS, B. H. WIIK, Phys. Rev. Letters 21,1271 (1968).

34. C. E. FRIEDBERG, D. F. BARTLETT, K. GOULIANOS, I. S. HAMMERMAN, D. P. HUTCHINSON, Bull. Am. Phys. Soc. 14, 76 (1969).

35. B. G. EROZOLIMSKY, L. N. BONDARENKO, Yu. A. MOSTOVOY, B. A.OBINYAKOV, V. P. ZACHAROVA, V. A. TITOV, Phys. Letters 27B, 557 (1968).

36. J. A. HELLAND, M. J. LONGO and K. K. YOUNG, Phys. Rev. Letters 21,257 (1968).

37. For more details see: G. CONFORTO, Λ decay, Preprint EFINS No. 67-101 to be published in Methods in Subnuclear Physics. M. NIKOLIĆ, Editor, Gordon and Breach New York, London, Paris, Vol.III (1969).

38. L. LANDAU, Nucl. Phys. 3, 127 (1957).

39. J. J. SAKURAI, Invariance Principles and Elementary Particles, Princeton University Press, Princeton, New Jersey.

40. D. G. BOULWARD, Nuovo Cimento 40, 1041 (1965).
41. L. WOLFENSTEIN, Phys. Rev. Letters 13, 562 (1964).
42. For a rather recent review of upper limits of electric dipole moments of elementary particles, nuclei and atoms see: P. G. H. SANDARS and E. LIPWORTH, Phys. Rev. Letters 13,718 (1964).
For a new limit to the electric dipole moment of the electron, $|d_e| < 3 \times 10^{-24}$ e cm see: M. C. WEISSKOPF, J. P. CARRICO, H. GOULD, E. LIPWORTH, and T. S. STEIN, Phys. Rev. Letters 21, 1645 (1968).

COMPOSITE PARTICLE COLLISIONS IN THE NON-RELATIVISTIC THREE-BODY THEORY†

By

Werner SANDHAS

Physikalisches Institut, Universität
Bonn, BRD

1. INTRODUCTION

The composite particle collision problem is evidently of great importance in many fields of quantum physics. In the last eight years or so this problem has been studied with increasing interest, in particular for the case of three nonrelativistic elementary particles, i.e., for the scattering of an elementary particle by a two-particle bound-state (for instance, a nucleon by a deuteron).

Most of the recent investigations have been initiated by the mathematical work of Faddeev [1], [2] on the one hand, and by the practical success of three-body calculations in the separable potential model on the other hand. (Such calculations were performed, in particular, by Mitra [3] and by the group: Aaron, Amado and Yam [5], [4] . It was Lovelace [6],[7] who first applied the separable potential model to the Faddeev equations. Thereby many aspects of the problem became rather transparent.

From a general point of view the three-body problem re-

---

† Seminar given at the VIII. Internationalen Universitätswochen für Kernphysik, Schladming, Feb.24 - March 8, 1969.

presents a first non-trivial (but still sufficiently simple) example of strong interaction dynamics. For, incoming and outgoing particles of complex structure can be studied which during the collision time may be rearranged to various other bound-states and resonances. A detailed study of the problem should, therefore, lead to an improvement of nuclear reaction theories, but it might also serve as a guide to a deeper understanding of some typical structures of elementary particle physics. In fact, a great part of the present interest in this field is induced by this hope. But let me say that in the latter respect the results are not very convincing up to now.

2. THE SINGLE-CHANNEL CASE

Before going to the three-body theory we have to recall some aspects of the single-channel problem. Here one studies the collision of two elementary particles with relative momenta $\vec{p}$ and $\vec{p}'$ (before and after the collision):

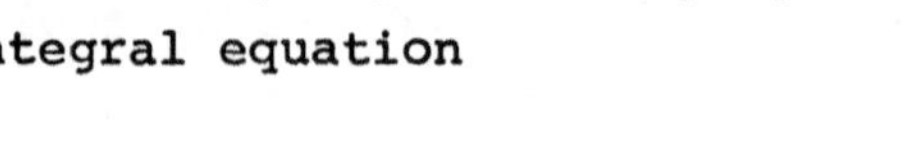

Fig. 1

What we need for the description of such a process is, of course, the scattering amplitude $T(\vec{p}',\vec{p};E)$, which is given by the integral equation

$$T(\vec{p}',\vec{p};E) = V(\vec{p}',\vec{p}) - \int d^3k\, V(\vec{p}',\vec{k}) \frac{1}{\frac{k^2}{2\mu} - E - io} T(\vec{k},\vec{p};E) \tag{2.1}$$

known as the Lippmann-Schwinger (L-S) equation for the T-matrix. (The "-io" in the denominator of the free

Green function indicates how we have to integrate around the pole at $k^2/2\mu = E$.) In operator form this equation simply reads (see also Fig. 2) :

$$T(E) = V - V\, G_o(E+io)\; T(E) \tag{2.2}$$

with

$$\langle \vec{p}\,' | T(E) | \vec{p} \rangle = T(\vec{p}\,', p;\; E) \tag{2.3}$$

and with the "free Green function"

$$G_o(E + io) = \frac{1}{H_o - E - io} \tag{2.4}$$

Fig. 2

The solution of the above L-S equation leads to the cross section

$$\frac{d\sigma}{d\Omega} = (2\pi)^4\; \mu^2 |T(\vec{p}\,', \vec{p};\; E)|^2 \quad . \tag{2.5}$$

Due to energy conservation we have the ("on shell") condition

$$\frac{p'^2}{2\mu} = p^2/2\mu = E \quad . \tag{2.6}$$

This means, that for the calculation of $d\sigma/d\Omega$ we only need the scattering amplitude "on-the-energy-shell", while the L-S equation yields an "off-shell" extension of this physical amplitude (this follows easily from the fact that in the L-S equation (2.1) we have to integrate over all values of $\vec{k}$, without any restriction of the form $k^2=p^2=2\mu E$). In other words, from the above equations (2.1) or (2.2) we find the full operator T, not only the special matrix elements which alone determine dσ/dΩ. Indeed the L-S equation allows to calculate T(E) from the potential V, but conversely the potential V is given by the same equation if T(E) is known (we have only to interchange in (2.2) the role of T and V). Consequently the off-shell amplitude contains the full information on the potential.

Note that we have not discussed more conventional methods for the determination of the scattering amplitude as, for instance, the solution of the Schrödinger equation with the correct asymptotic condition (plane wave plus outgoing spherical wave). Such elementary techniques are too restricted for our general program.

3. THE THREE-BODY PROBLEM

In the three-body case we now proceed in complete analogy to the two-particle problem. We again introduce transition amplitudes. Let us consider, for instance, the following process

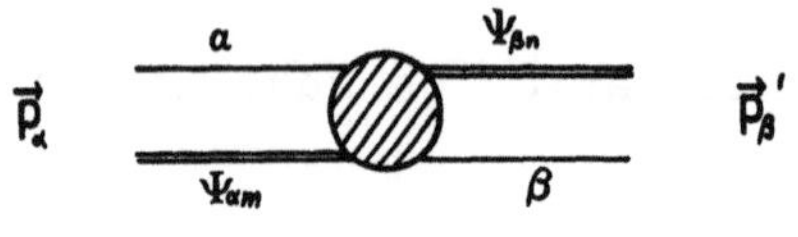

Fig. 3

Here, evidently, the transition amplitude $T_{\beta n,\alpha m}(\vec{p}_\beta\,',\vec{p}_\alpha)$ has to be labelled by

(i) an index α(β) which indicates the <u>unbound</u> particle (α = 1,2,3) in the initial (final) state,

(ii) an index m(n) which collectively denotes all the quantum numbers of the two-particle bound-state,

(iii) the relative momentum $\vec{p}_\alpha$ $(\vec{p}_\beta\,')$ of the colliding (composite) particles. (Translation invariance allows us to drop the total momentum).

It can be shown [8] that this amplitude is given again as a matrix element of a transition operator $U_{\beta\alpha}$:

$$T_{\beta n,\alpha m}(\vec{p}\,'_\beta,\vec{p}_\alpha) = \langle\vec{p}\,'_\beta|\langle\psi_{\beta n}|U_{\beta\alpha}|\psi_{\alpha m}\rangle|\vec{p}_\alpha\rangle \quad (3.1)$$

This representation shows some general aspects of our problem:

(i) To get the various amplitudes $T_{\beta n,\alpha m}$ we need several operators $U_{\beta\alpha}$. For example

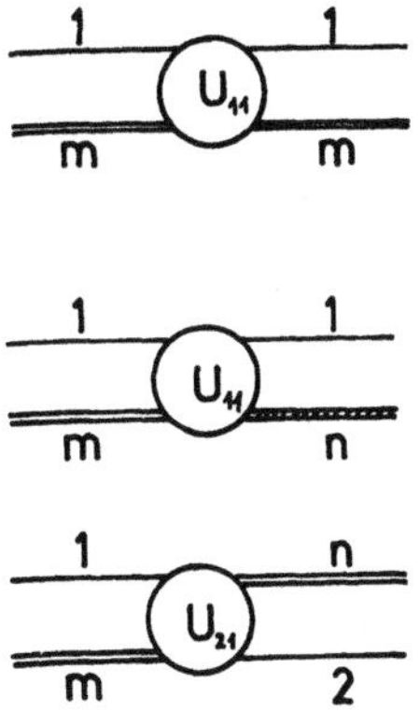

elastic scattering

"inelastic" scattering

rearrangement process

Fig. 4

Moreover an operator $U_{o\alpha}$ can also be introduced which leads to the break-up amplitude:

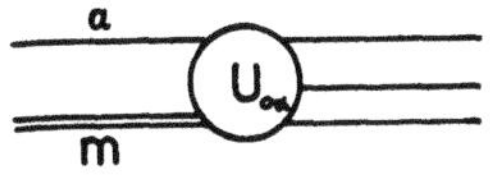

break-up process.

(ii) $U_{\beta\alpha}$ is, of course, an operator in the three-particle space - in accordance with the fact that we study a three-(elementary) particle problem [9]. What we really need for the calculation of the cross sections are, however, the $U_{\beta\alpha}$ sandwiched between the two-particle bound states,

$$\langle \psi_{\beta n} | U_{\beta\alpha} | \psi_{\alpha m} \rangle \tag{3.2}$$

i.e., effective two-particle operators which still act on the relative momenta of the colliding two particles. In more intuitive approaches one actually starts from equations for such "two-particle" amplitudes with effective potentials introduced phenomenologically. An exact theory has to be based, of course, on equations for the original three-particle operators $U_{\beta\alpha}$. Such equations turn out to be rather complicated (due to the multiplicity of variables). But we will show that it is possible to reduce the original three-particle equations to effective two-particle ones, which are as simple as in the intuitive approach but with "potentials" being now defined precisely by the original interactions.

4. DIFFICULTIES WITH THREE-BODY EQUATIONS

As discussed above we have now to write down integral equations for the operators $U_{\beta\alpha}$. In analogy to the two-particle case we find (for $\beta = \alpha$)

$$U_{\alpha\alpha} = \bar{V}_{\alpha} - \bar{V}_{\alpha} G_{\alpha} U_{\alpha\alpha} \quad . \tag{4.1}$$

$\bar{V}_{\alpha}$ is the interaction of particle α with the two particles in the bound system, i.e., the sum of potentials depicted in Fig. 5:

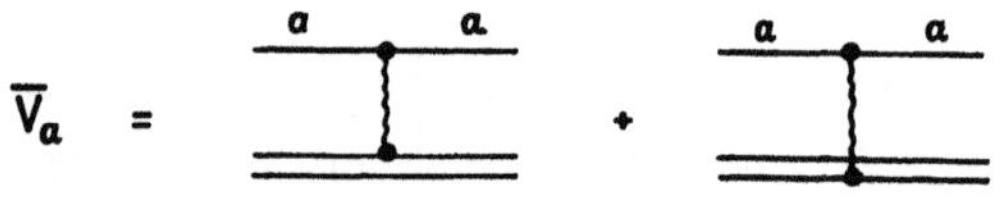

Fig. 5

Let us first consider the iteration of the two- and

three-particle equations. From (2.2) it follows

$$T = V - VG_oV + VG_oVG_oV - \dots , \qquad (4.2)$$

i.e., the Born series which in diagrams looks like:

Fig. 6

Correspondingly we have in the three-particle case the following picture :

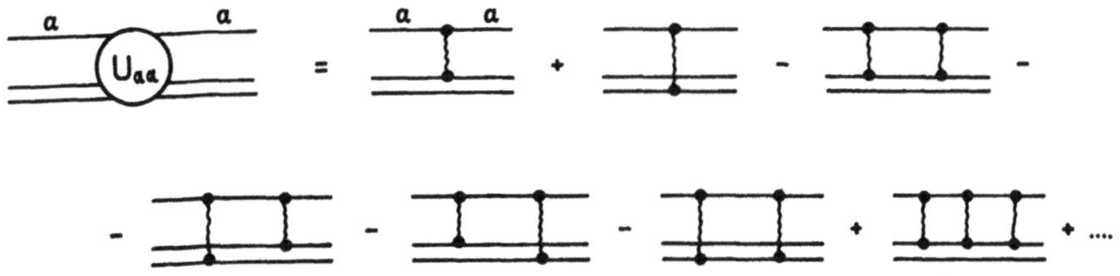

Fig. 7

We see, that the latter expansion contains partial series which coincide with the above two-particle Born series, apart from a third noninteracting particle: see, e.g., the terms 1, 3 and 7 in this expansion. Now it is well known that the Born series diverges for energies, which are in the neighbourhood of bound-states or resonances. Correspondingly the subsystem series of the three-particle problem will in general diverge since our interactions are strong enough to lead to two-particle bound states or resonances [10]. Consequently, if we work with a pure Born approximation of (4.1) we <u>ignore</u> the dominant structures of the subsystems.

In the two-particle case this difficulty is at once removed by application of the Fredholm theory (or of numerical treatments) which works even if iterative methods fail. This, however, remains not valid for the above three-particle equations. More precisely, the kernel $\bar{V}_\alpha G_\alpha$ does not fulfill the Fredholm condition. For, due to its disconnected parts, corresponding to one noninteracting particle in the three-body system (compare Fig. 5), it contains momentum conservation δ-functions which prevent the kernel from being of the Hilbert-Schmidt type.

This observation led Faddeev to look for equations which cure the latter shortcoming. Equations of this type will be studied in the following. But before ending the present discussion let us given a final comment. It is often claimed that we are forced to work with Faddeev's equations, since the above ones were not unique. This false statement originated by carrying over to the operator equations an argument due to Faddeev which only holds for the L-S equations for the scattering states.

5. FADDEEV EQUATIONS

To overcome the difficulties mentioned, Faddeev proposed a system of integral equations [2] of the form [11]

$$U_{\beta\alpha} = -(1-\delta_{\beta\alpha})G_o^{-1} - \sum_{\gamma\neq\beta} T_\gamma G_o U_{\gamma\alpha} \; . \qquad (5.1)$$

Let us summarize some typical properties of these equations.

(i) They represent a system of equations by which all the transition operators $U_{\beta\alpha}$ are treated simultaneously.

(ii) Eq. (5.1) contains in its kernel the two-particle scattering amplitudes T_γ instead of the potentials. Since we are now in a three-body problem we only have to label it by an index γ, in order to indicate the two-particle system where it acts. This is demonstrated in Fig. 8.

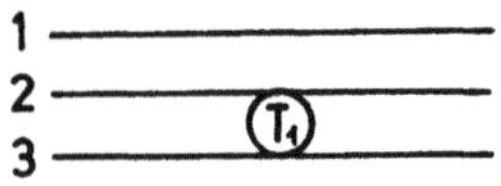

Fig. 8

It has to be stressed, however, that T_γ is not the two-particle amplitude on-the-energy-shell but the full two-particle operator (the off-shell amplitude!) studied in Section 2. Recalling the discussion given there we see that the complete information on the potential is contained and needed in the Faddeev equations - as in other three-particle equations. It will be shown that it is of great technical importance, that equations of the symmetric form (5.1) can be written down, which only implicitly (via T_γ) contain the potentials. But we have to note that this property does not imply that Faddeev's theory makes the knowledge of the potentials avoidable. Moreover, since the three-body problem requires the full knowledge of the potentials, it consequently allows to test various assumptions on it, which can not be decided solely by the two-particle data.

(iii) Further properties of the Faddeev equations become transparent by iterating them. For U_{11}, e.g. we arrive at

$$U_{11} = T_2 + T_3 - T_2G_oT_3 - T_3G_oT_2 + \dots , \tag{5.2}$$

that is, at a series shown in the following figure:

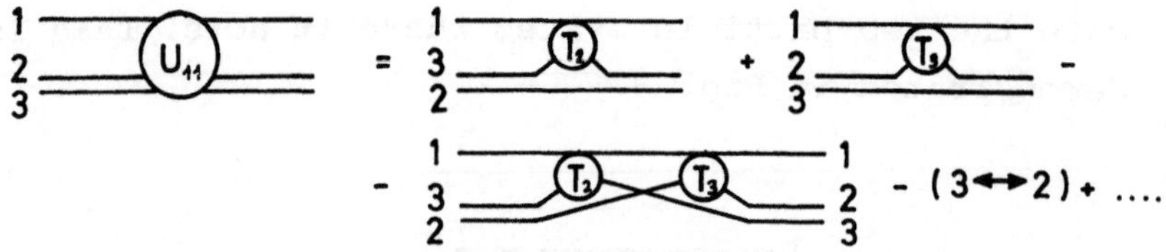

Fig. 9

Comparison with the expansion of (4.1), depicted in Fig. 7, shows that in the Faddeev equations the original subsystem series are summed [12] into the T_γ . This means that bound-states or resonances in the subsystems lead no longer to divergent series. We, therefore, expect better convergence of the expansion (5.2).

In this connection it is interesting to note that (5.2) represents Watson's multiple scattering series [13], proposed already in 1953. In particular it is closely related to Glauber's theory [14] which works so well in elementary particle - nucleus collision processes.

(iv) Despite of the fact that the divergence of the subsystem series is cured by the Faddeev equations, it is not true that the multiple scattering expansion is always convergent. In the neighbourhood of three-particle bound-states and resonances this series diverges, too. But now a second advantage of these equations turns up. Fig. 9 clearly shows that beginning with the second order of iteration the Faddeev equations only yield connected terms. This means that standard methods, as the Fredholm theory and, therefore, conventional numerical treatments, can be applied to them [2]. In other words, we are now in the same position as in single-channel scattering theory, where the Fredholm theory works even if perturbation theory fails. It was exactly this property which led to the statement that by Faddeev's theory the three-particle problem was completely solved.

Of course, such a statement is true. In practice, however, the equations (5.1) are too complicated. This is due to the fact that they are still two-dimensional after angular momentum decomposition. I.e., the typical difficulty of multiparticle problems, namely the high multiplicity of variables is <u>not</u> cured in Faddeev's theory. Thus, the formal advantage of this method is practically compensated by its difficult structure in detail.

6. SUBSYSTEM PROPERTIES AND REDUCTION OF THE THREE-PARTICLE PROBLEM

In this situation we ask: what is the real advantage of the Faddeev equations, if we intend to develop a practical composite particle theory? In fact, it turns out that neither its iteration properties nor the connectedness of the squared kernel are decisive. But, as already discussed, it is of considerable importance that instead of the potentials the two-particle scattering amplitudes T_γ occur. This allows to incorporate the dominant subsystem structures (bound-states and resonances) explicitly.

In the non-relativistic case we know (and in the relativistic case we have some opinions) how T_γ behaves near a bound-state energy $E = E_{\gamma r}$. Here we have a pole of the form

$$G_o T_\gamma (E) G_o \sim - |\psi_{\gamma r}\rangle \frac{1}{E_{\gamma r} - E} \langle\psi_{\gamma r}| =$$

$$= - |\psi_{\gamma r}\rangle \; t_{\gamma r}(E) \langle\psi_{\gamma r}| \quad . \tag{6.1}$$

Approximating in the Faddeev equations the T_γ by a sum of bound-state (or resonance) poles,

$$U_{\beta\alpha} = -(1-\delta_{\beta\alpha})G_o^{-1} + \sum_{\gamma} -(1-\delta_{\beta\gamma})G_o^{-1} \underbrace{G_o T_\gamma G_o}_{-\sum_r |\psi_{\gamma r}\rangle t_{\gamma r}\langle\psi_{\gamma r}|} U_{\gamma\alpha} , \quad (6.2)$$

and sandwiching the whole equation between the bound-states $\langle\psi_{\beta n}|$ and $|\psi_{\alpha m}\rangle$ we get

$$\langle\psi_{\beta n}|U_{\beta\alpha}|\psi_{\alpha m}\rangle = -(1-\delta_{\beta\alpha})\langle\psi_{\beta n}|G_o^{-1}|\psi_{\alpha m}\rangle - \quad (6.3)$$

$$- \sum_{\gamma,r} -(1-\delta_{\beta\gamma})\langle\psi_{\beta n}|G_o^{-1}|\psi_{\gamma r}\rangle t_{\gamma r}\langle\psi_{\gamma r}|U_{\gamma\alpha}|\psi_{\alpha m}\rangle .$$

We thus arrived at an equation for the effective two-particle operators $T_{\beta n,\alpha m}$, introduced in Section 3. In fact, (6.3) reads

$$T_{\beta n,\alpha m} = V_{\beta n,\alpha m} - \sum_{\gamma,r} V_{\beta n,\gamma r} t_{\gamma r} T_{\gamma r,\alpha m} , \quad (6.4)$$

or in matrix notation

$$T = V - V\, G_o\, T \quad . \quad (6.5)$$

This means that, at least in the pole approximation, we found an effective two-particle Lippmann-Schwinger type equation (compare (2.2)). According to (6.3) and (6.4) the "potential" is given by

$$V_{\beta n,\alpha m} = -(1-\delta_{\beta\alpha})\langle\psi_{\beta n}|G_o^{-1}|\psi_{\alpha m}\rangle =$$

$$= -(1-\delta_{\beta\alpha})\langle\psi_{\beta n}|G_o^{-1}\, G_o G_o^{-1}|\psi_{\alpha m}\rangle =$$

$$= -(1-\delta_{\beta\alpha})\langle\beta n|G_o|\alpha m\rangle \quad . \quad (6.6)$$

This expression represents an exchange potential of the form

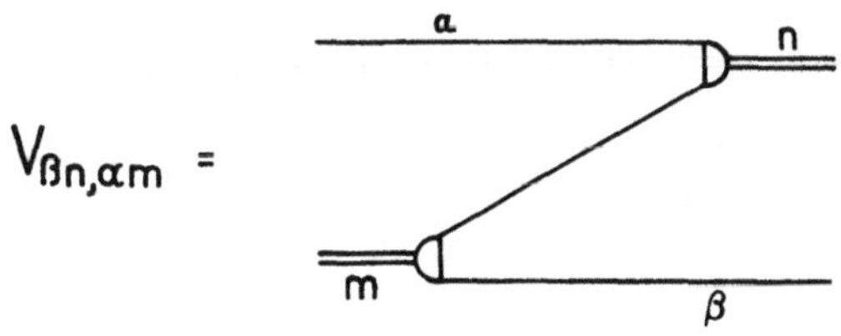

Fig. 10

The whole equation (6.4) has, therefore, the structure (compare also Fig. 2) :

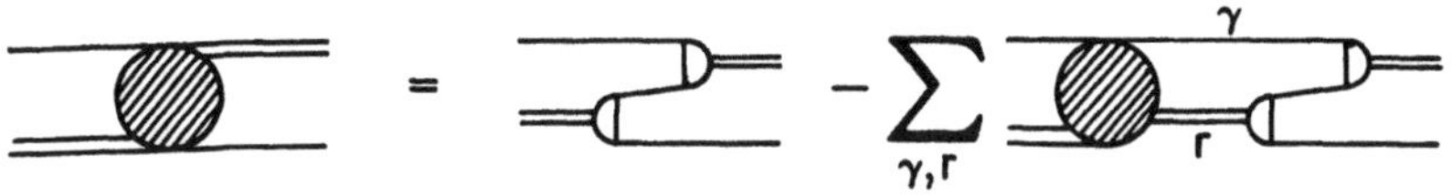

Fig. 11

By explicit incorporation of the dominant subsystem properties we succeeded in deriving effective two-particle equations which are practical as well as physically interpretable (compare the discussion in Section 3). We should stress once more that the possibility of incorporating directly the two-particle structures, and the resulting reduction to effective two-particle equations, is rather typical for Faddeev's theory and, in our opinion, its essential advantage. It was Lovelace [6] , who first emphasized this point and already arrived at equations of the form (6.4).

7. GENERALIZATION

In the preceding section we have shown, how the three-particle problem reduces to an effective two-particle theory in the pole approximation of the subsystem operators. This result is not only important for practical applications (after angular momentum decomposition we have one-dimensional equations, as in the genuine two-body case), it is also rather tempting from the physical point of view. For, as discussed in Section 3, such a structure is expected, a priori, in our problem where, in fact, two (composite) particles collide. We therefore expect, and we will show in the following, that our reduction procedure can also be performed <u>without</u> any approximation. For this purpose let us split T_γ according to

$$T_\gamma = \sum_r T_{\gamma,r}^{pole} + T'_\gamma \quad . \tag{7.1}$$

Here $T_{\gamma,r}^{pole}$ is given by (6.1):

$$T_{\gamma,r}^{pole}(E) = - G_o^{-1} | \psi_{\gamma r} \rangle \; t_{\gamma r}(E) \langle \psi_{\gamma r} | G_o^{-1} = \tag{7.2}$$

$$= - |\gamma r\rangle \; t_{\gamma r}(E) \langle \gamma r| \quad .$$

We assume that in the sum $\sum_{\gamma,r} T_{\gamma,r}^{pole}$ all bound-state poles (and, more generally, all resonance poles) are contained [15]. In other words, T'_γ is assumed to represent a "small" non-polar rest term. By a generalization [16] of the techniques used in Section 6 we derive again an equation of the form (6.4):

$$T_{\beta n,\alpha m} = V_{\beta n,\alpha m} - \sum_{\gamma,r} V_{\beta n,\gamma r} \; t_{\gamma r} \; T_{\gamma r,\alpha m} \; . \tag{7.3}$$

Here, $T_{\beta n,\alpha m}$ is given as in (6.4) or (6.3), representing exactly the effective two-particle amplitude:

$$T_{\beta n,\alpha m} = \langle \psi_{\beta n} | U_{\beta\alpha} | \psi_{\alpha m} \rangle \quad . \tag{7.4}$$

According to (3.1) this is sufficient for the calculation of the considered processes. The only difference between the present result and the approximate one of Section 6 is, that the "potentials" have now a more general form. They are given (similar to the amplitudes (7.4)) by

$$V_{\beta n,\alpha m} = \langle \psi_{\beta n} | U'_{\beta\alpha} | \psi_{\alpha m} \rangle \quad . \qquad (7.5)$$

$U'_{\beta\alpha}$ obeys the same Faddeev-type equation (5.1) as $U_{\beta\alpha}$, but with the small rest terms T'_{γ} instead of the full amplitudes T_{γ}. We, therefore, expect that a determination of $U'_{\beta\alpha}$ by iteration is justified. I.e., the Watson series with T_{γ} replaced by T'_{γ} should well converge.

In lowest order we reproduce the result of the preceding section. The first order approximation in T'_{γ} is shown, for $\beta \neq \alpha$, in Fig. 12.

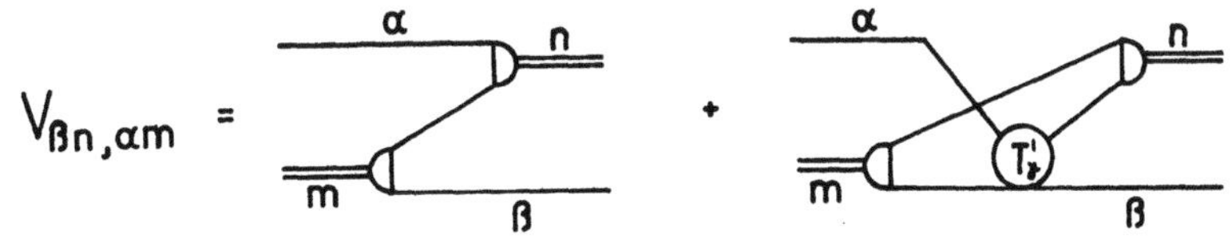

Fig. 12

For $\beta = \alpha$ no contribution exists in the pure pole approximation (compare Eq. (6.6)), i.e., only exchange diagrams occur for $T'_{\gamma} = 0$. But now we have the first order diagrams (compare Fig. 9):

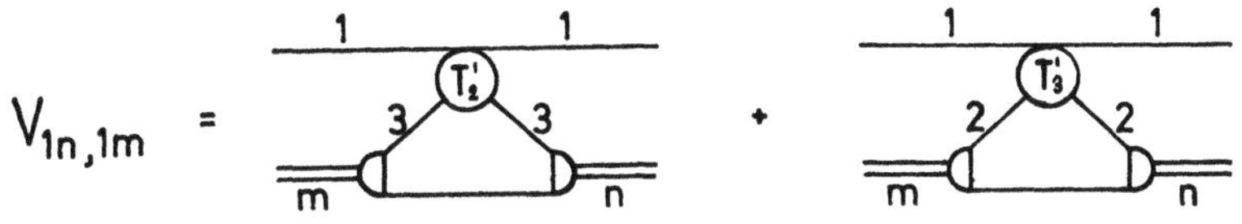

Fig. 13

Before ending this section we should note, that the equations (5.1) are not the original ones of Faddeev. Having their typical structure (connectedness, appearance of the T_γ instead of the potentials) they are written down for the operators $U_{\beta\alpha}$ which are more closely related to physically observable quantities than the operators introduced by Faddeev. In fact, these equations have been introduced by Alt, Grassberger and myself [11] in order to derive the exact effective two-particle theory given here [17].

8. BREAK-UP PROCESSES

The techniques which we have discussed for the case of two (composite) particles in the initial <u>and</u> in the final state, are also applicable to processes with three elementary particles in the final state (break-up). In low order of iteration in T'_γ we find expressions of a structure as shown in the following diagrams:

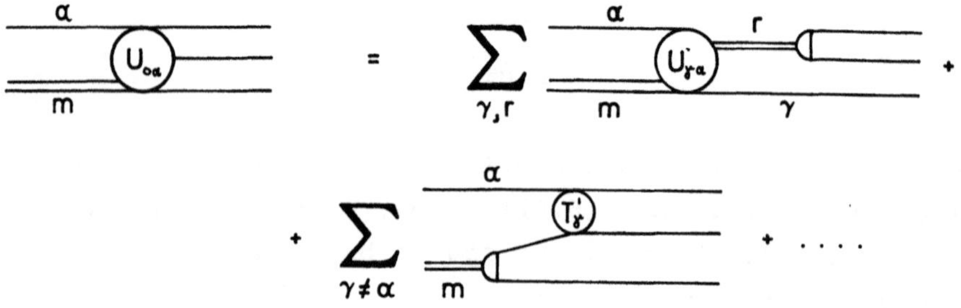

Fig. 14

We see that the result is <u>not</u> an integral equation. In fact, once we have calculated the occurring "two-particle" amplitudes $T_{\gamma r,\alpha m}$ (beyond the three-particle threshold!) by means of the integral equations (6.4) or (7.3), all expressions shown in Fig. 14 are calculable.

The first term which corresponds to the pure pole

approximation represents, of course, the isobar model (final state interaction). This term was already derived by Lovelace [6].

The second diagram has the form of the impulse approximation (spectator model). Of course, the isobar term is only large, if the relative energy of two of the produced particles is near to the energy of the intermediate bound-states or resonances $|\psi_{\gamma r}>$. Going away from the corresponding pole the spectator term becomes more and more important. I.e., its incorporation by means of the above exact theory is certainly not only a correction. Thus, we are in the position to test the applicability of the mentioned models in certain energy regions. Moreover, those regions can be studied where both mechanisms contribute.

9. ADDITIONAL COMMENTS

Some remarks on the above treatment might be helpful:

a) What we have used for our derivation is only the special way in which the Faddeev equations allow to consider the subsystem properties, but not any connectedness behaviour (Fredholm condition) of the Faddeev kernel. This observation is contrary to claims, often made with respect to the approximate treatment of Section 6.

b) The result (7.3) with the potentials (7.5) is exact. It is also practical, at least for the short range potentials of strong interaction theories, because in such cases we have only few bound-states or resonances. That is, few pole terms are sufficient to arrive at a small rest T'_{γ} .

c) Up to now, we nearly exclusively argued by means of physical intuition. However, the statements of b) can be formulated also in a more mathematical language: for short-range potentials it turns out that the Schmidt norm of $T_\gamma G_o$, read as an operator of the two-particle space, exists [18], [19]. Then, from functional analysis we know that this operator can be well approximated by a finite number of separable terms of the form $T_\gamma^{pole} G_o$. I.e., $T'_\gamma G_o$ becomes indeed sufficiently small to allow iteration [20]. We stress once more, that it is not any Hilbert-Schmidt (Fredholm) property of the whole (squared) Faddeev kernel which has been used. In this connection it should be noted that the subsystem operators $V_\gamma G_o$, occurring in the three-particle L-S equation, are also of the Hilbert-Schmidt type, (if they are read in the two-particle space). This gives a first hint, that the above general method is not so intimately related to Faddeev's theory (compare in this respect the final discussion in Section 10). We do not intend to go into further mathematical details. But we should be aware that the real application of the described methods requires some experience in functional analysis techniques. In particular, splittings of the form (7.1) with more refined expressions for T_γ^{pole} than the one given in (7.2) can thereby be found.

d) Since (7. 5) is an exact expression for the effective potential $V_{\beta n,\alpha m}$ we are not necessarily forced to determine it by perturbation theory. A variational method has been applied by Carew and Rosenberg [21] to these expressions which seems to be promising. Generalizing and simplifying their proposal it can be shown that the "potentials" are given by low-order iterative expressions and additional variational terms, all of simple structure. (This will be studied in detail in a publication which is under preparation).

e) It was the essential point of the method described

in Section 7, that the original multiparticle equations are reduced to two-particle ones. This goal is also achieved in Feshbach's projection operator formalism [22]. It is interesting that his method, as applied to the (in-) elastic case, turns out to be a special case of the above theory [23]. In fact, the projection on to certain channel eigenfunctions corresponds to explicitly introducing bound-state poles in T_γ. However, the generalization to rearrangement and break-up channels, which is rather cumbersome in Feshbach's theory, requires no additional effort in our Faddeev-induced result which incorporates all channels simultaneously. Furthermore, this is accomplished in a symmetrical way, i.e., the inclusion of the Pauli principle - a complicated problem in Feshbach's theory - is achieved at once.

f) Finally we note that the validity of such an important model as the distorted wave Born approximation can be studied quite well in the above exact "two-particle" theory [24].

10. SEPARABLE POTENTIALS

We should say some words about the separable potential model. Instead of our original interaction V,which we assume (at the moment) to lead only to one two-body bound-state $|\psi\rangle$, we consider the separable potential [25] (as in Section 2 we drop the index γ denoting the respective two-particle subsystem):

$$V^s = \frac{V|\psi\rangle\langle\psi|V}{\langle\psi|V|\psi\rangle} = |\chi\rangle \; \lambda \; \langle\chi| = \tag{10.1}$$

By the replacement $V \to V^s$, we arrive at an explicitly

solvable model. This is most easily seen in diagrams. Instead of the original expansion

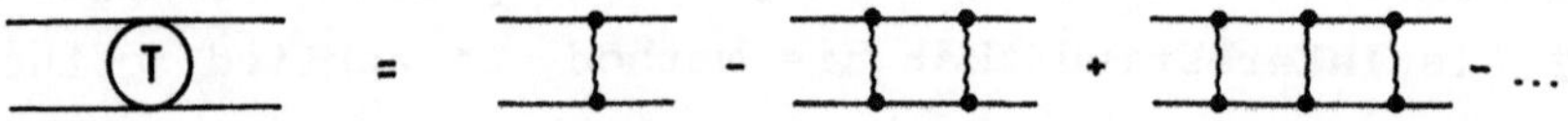

Fig. 15

we have now

Fig. 16

Fig. 16 represents a geometrical series of chain diagrams [26] (note that  = $\langle\chi|G_o|\chi\rangle$ is not an operator!) which can be summed into

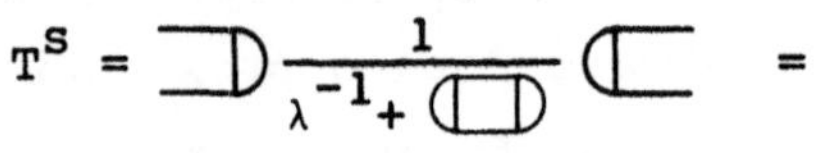

$$= V|\psi\rangle \frac{1}{\langle\psi|V|\psi\rangle+\langle\psi|VG_oV|\psi\rangle}\langle\psi|\ V\ . \qquad (10.2)$$

We see that the result is again of a separable form. The most interesting point, however, is that the explicit expression (10.2) has the same pole structure (corresponding to the bound-state $|\psi\rangle$) as the original amplitude T. Inserting T^s in the Faddeev equations, effective two-particle equations are derived by the same method as demonstrated in Section 6 for the pure pole approximation.

Such equations have been found already by Mitra [3], [27] without any reference to Faddeev's theory. But as emphasized by Lovelace [6], Faddeev's theory shows rather suggestively why the separable potential model is a good approximation. For, replacing V by V^s we do not change the dominant subsystem pole structure of the problem and this should be essential for a good approximation.

Generalizing this approximate method we can again derive exact "two-particle" equations, if we work with the splitting [28]

$$V = V^{S} + V' \qquad (10.3)$$

The effective potentials are determined by series which only contain the "small" nonseparable rest terms V' (compare the analogous results of Section 7).

As in the pure separable model the derivation can also be performed without reference to Faddeev's theory [29]. This shows transparently that a reduction procedure as described in Section 7 is quite independent of the special version of the original three-body equations. Since, furthermore, the results are exact in any case we are not even forced to use Faddeev's equations for plausibility arguments, as done by Lovelace and as discussed above. Moreover, by the non-Faddeev approach we find at once a particularly simple form of the "potentials". For $\beta \neq \alpha$, e.g., we have [29]

$$V_{\beta n,\alpha m} = - \langle \chi_{\beta n} | \frac{1}{H_o + \sum_{\gamma} V'_{\gamma} - E - io} | \chi_{\alpha m} \rangle \, . \qquad (10.4)$$

11. NUMERICAL RESULTS

Many calculations of the three-nucleon problem have been performed in the separable model with various degrees of sophistication in the choice of the separable potentials (the parameters are adjusted to yield the deuteron binding energy and the low energy scattering data of the two-nucleon system). The results obtained are in good agreement with experiment [30], [31]. This is consistent with the main conclusion of the preceding sections, namely that it is sufficient for a good description of the three-particle problem if the dominant subsystem properties, i.e., the essential structures of the internal dynamics, are suitably taken into account-that is just what separable potentials allow to do.

In detail, however, deviations are expected and, in fact, appear. In order to get some feeling for the effect of the non-separable rest parts $V'_{\gamma}=V_{\gamma}-V^{s}_{\gamma}$ (see Equ. (10.3)) we have studied a model problem of three identical spinless particles interacting via Yukawa potentials (with coupling constant = g).

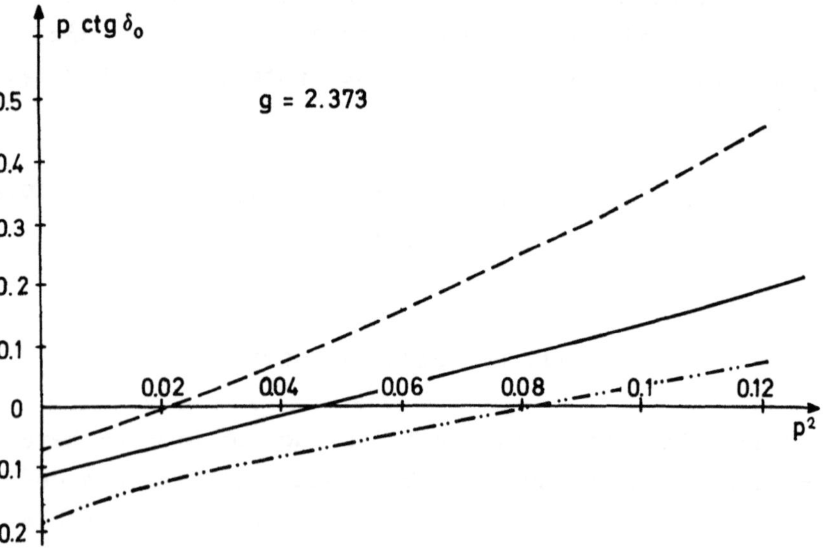

Fig. 17

The dashed line in Fig. 17 shows p.ctgδ_o (δ_o=s-wave phase shift) calculated in zeroth order (separable potential approximation) of our general iteration scheme described in Sections 7 and 10. The first order result in V'_γ is indicated by the solid line. Since it can be shown that the latter curve gives a rigorous upper bound for the exact values, we conclude that the first order correction represents a considerable step in the right direction [32]. Further improvement is expected by adding the variational terms mentioned in Section 9 d). Hereby the discrepancy between our results and nearly exact ones of Ball and Wong [33] (the lowest curve in Fig. 17 : –··–·–··–··) should still diminish. This would provide us with a practical tool for calculating in a sufficiently accurate way the effective potentials (and consequently the whole collision problem) in more realistic cases, too. Such methods are, of course, of decisive importance if we want to get any answer on the detailed form of the two-nucleon interaction.

Coulomb effects can be taken into account correctly in our formalism. We only have to incorporate the Coulomb potentials in the nonseparable rest terms V'_γ. This permits, e.g., to calculate the corresponding corrections in the proton-deuteron process compared with the neutron-deuteron case. Particularly simple expressions are obtained for separable nuclear potentials. Then, besides Coulomb corrected proton-proton form factors and propagators $t_{\gamma r}$, we have effective potentials which are given, without any approximation, by expressions of the form shown in Figs. 12 and 13: T'_γ is now the pure Coulomb T-matrix (compare Section 5 of the reference given in [11] and the discussion in [34]).

This method has been used by Alessandrini and collaborators [34] to calculate the difference, ΔE, of the binding energies of triton and He^3 due to Coulomb effects. Two choices of the separable nucleon-nucleon potential led to $\Delta E = 0.65$ and 0.81 MeV, while the experimental value is $\Delta E_{exp} = 0.764$ MeV. In view of a possibly existing charge asymmetry (which could not be excluded by variational treatments [35] an extension of the calculations to more realistic nuclear interactions would be of fundamental importance.

12. THE N-BODY CASE

The n-body collision problem has been treated by several authors in analogy to Faddeev's approach. That is, equations with connected kernels were derived which are, therefore, solvable by standard methods - at least in principle [36]. However these derivations consist in purely algebraic transformations of the full n-particle operators. This means that they lead to equations of high dimension which are far beyond any practical applicability (recall that already in the three-body case a direct solution of the Faddeev equations is nearly impossible due to their high dimension).

On the other hand manageable effective two-particle equations can be derived also in the general case [37], [29] if, step by step, the essential subsystem structures are taken into account [38] by the methods described in Sections 6 to 10. In fact, the same treatment by which the three-body problem was reduced to an effective two-particle one (namely the explicit incorporation of the two-body poles) allowes to reduce the four-body problem to a formal three-particle theory (we already know how to handle this!). By extracting again the "two-particle" subsystem poles, which now, in fact, correspond to three-body bound-states and resonances of the original problem, we end up with a system of effective two-particle equations of the form shown in Fig. 18 [39].

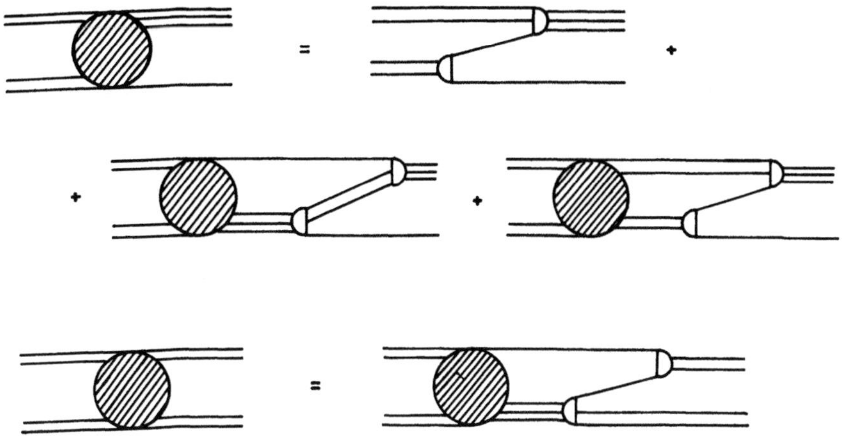

Fig. 18

These equations allow to calculate, among others, the differential cross section for the process deuteron + + deuteron → proton + triton [40]. The results obtained for the energy E_{Lab} = 13.8 MeV are compared with experimental values in Fig. 19.

We note, that in the calculation no open parameters exist which could be varied in order to reproduce the experimental data. Only the low energy parameters of the two-body interactions are taken from experiment, as discussed in Section 11 in connection with the three-body problem. The occurring transition potentials, shown in Fig. 18, have been determined carefully while, as a first crude attempt, we only used a K-matrix Born approximation of our final "two-particle" equation instead of solving the coupled integral equations. This is, probably, responsible for most of the deviations seen in Fig. 19. We

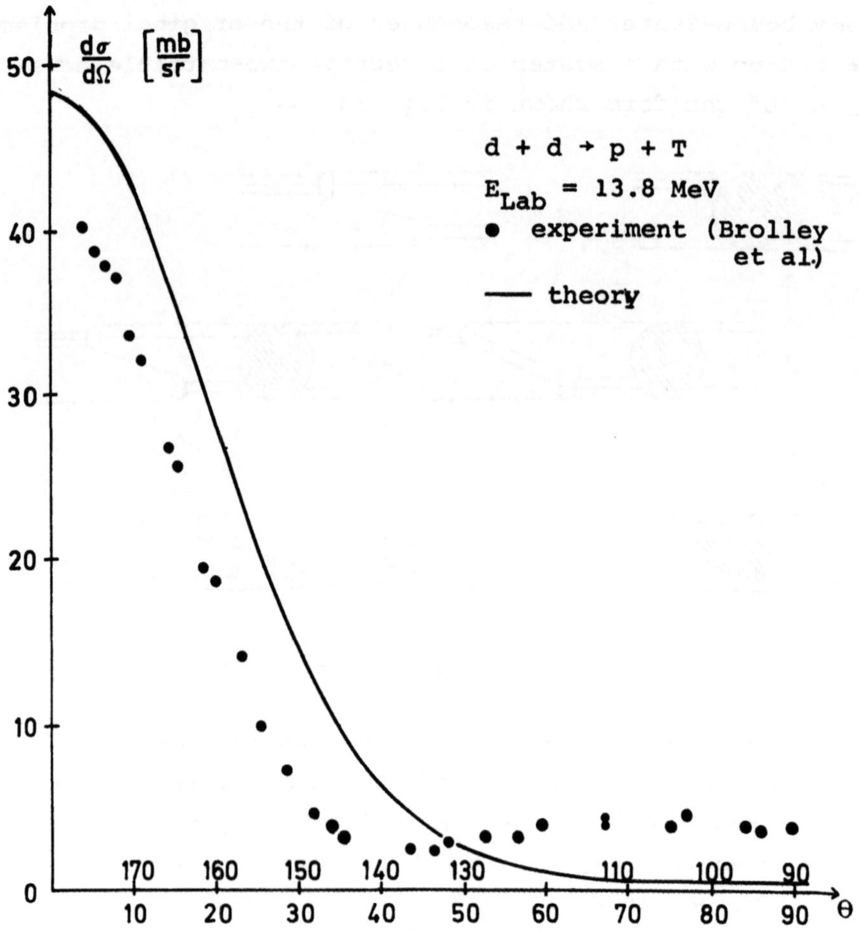

Fig. 19

stress, however, that the final equations, being one dimensional as in the genuine two-particle case, are amenable to modern computer possibilities. Thus, more computational effort along these lines seems to be suggested by the above results.

13. CONCLUSIONS

We have described a method which reduces the three-body collision problem (and even the n-bodycase) to two-particle equations with effective potentials. That is, exact equations are obtained of the form as expected, a priori, by physical intuition. From the general point of view this treatment shows how perturbation theory can be applied to strong interaction multiparticle problems, namely, by an iterative determination of the effective potentials. As in any perturbation theory we are, thereby, in the position to separate main effects from higher order corrections. Dominant structures and, therefore, the basic mechanisms of the processes can be studied in such an approach.

As the starting point we used Faddeev's equations. While they were not necessary for our derivations they made many arguments very suggestive. We have already mentioned, that one could also try to solve these equations directly, at least as soon as sufficiently fast computers become available. However, by putting such complicated (high dimensional) relations into a computer - which at the end (hopefully) yields the right numbers - we would learn less about the internal dynamics of multiparticle systems than in the approach discussed here.

REFERENCES AND FOOTNOTES

1. L. D. FADDEEV, Soviet Phys.-JETP 12, 1014 (1961),Soviet Phys. Dokl. 6, 384 (1961); 7, 600 (1963).

2. L. D. FADDEEV, Mathematical Aspects of the Three-Body Problem in the Quantum Scattering Theory (Israel Program for Scientific Translation, Jerusalem 1965).

3. A. N. MITRA, Nucl. Phys. 32, 529 (1962); Phys. Rev. 139, B1472 (1965).

Compare also: A. N. MITRA and V. S. BHASIN, Phys. Rev. 131, 1265 (1963).

4. R. D. AMADO, Phys. Rev. 132, 486 (1963); 141, 902 (1966);

R. AARON, R. D. AMADO and Y. Y.YAM, Phys. Rev. 136, B650 (1964); 140, B1291 (1965); Phys. Rev. Lett. 13, 574 (1964).

5. Compare also: P. E. SHANLEY and R. AARON, Annals of Physics 44,363 (1967) where many relevant references are given.

6. C. LOVELACE, Phys. Rev. 135, B1225 (1964).

7. See also: A. C. PHILLIPS, Phys. Rev. 142, 984 (1966); 145, 733 (1966).

8. H. EKSTEIN, Phys. Rev. 101, 880 (1956);

Compare also [6] and Section 2 of the reference given in[11].

9. $U_{\beta\alpha}$ can be defined explicitly. Instead of doing this we give in the following the integral equations which it fulfills.

10. R. AARON, R. D. AMADO and B. W. LEE, Phys. Rev. 121, 319 (1961).

11. The Faddeev-type equations (5.1) were given by E. O. ALT, P. GRASSBERGER and W. SANDHAS, Nucl. Phys. B2, 167 (1967).

Compare in this respect the discussion at the end of Section 7, and footnote [17].

12. Faddeev originally constructed his equations by summing up the subsystem series [1].

13. K. M. WATSON, Phys. Rev. 89, 575 (1953).

14. R. J. GLAUBER, in: High Energy Physics and Nuclear Structure,ed.: G.ALEXANDER (North Holland, Amsterdam (1967)).

15. Instead of $|\gamma r\rangle = -G_o^{-1}|\psi_{\gamma r}\rangle$ more general expressions for the "form factors" $|\gamma r\rangle$ can also be used.

16. We apply, in fact, a suitable version of the quasiparticle method developed for the two-body case by

S. WEINBERG: Phys. Rev. 130, 776 (1963); 131, 440 (1963);

see also K. MEETZ, J. Math. Phys. 3, 690 (1962).

It should be noted that Weinberg himself proposed an extension of this treatment to the three-(n-)body-case (Phys. Rev. 133, B232 (1964)). However, his very complicated version seems to be beyond any practical applicability.

17. Faddeev-type equations for transition operators $U^{(\pm)}_{\beta\alpha}$ which are different from our $U_{\beta\alpha}$ have previously been introduced by Lovelace who proposed many of the discussed ideas. Since his equations, however, contain the T_γ and the potentials V_γ it is even cumbersome to apply the approximation of Section 6. For the performance of the general treatment of Section 7 the replacement of his relations by the equations (5.1) was decisive.

18. C. LOVELACE, in: Strong Interactions and High Energy Physics, ed.: R. G. MOORHOUSE (Oliver and Boyd, London 1964).

S. WEINBERG, Phys. Rev. 131, 440 (1963).

19. To be precise: this property holds for complex energies. Some mathematical effort is required in going to the real axis.

20. In the Faddeev equations $T_\gamma G_o$ or $T'_\gamma G_o$ is a two-particle operator acting in the three-particle space. The above considerations on the Schmidt norms are only valid for genuine two-particle operators. But they lead to similar statements with respect to the operator norms of $T_\gamma G_o$ as read in the three-particle space. Such properties are sufficient for our argumentation.

21. J. CAREW and L. ROSENBERG, preprint.

22. H. FESHBACH, Annals of Phys. 5, 357 (1958); 19, 287 (1962).

23. P. GRASSBERGER and W. SANDHAS, Zeitschr. f. Physik

220, 29 (1969).

24. E. O. ALT, P. GRASSBERGER and W. SANDHAS, Bonn preprint 2-37, See also the relevant references given there and in [5].

25. Y. YAMAGUCHI, Phys. Rev. 95, 1628 (1954).

26. Recall that solvable field theoretical models, as, e.g., the Zachariasen model (F. ZACHARIASEN, Phys. Rev. 121, 1851 (1961), W. THIRRING, Nuovo Cimento 23,1064 (1962)) have usually such a structure.

27. An equivalent field theoretical (Lee-type) model has been developed by R. D. AMADO [4].

28. The separation (10.3) leads to a splitting of T of the structure (7.1) [16].

29. P. GRASSBERGER and W. SANDHAS, Zeitschr. f. Physik 217, 9 (1968).

30. Compare the relevant papers cited in [3], [4], [7].

31. For further references see H. P. NOYES and H. FIELDELDEY, in: Three-Particle Scattering in Quantum Mechanics, ed.: J. GILLESPIE and J. NUTTALL (Benjamin, New York 1968).

32. The details are given in: E. ALT, P. GRASSBERGER and W. SANDHAS, Bonn-preprint 2-55 (1969), where also the three-body bound-state problem is studied.

33. J. S. BALL and D. Y. WONG, Phys. Rev. 169, 1362 (1968) Compare also: T. A. OSBORN, SLAC-report 79 (1967).

34. V. A. ALESSANDRINI, H. FANCHIOTTI and C. A. GARCIA, Phys. Rev. 170, 935 (1968).

35. Note, that in contrast to the above results, variational calculations usually yield too small values of ΔE. Compare e.g., the discussion of the whole problem in [31], Section V. 1, where further methods and results are reviewed.

36. Compare the references 2 - 7 given in [37] and furthermore:

R. G. NEWTON, J. Math. Phys. 8, 851 (1967);
R. OMNES, Phys. Rev. 165, 1265 (1968).

37. P. GRASSBERGER and W. SANDHAS, Nucl. Phys. B2, 181 (1967).

38. Applying such treatments, it is not necessary to start by first writing integral equations with connected kernels. This is in contrast to claims often made (compare also the discussion of this point in [29] and in Section 9).

39. More general expressions for the "potential" occur if nonpolar rest terms are retained in the subsystems (compare the generalizations of the treatment of Section 6 given in Section 7).

40. E. O. ALT, P. GRASSBERGER and W. SANDHAS, Bonn-preprint 2-48(1968).

CP VIOLATION AND COSMOLOGICAL FIELDS†

By

O. NACHTMANN

CERN, Geneva

1. INTRODUCTION

The rather surprising fact that the decay properties of the neutral K mesons can provide information on the gravitational interaction of these particles was first discovered by Good [1] long before CP violation. He realized that the very observation of a short and long lived component implied that the gravitational energies of K^0 and $\bar{K}^0$ could not be very different. So the $\bar{K}^0$ cannot fall upwards as was proposed sometimes for antiparticles. The discovery of CP violation [2] brought increased interest to this field. A number of suggestions were made to explain the observed CP violation by the action of some unknown cosmological field [3]- [6]. Most of these explanations have now been ruled out by experiment. But we can take another point of view, as Professor Thirring did last year here in Schladming [7], namely to ask what information we can get from existing experiments on cosmological fields. We will explore this below in a simple field theoretical model. But before we will review the propagator method for defining the mass matrix as given by Sachs [8], [9].

† Seminar given at the VIII. Internationale Universitätswochen für Kernphysik, Schladming, Feb. 24 - March 8, 1969.

We use units $\hbar = c = 1$, metric $g_{00} = -g_{11} = -g_{22} = -g_{33} = 1$.

2. THE MASS MATRIX AND ITS INVARIANCE PROPERTIES

In the approach of Sachs, one assumes the existence of fields describing the K mesons. Now we find it convenient to work with Hermitian fields ϕ_1, ϕ_2 corresponding to the CP eigenstates K_1, K_2 which are the following linear combinations of K^o and $\bar{K}^o$

$$K_1 = \frac{1}{\sqrt{2}} (K^o - \bar{K}^o)$$

$$K_2 = \frac{1}{\sqrt{2}} (K^o + \bar{K}^o)$$

An expansion of these fields in terms of creation and annihilation operators would be

$$\phi_1(x) = \sum \frac{i}{\sqrt{2\omega V}} (e^{ikx} a_1^+(k) - e^{-ikx} a_1(k))$$

$$\phi_2(x) = \sum \frac{1}{\sqrt{2\omega V}} (e^{ikx} a_2^+(k) + e^{-ikx} a_2(k)) \quad (2.1)$$

where $a_{1,2}^+$ create the states $K_{1,2}$ respectively* .

It is useful to list the transformation properties of these fields under the operations CP, T and Θ = CPT (Table 1).

The essential tool in the further analysis is the propagator, defined by

$$\Delta_{ij}(x-x') = i\langle 0|T(\phi_i(x),\phi_j(x'))|0\rangle = \qquad i,j=1,2$$

$$= \int \frac{dk}{(2\pi)^4} e^{-ik(x-x')} \Delta_{ij}(k^2) \quad (2.2)$$

---

* The factor i in the expansion for ϕ_1 corresponds to the phase conventions given in [10].

A	$CPA(CP)^{-1}$	$(TAT^{-1})^{+}$	$(\Theta A\Theta^{-1})^{+}$
$\phi_1(\vec{x},t)$	$\phi_1(-\vec{x},t)$	$\phi_1(\vec{x},-t)$	$\phi_1(-\vec{x},-t)$
$\phi_2(\vec{x},t)$	$-\phi_2(-\vec{x},t)$	$-\phi_2(\vec{x},-t)$	$\phi_2(-\vec{x},-t)$
$a_1(\vec{k})$	$a_1(-\vec{k})$	$-a_1^+(-\vec{k})$	$-a_1^+(\vec{k})$
$a_2(\vec{k})$	$-a_2(-\vec{k})$	$-a_2^+(-\vec{k})$	$a_2^+(\vec{k})$

Table 1

If we turned off all interactions, the propagator would have the simple structure

$$\Delta(k^2) = \frac{1}{m_o^2 - k^2} \tag{2.3}$$

where m_o is the common mass of $K_{1,2}$ and Δ is understood to be a multiple of the unit 2×2 matrix.

The propagator in presence of interaction can be calculated in principle [11] and is given by

$$\Delta(k^2) = [m_o^2 + \Pi^*(k^2) - k^2]^{-1} \tag{2.4}$$

$\Pi^*(k^2)$ is the sum of all proper self-energy diagrams. Since in our case the coupling is weak, it is sufficient to approximate $\Pi^*(k^2)$ by a constant: $\Pi^*(m_o^2+i\varepsilon)$. So the propagator reads finally* :

$$\Delta(k^2) = [W - k^2]^{-1} \tag{2.5}$$

$$W = m_o^2 + \Pi^*(m_o^2 + i\varepsilon) \tag{2.6}$$

W, of course, is a 2×2 matrix containing transitions

* If we take into account strong interactions, $\Delta(k^2)$ of (2.5) is understood to be the pole term of the propagator.

from K_1 to K_2.

It has been shown in Refs. [8], [9] that the mass matrix $\mathcal{M}$ governing the decay of $K_{1,2}$ is given by the square root of W

$$\mathcal{M} = W^{\frac{1}{2}} = m_o + \frac{1}{2m_o} \Pi^*(m_o^2 + i\varepsilon) \qquad (2.7)$$

If we construct the propagator from the fields ϕ_1, ϕ_2 we will get $\mathcal{M}$ in the following basis:

$$\mathcal{M} = \begin{pmatrix} \langle iK_1|\mathcal{M}|iK_1\rangle & \langle iK_1|\mathcal{M}|K_2\rangle \\ \langle K_2|\mathcal{M}|iK_1\rangle & \langle K_2|\mathcal{M}|K_2\rangle \end{pmatrix} \qquad (2.8)$$

Any invariance of the theory will reflect itself in an invariance of the propagator and therefore also of the mass matrix $\mathcal{M}$. Let us for example assume CPT invariance. Then Table 1 tells us that

$$\begin{aligned} \Delta_{ij}(x-x') &= \Theta(t-t')i\langle 0|\phi_i(x)\phi_j(x')|0\rangle + \\ &\quad + \Theta(t'-t)i\langle 0|\phi_j(x')\phi_i(x)|0\rangle = \\ &= \Theta(t-t')i\langle 0|\Theta\phi_i(-x)\phi_j(-x')\Theta^{-1}|0\rangle + \\ &\quad + \Theta(t'-t)\ i\ \langle 0|\Theta\phi_j(-x')\phi_i(-x)\Theta^{-1}|0\rangle = \\ &= \Theta(t-t')i\langle 0|\phi_j(-x')\phi_i(-x)|0\rangle + \\ &\quad + \Theta(t'-t)i\langle 0|\phi_i(-x)\phi_j(-x')|0\rangle = \Delta_{ji}(x-x') \end{aligned}$$

From this, and from corresponding arguments for CP and T, follow the relations for the mass matrix given below in Table 2.

Invariance	Relation for $\mathcal{M}$
CP	$\langle iK_1\|\mathcal{M}\|K_2\rangle = \langle K_2\|\mathcal{M}\|iK_1\rangle = 0$
T	$\langle iK_1\|\mathcal{M}\|K_2\rangle = -\langle K_2\|\mathcal{M}\|iK_1\rangle$
Θ	$\langle iK_1\|\mathcal{M}\|K_2\rangle = \langle K_2\|\mathcal{M}\|iK_1\rangle$

Table 2

3. LAGRANGIAN MODEL FOR INTERACTION OF K MESONS WITH COSMOLOGICAL FIELDS

Let us write down the most general Lagrangian for the interaction of two K mesons with scalar, vector and tensor fields

$$\begin{aligned} \mathcal{L}(x) = \; & \frac{1}{2} f_{ik} : \phi_i(x) \; \phi_k(x) :+ \\ & + \frac{1}{2} f^{\mu}_{ik} : \phi_{i,\mu}(x) \; \phi_k(x) : + \\ & + \frac{1}{2} f^{\mu\nu}_{ik} : \phi_{i,\mu}(x) \; \phi_{k,\nu}(x) : \end{aligned} \quad (3.1)$$

The double points mean as usual normal ordering. Summation over i,k = 1,2 is implied. The f's are supposed to be cosmological fields generated by the surrounding heavy objects as the earth, the sun or the galaxy and are taken as constants. It must, however, be emphasized that it is difficult to go beyond the external field approximation. It was shown in Ref. [6] that quantization of a long range vector field will lead to an enormous production rate if the field is not coupled to a conserved current.The same argument can be used to conclude that the gravitational coupling must be universal. Still it may be useful to ask how well this is borne out by experiment, disregarding paradoxes which arise on the quantum level.

Now, since $\mathcal{L}$ must be Hermitian, the f's have to be real. We get further restrictions on the f's by using the symmetry properties of the normal ordered products and by discarding total divergence expressions from the Lagrangian (3.1).

Since $:\phi_i(x)\phi_k(x):$ is symmetric in i,k,f_{ik} can also be taken to be symmetric

$$f_{ik} = f_{ki} = f^*_{ik} \tag{3.2}$$

In the second term of (3.1), we can use the identity

$$:\phi_{i,\mu}(x)\phi_k(x): = \tag{3.3}$$

$$= \frac{1}{2}:[\phi_{i,\mu}(x)\phi_k(x)-\phi_{k,\mu}(x)\phi_i(x)]: + \frac{1}{2}\frac{\partial}{\partial x^\mu}:\phi_i(x)\phi_k(x):$$

The divergence can be discarded. Because of the antisymmetry in i,k of the first term in (3.3), we can choose f^μ_{ik} to satisfy

$$f^\mu_{ik} = - f^\mu_{ki} = f^{\mu *}_{ik} \tag{3.4}$$

By analogous reasoning, we find, for $f^{\mu\nu}_{ik}$

$$f^{\mu\nu}_{ik} = f^{\nu\mu}_{ik} = f^{\mu\nu}_{ki} = f^{\mu\nu *}_{ik} \tag{3.5}$$

In this model the Feynman diagrams are of course very simple and we find for the self-energy matrix Π^*

$- i\, \Pi^*_{ik}$

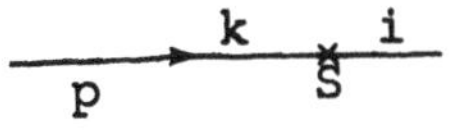

$i\, f_{ik}$

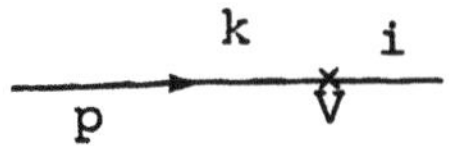

$- f^\mu_{ik}p_\mu$

$i f^{\mu\nu}_{ik}p_\mu p_\nu$ (3.6)

where S, V, T stand for scalar, vector and tensor.

The mass matrix is according to (2.7)

$$\mathcal{M} = \begin{pmatrix} M_1^{(o)} + \varepsilon_{11} & \varepsilon_{12} \\ \varepsilon_{21} & M_2^{(o)} + \varepsilon_{22} \end{pmatrix} \qquad (3.6')$$

$M_1^{(o)}$ and $M_2^{(o)}$ are the eigenvalues of the mass matrix when the cosmological fields are turned off. For simplicity we have assumed CP invariance in this case. The ε_{ik} are given below

$$\varepsilon_{11} = -\frac{1}{2m_o} f_{11}^{\mu\nu} p_\mu p_\nu$$

$$\varepsilon_{22} = -\frac{1}{2m_o} f_{22}^{\mu\nu} p_\mu p_\nu$$

$$\varepsilon_{12} = -\frac{1}{2m_o} \{f_{12} + if_{12}^{\mu} p_\mu + f_{12}^{\mu\nu} p_\mu p_\nu\}$$

$$\varepsilon_{21} = -\frac{1}{2m_o} \{f_{21} + if_{21}^{\mu} p_\mu + f_{21}^{\mu\nu} p_\mu p_\nu\} \qquad (3.7)$$

The terms with f_{11} and f_{22} have been absorbed in $M_1^{(o)}$ and $M_2^{(o)}$. p^μ is the four-momentum of the K meson in the laboratory system.

If we now compare the symmetry properties of the f's (3.2), (3.4), (3.5) with the invariance requirements for $\mathcal{M}$ under T and CPT (Table 2), we find that scalar and tensor fulfil CPT but violate T whereas vector coupling fulfils T but violates CPT. Translating the relations of Table 2 to the basis K^o, $\bar{K}^o$ this means that in the scalar and tensor case we get

$$\langle K^o|\mathcal{M}|K^o\rangle = \langle \bar{K}^o|\mathcal{M}|\bar{K}^o\rangle$$

So these interactions do not give rise to different masses for particle and antiparticle.

To get the states with exponential decay law, we have to construct the (non-orthogonal) eigenstates of $\mathcal{M}$

$$\mathcal{M}|K_S\rangle = M_S|K_S\rangle$$
$$\mathcal{M}|K_L\rangle = M_L|K_L\rangle \quad (3.8)$$

$$|K_S\rangle = |K_1\rangle + \varepsilon_1|K_2\rangle$$
$$|K_L\rangle = |K_2\rangle + \varepsilon_2|K_1\rangle \quad (3.9)$$

$$M_S = M_1^{(o)} + \varepsilon_{11}$$
$$M_L = M_2^{(o)} + \varepsilon_{22} \quad (3.10)$$

$$\varepsilon_1 = \frac{i\varepsilon_{21}}{M_S - M_L}$$
$$\varepsilon_2 = \frac{i\varepsilon_{12}}{M_S - M_L} \quad (3.11)$$

Let us recall the definition of the observable quantities η_{+-}, η_{oo} (see Ref. [10])

$$\eta_{\substack{+-\\oo}} = \frac{A(K_L \to \pi^{\substack{+\\o}}\,\pi^{\substack{-\\o}})}{A(K_S \to \pi^{\substack{+\\o}}\,\pi^{\substack{-\\o}})} \quad (3.12)$$

Since we have assumed CP invariance in the absence of cosmological fields, only the K_1 component of K_L can decay into 2π's and we get

$$\eta_{+-} = \eta_{oo} = \varepsilon_2 \quad (3.13)$$

If there are other sources of CP violation the resulting η will be the sum of all contributions. Below, we list $\eta_{\substack{+-\\oo}}$ and $\Delta m = \mathrm{Re}(M_L - M_S)$ for S,V,T

$$\eta_{\substack{+-\\oo}} = |\eta_{\substack{+-\\oo}}| \exp\{i\phi_{\substack{+-\\oo}}\}$$

	S	V	T
$\eta_{+-}=\eta_{oo}$	$\frac{f_{12}}{m_o(\Gamma_s-2i\Delta m)}$	$\frac{if^{\mu}_{12}p_{\mu}}{m_o(\Gamma_s-2i\Delta m)}$	$\frac{f^{\mu\nu}_{12}\, p_{\mu}\, p_{\nu}}{m_o(\Gamma_s-2i\Delta m)}$
$\phi_{+-}=\phi_{oo}$	$\pi/4,\ -3\pi/4$	$3\pi/4,\ -\pi/4$	$\pi/4,\ -3\pi/4$
$\Delta\ m$	$\Delta m^{(o)}$	$\Delta\ m^{(o)}$	$\Delta m^{(o)}+\frac{f^{\mu\nu}_{11}-f^{\mu\nu}_{22}}{2m_o}p_{\mu}p_{\nu}$

Table 3

In calculating $\phi_{+-} = \phi_{oo}$ we have used $\Delta m = \frac{1}{2}\,\Gamma_s$.
Γ_s : decay rate of the short lived component.

4.UPPER LIMITS FOR COSMOLOGICAL FIELDS

According to Section 3, we will regard the f's as static fields generated by cosmological objects. It will be convenient to compare these fields with the gravitational potential generated by the earth, sun, galaxy and universe, respectively.

Φ : gravitational potential $(\sim\frac{\kappa M}{r})$

f : gravitational coupling constant $f^2/8\pi = \kappa$

κ : Newton's gravitational constant

	$m_o\Phi$
earth	0.4 eV $\simeq$ $5.10^4\ \Gamma_s$
sun	6 eV $\simeq$ $8.10^5\ \Gamma_s$
galaxy	300 eV $\simeq$ $4.10^7\ \Gamma_s$
universe	500 MeV $\simeq$ $6.10^{13}\Gamma_s$

Table 4

a - The Scalar Case

Looking at the Lagrangian (3.1), we realize that we could as well interpret f_{ik} as coupling constants. This would then be a superweak model in the sense of Wolfenstein [12], [13]. Therefore, we see that a superweak model and a cosmological model with scalar coupling could only be distinguished if emission of real quanta of the cosmological field were observed.

From the experimental point of view, a superweak model is not ruled out by the latest experiments.

Reference	$\vert\eta_{+-}\vert$	$\vert\phi_{+-}\vert$
14	$(1.92 \pm 0.04).10^{-3}$	$(50 \pm 8)^{o}$
	$\vert\eta_{oo}\vert$	$\vert\phi_{oo}\vert$
15	$(2.3 \pm 0.3).10^{-3}$	
16	$(2.2 \pm 0.4).10^{-3}$	
17	$(3.2 \pm 0.7).10^{-3}$	$(37 \pm 30)^{o}$

Table 5

If we want over-all CP conservation, the cosmic field f_{12} must be odd under CP and also T. Therefore also its source will have to be odd under T, which leads to a vanishing thermodynamical expectation value for a stationary system. In other words, one cannot generate a T odd scalar field by a static distribution of matter in the universe. However, it is claimed in Ref.[5] that, by considering more general types of universes, one can overcome this difficulty* .

* I am grateful to Professor J. S. Bell for having drawn my attention to this point .

Let us compare anyway the strength of a superweak interaction, if it is generated by a cosmic potential to the strength of gravitation. We set, by dimensional arguments (f is again the gravitational coupling constant):

$$f_{12} = \frac{f_s}{f} \; \Phi \, m_o^2 \tag{4.1}$$

Inserting this in the expression for η_{+-} of Table 3, we get

$$\left| \frac{f_s}{f} \right| = 2.\sqrt{2} \; . \; 10^{-3} \; \frac{\Gamma_s}{m_o \, \Phi} \tag{4.2}$$

Assuming that only the earth, sun, etc., contribute to the scalar field we find the following orders of magnitude.

	$\lvert f_s/f \rvert$
earth	$\simeq 1{,}8.10^{-7}$
sun	$\simeq 3{,}5.10^{-9}$
galaxy	$\simeq 0{,}7.10^{-10}$
universe	$\simeq 5 \; .10^{-14}$

Table 6

We see that such a coupling would be even much weaker than gravitation, but would still have observable consequences on the elementary particle level.

b - The Vector Case

Such a coupling was proposed in Ref. [3] to explain the observed CP violation. As source of the vector field, the hypercharge current was used. This theory, however, would give an $|\eta_{+-}|$ dependent on the velocity of the K mesons in the laboratory system. Such a velocity dependence was looked for experimentally but was not found.

Moreover we find from Table 3 that ϕ_{+-} for this case does not agree with the experimental value (Table 5). So certainly this interaction is not responsible for the main part of observed CP violation. Experiments done on K mesons with laboratory energy up to 10 GeV/c [18] show a constant $|\eta_{+-}|$ to within 10%. From this we conclude

$$\left|\frac{if^{\mu}_{12}\, p_{\mu}}{m_o\sqrt{2}\,\Gamma_s}\right| \leq 2.10^{-4} \tag{4.3}$$

For f^{μ}_{12} we take analogously to (4.1)

$$f^{\mu}_{12} = \frac{f_v}{f}\, m_o\Phi \begin{pmatrix}1\\0\\0\\0\end{pmatrix} \tag{4.4}$$

We find, from (4.3), the following upper limits for $|f_v/f|$

$$\left|\frac{f_v}{f}\right| \lesssim 2.\sqrt{2}\ .\ 10^{-4}\,\frac{\Gamma_s}{m_o\Phi\gamma} \tag{4.5}$$

$$\gamma = \frac{p_o}{m_o} = \frac{1}{\sqrt{1-v^2}}$$

	$\lvert f_v/f\rvert$
earth	$\lesssim 3.10^{-10}$
sun	$\lesssim 2.10^{-11}$
galaxy	$\lesssim 4.10^{-13}$
universe	$\lesssim 2.10^{-19}$

$\gamma \simeq 20$ for 10 GeV/c momentum.

Table 7

We see that we can be quite sure of the absence of a long range vector field.

c - The Tensor Case

From the point of view of gravitation, a tensor interaction is of course of the greatest interest. We may note that the interaction usually written down in gravitation [19]

$$f \, \psi_{\mu\nu} \, T^{\mu\nu} \tag{4.6}$$

with $T^{\mu\nu}$ the energy momentum tensor of the particles

$$T^{\mu\nu} = \Phi'^{\mu} \, \Phi'^{\nu} - \frac{g^{\mu\nu}}{2} (\Phi_{,\lambda} \, \Phi'^{\lambda} - m_o^2 \, \Phi^2) \tag{4.7}$$

is completely equivalent to the tensor part of (3.1) as far as the mass matrix is concerned. This comes about because the part with $g^{\mu\nu}$ in (4.7) becomes proportional to $p^2-m_o^2$ which gives 0 on the mass shell.

As in the previous cases, we can derive upper bounds for the non-diagonal coupling $f_{12}^{\mu\nu}$ if we assume it proportional to the gravitational field (see Ref. [19] for the specific form)

$$f_{12}^{\mu\nu} = \frac{f_T}{f} \Phi \begin{pmatrix} 1 & & & \\ & 1 & & \\ & & 1 & \\ & & & 1 \end{pmatrix} \tag{4.8}$$

η_{+-} $_{oo}$ is now even more strongly dependent on γ , so the same argument as for the vector case gives:

$$\left| \frac{f_T}{f} \right| \lesssim 2 \sqrt{2} \, . \, 10^{-4} \, \frac{\Gamma_S}{2 m_o \Phi \gamma^2} \tag{4.9}$$

Now we can also test the diagonal couplings $f_{11}^{\mu\nu}$ and $f_{22}^{\mu\nu}$ by observing the dependence of the mass difference of K_L and K_S on the laboratory momentum. Table 8 below lists some experiments on Δm. We see that Δm is constant to about 10 % for γ up to 9.

Reference	Lab.momentum GeV/c	γ	Δm in units Γ_s
20	$\sim$ 2.7	5.5	0.44 ± 0.03
21	$\sim$ 0.8	1.9	0.42 ± 0.2
22	$\sim$ 4.5	9.0	0.48 ± 0.02
23	$\sim$ 1.0	2.2	0.42 ± 0.04
24			0.47 ± 0.02

Table 8

We now set

$$f_{11}^{\mu\nu} = \frac{f_1}{f}\Phi \begin{pmatrix} 1 & & & \\ & 1 & & \\ & & 1 & \\ & & & 1 \end{pmatrix} \tag{4.10}$$

$$f_{22}^{\mu\nu} = \frac{f_2}{f}\Phi \begin{pmatrix} 1 & & & \\ & 1 & & \\ & & 1 & \\ & & & 1 \end{pmatrix} \tag{4.11}$$

where f_1 and f_2 can be interpreted as the deviations of the gravitational couplings of K_1 and K_2 from the universal coupling. This gives the following upper limit for $(f_1-f_2)/f$:

$$\left|\frac{f_1-f_2}{f}\right| \lesssim \frac{5.10^{-2}\Gamma_s}{m_o\ \Phi\ \gamma^2} \tag{4.12}$$

The upper bounds for the off-diagonal and the difference of the diagonal couplings which one gets from (4.9) and (4.12) respectively, are listed below in Table 9. These values have to be compared with the Eötvös-Dicke experiment [25] which gives

$$\left|\frac{\delta f}{f}\right| \leq 10^{-11} \tag{4.13}$$

for the difference of the gravitational couplings of protons and neutrons. Since the gravitational field extends

	f_T/f	$\frac{f_1-f_2}{f}$
earth	$\lesssim 5.10^{-12}$	$\lesssim 1{,}2.10^{-8}$
sun	$\lesssim 3.10^{-13}$	$\lesssim 0{,}8.10^{-9}$
galaxy	$\lesssim 6.10^{-15}$	$\lesssim 1{,}6.10^{-11}$
universe	$\lesssim 4.10^{-21}$	$\lesssim 1{,}0.10^{-17}$

Table 9

certainly to the galaxy, we find that the universal gravitational coupling of K mesons is verified to the same accuracy as in the classical experiment for the diagonal coupling and four orders of magnitude better for the off-diagonal. We must, however, note that the Eötvös-experiment is sensitive to the field strengths whereas in the analysis of the K system the potential itself enters. Finally we have to say that, in the language of gravitational theory in both experiments, the so-called weak equivalence principle is tested.

ACKNOWLEDGMENTS

The author is grateful to Professor J. S. Bell and to Professor W. Thirring for illuminating discussions and critical reading of the manuscript.

REFERENCES

1. M. L.GOOD, Phys. Rev. 121, 311 (1961).
2. J. H. CHRISTENSON, J. W. CRONIN, V. L. FITCH and R. TURLAY, Phys. Rev. Letters 13, 138 (1964).
3. J. S. BELL and J. K. PERRING, Phys. Rev. Letters 13, 348 (1964).

4. J. BERNSTEIN, N. CABIBBO and T. D. LEE, Phys. Letters 12, 146 (1964); T. D. LEE, Phys. Rev. 137, B1621(1965).
5. F. GÜRSEY and A. PAIS, unpublished.
6. S. WEINBERG, Phys. Rev. Letters 13, 495 (1964).
7. W. THIRRING, Acta Phys. Austriaca, Suppl. V, (1968).
8. R. JACOB and R. G. SACHS, Phys. Rev. 121, 350 (1961).
9. R. G. SACHS, Ann. Phys. 22, 239 (1963).
10. J. S. BELL and J. STEINBERGER - Proceedings Oxford Conference (1965).
11. F. J. DYSON, Phys. Rev. 75, 1736 (1949).
12. L. WOLFENSTEIN and T. D. LEE, Phys. Rev. 138, B1490 (1965).
13. L. WOLFENSTEIN, Nuovo Cimento 42A, 17 (1966).
14. Review of Particle Properties, Revs. Modern Phys. (January 1969).
15. M. BANNER et al., Phys. Rev. Letters 21, 1107 (1968).
16. I. A. BUDAGOV et al., Phys. Letters 28B, 215 (1968).
17. J. CHOLLET et al., CERN Preprint, (Jan. 1969). This experiment was designe principally to measure ϕ_{oo}.
18. X. de BOUARD et al., Phys. Letters 15, 58 (1965).
19. W. THIRRING, Fortschr. d. Phys. 7, 79 (1959) and Ann. Phys. 16, 96 (1961).
20. C. ALFF-STEINBERGER et al., Phys. Letters 21, 595 (1966).
21. M. BALDO-CEOLIN et al., Nuovo Cimento 45A, 733 (1966).
22. M. BOTT-BODENHAUSEN et al., Phys. Letters 23, 277 (1966).
23. W. A. MEHLHOP et al., Phys. Review 172, 1613 (1968).
24. World average, taken from Ref. [14].
25. R. H. DICKE in: "Relativity, Groups and Topology", De Witt Ed., Les Houches (1963).

STOCHASTIC APPROACH TO QUANTUM MECHANICS AND TO QUANTUM FIELD THEORY†

By

W. GARCZYŃSKI

Institute for Theoretical Physics,
University of Wrócław, Poland

I. INTRODUCTION

In these notes I shall inform you about my recent work on the, what I call, quantum stochastic processes which has been proved to be the very natural language for expressing the contents of quantum mechanics and quantum field theory as well [1] -[7]. The theory of the quantum Markovian processes is especially useful there and permits rigorous mathematical formulation of the Dirac and Feynman heuristic approach to quantum theory. Since I already talked two weeks ago about the stochastic formulation of the nonrelativistic quantum mechanics at the VI-th Winter School for Theoretical Physics in Karpacz, organized by the University of Wroclaw [8], here I would like to emphasize the applications of the quantum Markovian processes to quantum field theory. Let me just mention that the main idea of nonrelativistic description of one particle moving in external field consists of introducing the density of the transition probability amplitude $(s,\underline{y}|t,\underline{x})$ from the point $\underline{y}$ at the time s to the point $\underline{x}$

† Seminar given at the VIII. Internationale Universitätswochen für Kernphysik, Schladming, Feb. 24 - March 8, 1969.

at the time $t > s$. Both $\underline{x}$ and $\underline{y}$ vary within the domain X which is a part of space R^3 accessible for a particle. The amplitude is a complex function satisfying certain conditions (confr. [1]) of which the most important one is the quantum causality condition

$$\int_X d\underline{z}(s,\underline{y}|\tau,\underline{z})(\tau,\underline{z}|t,\underline{x}) = (s,\underline{y}|t,\underline{x})$$

$$s < \tau < t$$

$$x,y \in X \quad . \qquad (1.1)$$

Using the well-known techniques from the theory of probabilities [9]-[17], we may construct corresponding to this amplitude the quantum Markovian process (x_t, N_t, M_y) and derive differential equations for the amplitude which are precisely the Schrödinger equations .

Simplest examples of transition amplitudes are:

A. The quantum Poisson process

$$(s,\underline{y}|t,\underline{x}) = \sum_{n=o}^{\infty} \frac{[i\lambda(t-s)]^n}{n!} \Pi_n(\underline{y},\underline{x}) \exp[-i\lambda(t-s)] \qquad (1.2)$$

where

$$\Pi_o(\underline{y},\underline{x}) = \delta(\underline{y}-\underline{x}) \quad ,$$

$$\Pi_n(\underline{y},\underline{x}) = \int_X d\underline{z}\; q(\underline{y},\underline{z})\Pi_{n-1}(\underline{z},\underline{x}) \,,\; n > o \qquad (1.3)$$

and $q(\underline{y},\underline{x})$ is a complex function satisfying the conditions

1) $q(\underline{y},\underline{x}) = q^*(\underline{x},\underline{y})$,

2) $|\Pi_n(\underline{y},\underline{x})| \le 1$.

B. The quantum Brownian motion

$$(s,\underline{y}|t,\underline{x}) = \left[\frac{m}{2\pi i\hbar(t-s)}\right]^{\frac{3}{2}} \exp\left[-\frac{m}{2i\hbar(t-s)}(\underline{y}-\underline{x})^2\right] \quad (1.4)$$

where m is the mass of a particle. This is simply the Green function of the free particle Schrödinger equation [19] - [21] . Corresponding to it the quantum Markovian process X_t will be called the Feynman process.

This process may be used for the definition of the Feynman integral in the same way as the Wiener process is used for the definition of the continual Wiener integrals, [7], [8], [22],

$$\int F[f]df = \int F[f_t] \prod_{t\in R_1} df_t = Q\, F[X_t] \quad (1.5)$$

where Q is the quantum expectation value (confr. e.g. [7]) and X_t is the Feynman process.

Similar in principle theory may be formulated for the relativistic case. One should replace $\underline{x}$ and $\underline{y}$ in (1.1) by the four-dimensional x and y elements of the Minkowski space lying on the space-like surfaces σ,σ' replacing times s,t. Integral over X should be replaced by the surface integral. Additionally, we assume that the amplitude $(\sigma,y|\sigma',x)$ vanishes when y-x become space-like. The Klein-Gordon equation may be then derived in the simplest case of one amplitude describing spinless relativistic particle.

Instead of discussing the particular relativistic cases let us start with the most general situation given by quantum field theory which includes as a special case the theory of many particle system (if one forbids the production processes and assumes only the Euclidean invariance of a theory).

As we know there are two different kinds of approaches to quantum field theory. The first, say classical one, starts with field operators satisfying several conditions (confr.,e.g., [26]) from which the S-matrix is constructed. The second approach proceeds in the inverse order. One starts with the S-matrix elements and formulates a theory entirely in terms of these elements [27]. However, one should emphasize the fact that so far it was not possible to formulate a theory in terms of observable quantities, i.e., in terms of S-matrix elements on the mass-shell only because of the lack of the causality postulate in this framework. The only place where the causality postulate was directly formulated is the so-called off-mass-shell formulation of S-matrix theory [29] which deals with unphysical scattering amplitudes. The causality postulate in this theory takes the form of the Bogoliubov condition [30], which corresponds essentially to the locality postulate in conventional field theory.

In this lecture we will discuss first the on-mass-shell formulation of S-matrix theory and give the causality postulate in terms of physical amplitudes. Then we shall go over into corresponding off-mass-shell formulation and derive the Bogoliubov causality postulate which was used extensively by Rzewuski in his functional formulation of quantum field theory [31]. Finally, we shall discuss the stochastic approach to the conventional formulation of quantum theory of fields.

2. ON-MASS-SHELL FORMULATION

In the simplest possible case of neutral spinless massive bosons one deals with the set of amplitudes

$$(\sigma;\underline{p}_1,\ldots,\underline{p}_m|\sigma';\underline{q}_1,\ldots,\underline{q}_n) = (\sigma;\underline{p}_m|\sigma';\underline{q}_n) \quad (2.1)$$

$$m,n = 0,1,\ldots, \quad \underline{p},\underline{q} \in R^3$$

describing the transitions of a system from the states, e.g., with m particles $\underline{p}_1,\dots,\underline{p}_m$ on the space-like surface σ to that with n particles with momenta $\underline{q}_1,\dots,\underline{q}_n$ on the surface σ' following σ : $\sigma < \sigma'$. The surfaces with finite σ,σ' appear in intermediate steps while physical amplitudes correspond to the case when $\sigma\to-\infty$, $\sigma'\to\infty$. In this case one thinks of transitions between different configurations of free particles. We impose the following conditions on the physical amplitudes (confr. [5]) :

(i) $$(\sigma;p_1,\dots,p_m|\sigma';q_1,\dots,q_n) =$$

$$= (\sigma;\underline{p}_{j_1},\dots,\underline{p}_{j_m}|\sigma';\underline{q}_{k_1},\dots,\underline{q}_{k_n})$$

where $(j_1,\dots,j_m)$, $(k_1,\dots,k_n)$ are any permutations of $(1,\dots,m)$ and $(1,\dots,n)$ numbers. This is the symmetry principle for bosons.

(ii) $$(\sigma;\underline{p}_m|\sigma';\underline{q}_n) = (\sigma';\underline{q}_n|\sigma;\underline{p}_m)^* ,$$

(time reversal invariance)

(iii) $$\lim_{\sigma\to\sigma'} (\sigma;\underline{p}_m|\sigma';\underline{q}_n) = \delta_{mn}\ \delta(\underline{p}_m;\underline{q}_m)$$

(time continuity property)

Here, $\delta(\underline{p}_m;\underline{q}_m)$ is the symmetrized δ-function

$$\delta(\underline{p}_m;\underline{q}_m) = \frac{1}{m!} \sum_{\pi\in S_m} \delta(\underline{p}_1-\underline{q}_{\pi(1)})\cdot\ldots\cdot\delta(\underline{p}_m-\underline{q}_{\pi(m)}) \tag{2.2}$$

(iv) $$\sum_{\ell=0}^{\infty} \int d\underline{k}_\ell (\sigma;\underline{p}_m|\sigma';\underline{k}_\ell)(\sigma';\underline{k}_\ell|\sigma;q_n) = \delta_{mn}\delta(\underline{p}_m;\underline{q}_m)$$

(unitarity condition) ,

(v) $$\sum_{\ell=0}^{\infty} \int d\underline{k}_\ell (\sigma';\underline{p}_m|\sigma;\underline{k}_\ell)(\sigma;\underline{k}_\ell|\sigma'';\underline{q}_n)=(\sigma';\underline{p}_m|\sigma'';\underline{q}_n) ,$$

(quantum causality condition), $\sigma' < \sigma < \sigma''$.

(vi-a) $$(\sigma|\sigma';\underline{q}_n) = 0 \ , \quad n > 0 \ ,$$

$$(\sigma|\sigma') = 1$$

(vacuum stability condition).

(vi-b) $$(\sigma;\underline{p}|\sigma';\underline{q}_n) = 0 \quad \text{for } n > 1$$

$$(\sigma;\underline{p}|\sigma';\underline{q}) = \delta(\underline{p}-\underline{q})$$

(one-particle stability condition)

(vii) $$(\sigma;\underline{p}_m|\sigma';\underline{q}_n) = (\sigma+a;\ L\underline{p}_m|\sigma'+a;\ L\underline{q}_n) \times$$

$$\times \exp ia \left(\sum_{j=1}^{m} p_j - \sum_{k=1}^{n} q_k\right)\Big|_{p_j^o=\omega_{p_j},\ q_k^o=\omega_{q_k}} ,$$

$$\omega_p = \sqrt{p^2+m^2} \ ; \quad L \in L_+^{\uparrow}(R)$$

(Lorentz invariance condition).

These are all postulates which we impose on the physical amplitudes. Clearly, the fifth one is most significant here since it determines the dynamics of a system. The unitarity postulate (iv) follows from it and the stochastic continuity condition (iii).

It is more convenient to introduce the generating functional $S[\sigma,\sigma';\alpha,\beta]$ instead of an infinite family of amplitudes and to express all basic assumptions (i-vii) in terms of it [31],

$$S[\sigma,\sigma';\alpha,\beta] = \tag{2.3}$$

$$= \sum_{m=o}^{\infty} \sum_{n=o}^{\infty} \frac{1}{\sqrt{m!n!}} \int d\underline{p}_m \int d\underline{q}_n (\sigma;\underline{p}_m|\sigma';\underline{q}_n) e[\alpha;\underline{p}_m] . e[\beta;\underline{q}_n]$$

where

$$\int d\underline{p}_m = \prod_{j=1}^{m} \int d\underline{p}_j \quad ,$$

$$e[\alpha;\underline{p}_m] = \prod_{j=1}^{m} \alpha(\underline{p}_j) \qquad (2.4)$$

$\alpha(\underline{p})$, $\beta(\underline{p})$ are real and smooth functions from $J_{\underline{p}}$ L. Schwartz space. A value $S[\sigma,\sigma';\beta,\alpha]$ at given α, β may be considered as the matrix element with indices α and β of continuous functional matrix $S[\sigma,\sigma']$. The postulates (i-vii) may be expressed in terms of this matrix as follows

(i) $S[\sigma,\sigma']$ - functional matrix

(ii) $S^+[\sigma,\sigma'] = S[\sigma',\sigma]$,

(iii) $\lim_{\sigma\to\sigma'} S[\sigma,\sigma'] = I$, $(\sigma < \sigma')$

(iv) $S[\sigma,\sigma'] * S^+[\sigma,\sigma'] = I$,

(v) $S[\sigma',\sigma] * S[\sigma,\sigma''] = S[\sigma',\sigma'']$, $\sigma' < \sigma < \sigma''$,

(vi-a) $\Omega[\sigma] * S[\sigma,\sigma'] = \Omega[\sigma]$,

(vi-b) $V_1[\sigma] * S[\sigma,\sigma'] = V_1[\sigma]$,

(vii) $\tilde{S}[\sigma,\sigma'] = \tilde{S}_{(a,L)}[\sigma,\sigma']$,

where $\tilde{S}[\sigma,\sigma';\alpha,\beta] = S[\sigma,\sigma'; \frac{\alpha}{\sqrt{2\omega}}, \frac{\beta}{\sqrt{2\omega}}]$,

$$\tilde{S}_{(a,L)}[\sigma,\sigma';\beta,\alpha] = \tilde{S}[\sigma+a,\sigma'+a; \alpha_{(a,L)},\beta_{(a,L)}] ,$$

$$\alpha_{(a,L)}(\underline{p}) = \alpha(L^{-1}p)\exp(iap)\Big|_{p^0=\omega_p}$$

$$\beta_{(a,L)}(\underline{p}) = \beta(L^{-1}p)\exp(-iap)\Big|_{p^0=\omega_p} . \qquad (2.5)$$

Any functional matrix $M[\alpha,\beta]$ is expressable (by definition) in form of the Volterra series

$$M[\alpha,\beta] = \tag{2.6}$$

$$= \sum_{m=o}^{\infty} \sum_{n=o}^{\infty} \frac{1}{\sqrt{m!n!}} \int d\underline{p}_m \int d\underline{q}_n \qquad M(\underline{p}_m;\underline{q}_n)\, e[\alpha;\underline{p}_m]\cdot e[\beta;\underline{q}_n]$$

The multiplication by numbers, sum of two functional matrices, the inverse matrix and hermitian conjugation operations are in natural algebraic way while the multiplication prescription "*" is the following

$$(M_1 * M_2)[\alpha,\beta] = M_1[\alpha,\frac{\delta}{\delta\rho}] M_2[\rho,\beta]\Big|_{\rho=o} \tag{2.7}$$

Correspondingly, the unit matrix I has elements

$$I[\alpha,\beta] = \exp(\alpha.\beta) \quad ,$$

$$\alpha.\beta = \int d\underline{p}\,\alpha(\underline{p})\ \beta(\underline{p}) \ . \tag{2.8}$$

The vacuum functional $\Omega[\sigma]$ is given by

$$\Omega[\sigma;\alpha] = 1 \quad \text{for any } \alpha \in J_{\underline{p}} \tag{2.9}$$

while one-particle functional $V_1[\sigma]$ is such that

$$V_1[\sigma;\alpha] = \int d\underline{p}\ v(\sigma;\underline{p})\,\alpha(\underline{p}) \tag{2.10}$$

From the causality condition (v) we get the following dynamical equations

$$i\hbar \frac{\delta S[\sigma,\sigma']}{\delta\sigma'(x)} = S[\sigma,\sigma'] * H[\sigma';x] \tag{2.11}$$

$$-i\hbar \frac{\delta S[\sigma,\sigma']}{\delta\,\sigma(y)} = H[\sigma;y] * S[\sigma,\sigma'] \tag{2.12}$$

where

$$H[\sigma';x] = i\hbar\ S^{+}[\sigma,\sigma'] * \frac{\delta S[\sigma,\sigma']}{\delta\sigma'(x)} \qquad (2.13)$$

By solving them for particular Hamiltonian $H[\sigma',x]$ we get the S-matrix generating functional after the limit $\sigma \to -\infty, \sigma' \ +\infty$ has been performed.

3. OFF-MASS-SHELL-FORMULATION

Going over into off-mass-shell formulation of quantum field theory we look for an analytic functional $J[\sigma,\sigma';q]$ where $q(x) \in J_x$ (x belongs to the Minkowski space M), such that

$$S[\sigma,\sigma';\alpha,\beta] = e^{\alpha\beta} J[\sigma,\sigma';q_o[\alpha,\beta]] \ , \qquad (3.1)$$

(confr. [6] and [31]), where $q_o[\alpha,\beta](x)$ is a general solution of the Klein-Gordon equation

$$q_o[\alpha,\beta](x) = \qquad (3.2)$$

$$= (2\pi)^{-3/2} \int dp\,\theta(p)\,\delta(p^2-m^2)\sqrt{2\omega_p}\{\alpha(\underline{p})e^{-ipx}+\beta(\underline{p})\ e^{ipx}\} \ .$$

The functional $J[\sigma,\sigma',q]$ is not fully determined by (3.1) because of the presence of $\delta(p^2-m^2)$ term in $q_o[\alpha,\beta]$. We assume however that $J[\sigma,\sigma';q]$ is an analytic functional, i.e.,

$$J[\sigma,\sigma';q] = \sum_{n=o}^{\infty} \frac{i^n}{n!} \int dx_n\ J(\sigma,\sigma';x_n)e[q;x_n] \ , \qquad (3.3)$$

$J(\sigma,\sigma';x_n)$ - symmetric functions.

The most natural off-mass-shell extension $J[\sigma,\sigma';q]$ of the functional S is determined by the following conditions:

(i) $\qquad J[\sigma,\sigma';q]$ - analytic functional,

(ii) $\{J[\sigma,\sigma';q]\}^{*} = J[\sigma',\sigma;q^{*}]$,

(iii) $\lim_{\sigma\to\sigma'} J[\sigma,\sigma';q] = 1 \quad (\sigma < \sigma')$

(iv) $J[\sigma,\sigma';q]\ (*)\ J^{*}[\sigma,\sigma';q] = 1$,

(v) $J[\sigma',\sigma;q]\ (*)\ J[\sigma,\sigma'';q] = J[\sigma',\sigma'';q]$

(vi-a) $e[f^{*}\frac{\delta}{\delta q};\underline{q}_n]\ J[\sigma,\sigma';q]\Big|_{q=o} = 0 \quad , n > o ,$

$$J[\sigma,\sigma';q]\Big|_{q=o} = 1 ,$$

(vi-b) $e[f^{*}\frac{\delta}{\delta q};\underline{q}_n]\, f\frac{\delta}{\delta q}(\underline{p})\, J[\sigma,\sigma';q]\Big|_{q=o} = 0 , \quad n\geq 1$

(vii) $J[\sigma,\sigma';q] = J[\sigma+a,\sigma'+a;q_{(a,L)}]$

$$q_{(a,L)}(x) = q[L^{-1}(x-a)] \quad , L\varepsilon L_{+}^{\uparrow}(R) \quad .$$

Here (*) means the following multiplication prescription

$$(*) = \exp\{i\ \frac{\overleftarrow{\delta}}{\delta q}\ \Delta^{+}\ \frac{\overrightarrow{\delta}}{\delta q}\} \tag{3.4}$$

where Δ^{+} is the usual Pauli-Jordan function .

The dynamical equations (2.11) and (2.12) are replaced here by the equations

$$i\hbar\ \frac{\delta}{\delta\sigma'(x)}\ J[\sigma,\sigma';q] = J[\sigma,\sigma';q]\ (*)\ \mathscr{H}\,[\sigma',x;q] \ , \tag{3.5}$$

$$-i\hbar\ \frac{\delta}{\delta\sigma(y)}\ J[\sigma,\sigma';q] = \mathscr{H}\,[\sigma,y;q\ ]\ (*)\ J[\sigma,\sigma';q] \tag{3.6}$$

where the Hamiltonian $\mathscr{H}$ gives the off-mass-shell extension of H-matrix,

$$H[\sigma;y;\alpha,\beta] = e^{\alpha\beta}\ \mathscr{H}\,[\sigma,y;q_o[\alpha,\beta]] \quad , \tag{3.7}$$

$$\mathcal{H}[\sigma;y;q] = i\hbar\, J^{+}[\sigma,\sigma';q]\ (*)\ \frac{\delta}{\delta\sigma(y)}\, J[\sigma,\sigma';q] \tag{3.8}$$

The surface dependence of $J[\sigma,\sigma';q]$ may be a priori very complicated. However, in the special case which leads to the Rzewuski formulation of quantum field theory [31] it is rather simple and given by

$$J[\sigma,\sigma';q] = \Omega\,[q_{\sigma\sigma'}] \tag{3.9}$$

where $\Omega[q]$ is some analytic functional and $q_{\sigma\sigma'}$ is the restriction of q to the domain between σ and σ' surfaces

$$q_{\sigma\sigma'}^{(x)} = \begin{cases} q(x) & \text{if } \sigma \leq x \leq \sigma' \ , \\ 0 & \text{outside} \end{cases} \tag{3.10}$$

Clearly, $\Omega[q]$ determines the S-matrix elements since

$$S[\alpha,\beta] = \lim_{\substack{\sigma\to-\infty \\ \sigma'\to\infty}} S[\sigma,\sigma';\alpha,\beta] = e^{\alpha\beta}\Omega[q_0[\alpha,\beta]] \tag{3.11}$$

The functional $\Omega[q]$ satisfies the following conditions which are implied by the postulates (i-vii) for $J[\sigma,\sigma';q]$ functional:

1) $\Omega[q]$ - analytic functional ,

2) $\Omega[o] = 1$,

$$e[f^{*}\tfrac{\delta}{\delta q};\underline{q}_n]\,\Omega[q]\Big|_{q=o} = 0 \ , \quad n > o \ ,$$

$$e[f^{*}\tfrac{\delta}{\delta q};\underline{q}_n]\, f\tfrac{\delta}{\delta q}(\underline{p})\,\Omega[q]\Big|_{q=o} = 0 \ , \quad n \geq o$$

3) $\Omega[q]\ (*)\ \Omega^{*}[q] = 1$, q real ,

4) $\Omega[q_{\sigma'\sigma}]\ (*)\,\Omega[q_{\sigma\sigma''}] = \Omega[q_{\sigma'\sigma''}]$, $\sigma'<\sigma<\sigma''$

5) $\Omega[q] = \Omega[q_{(a,L)}]$, $L \in L_+^+(R)$.

The main condition is the 4-th one which expresses the Markovian property of the dynamics. We may easily derive from it the following differential causality condition

$$\frac{\delta}{\delta q(y)}\{\Omega[q](*)\ \frac{\delta\Omega^*[q]}{\delta q(x)}\} = 0 \quad \text{for } x \lesssim y\ , \qquad (3.12)$$

which was postulated by Rzewuski [29] and studied extensively in his new forthcoming book [31] on the functional formulation of quantum field theory.

We shall close this section with the remark that it should be possible to derive from the quantum causality condition (iv) all the relevant analytic properties of amplitudes which are usually postulated in the S-matrix theory. Moreover, this condition together with the unitarity postulate may serve as a general guide in construction of phenomenological models which one gets by cutting off the sums over the number of particles involved in the intermediate states.

4. STOCHASTIC APPROACH TO QUANTUM FIELD THEORY

Now let us briefly describe the stochastic approach to the Wightman type theory of local quantum field [32]. We discuss here the simplest possible case of one, neutral, scalar self-interacting field which is defined as a generalized quantum stochastic field $A(f)$ [33],[34] satisfying certain conditions described below.

Let us assume that for any real functions $f_1(x),\ldots$ $\ldots,f_n(x)$ from the L.Schwartz space J_x , $x \in M$, there are given real random variables $\{A(f_1),\ldots,A(f_n)\}$ such that their common distribution amplitudes

$$m\begin{pmatrix}\xi_1,\dots,\xi_n\\ A(f_1),\dots,A(f_n)\end{pmatrix} = M\{A(f_1)<\xi_1,\dots,A(f_n)<\xi_n\} \qquad (4.1)$$

satisfy the selfconsistency requirements,i.e.,

$$m\begin{pmatrix}\xi_1,\dots,\xi_n\ ,\ \infty\\ A(f_1),\dots,A(f_n),A(f)\end{pmatrix} = m\begin{pmatrix}\xi_1,\dots,\xi_n\\ A(f_1),\dots,A(f_n)\end{pmatrix} \qquad (4.2)$$

for any real f and $f_1,\dots,f_n \in J_x$. Here $m(^{\xi}_{X})$ is the distribution function of the variable X,i.e.,

$$m(^{\xi}_{X}) = M\{\omega : X(\omega) < \xi\} \qquad (4.3)$$

where ω is an element of the space Ω and M is an amplitude on the σ-algebra $\mathcal{B}$ of subsets of Ω (i.e., it is a complex charge having the value zero for the empty set $\phi \in \mathcal{B}$ and the value one for Ω ,$|M(A)| \leq 1$ for any $A \in \mathcal{B}$ (confr. [7] and[8])) .

We assume further that A(f) is linear in f in a sense that the random variables $A(\alpha f+\beta g)$ and $\alpha A(f)+\beta A(g)$ are stochastically equivalent

$$A(\alpha f+\beta g) = \alpha A(f) + \beta A(g)\ ,\quad f,g \in J \qquad (4.4)$$

α,β - real numbers.

We shall assume also the continuity property of A(f)

$$\lim_{n\to\infty} A(f_n) = A(f) \qquad (4.5)$$

if

$$\lim_{n\to\infty} f_n = f$$

in topology of J_x.

This condition means that the distribution functions for $A(g_1),\dots,A(g_m)$, $A(f_n)$ variables are weakly convergent to the distribution functions of $A(g_1),\dots,A(g_m),A(f)$

variables for any $g_1,\dots,g_m \in J$ and on the space of bounded and continuous functions $\phi(\xi_1,\dots,\xi_m,\xi)$.

We assume moreover the Poincaré invariance of A(f), i. e., we assume that

$$A(f) = A(f_{(a,L)}) \qquad f \in J \tag{4.6}$$

in the sense of stochastical equivalence. In this case the moments of A(f) are Lorentz invariant and homogeneous distributions on J.

$$\kappa(f_1,\dots,f_n) = Q[A(f_1)\cdot\ldots\cdot A(f_n)] = \tag{4.7}$$

$$= \int \xi_1\cdot\ldots\cdot\xi_n \, dm\begin{pmatrix}\xi_1,\dots,\xi_n\\ A(f_1),\dots,A(f_n)\end{pmatrix} = \kappa(f_{1(a,L)},\dots,f_{n(a,L)})$$

We say that a theory is local if the variables $A(f_1)$ and $A(f_2)$ are stochastically independent for $f_1(x)$ and $f_2(y)$ space-like separated functions, i.e., if

$$f_1(x)f_2(y) = 0 \quad \text{for} \quad x-y \quad \text{time-like}$$

then

$$Q[A(f_1)A(f_2)] = Q[A(f_1)]\, Q[A(f_2)] \, . \tag{4.8}$$

In this case the characteristic functional

$$L(f) = Qe^{iA(f)} = \int e^{i\xi} dm\binom{\xi}{A(f)} \tag{4.9}$$

satisfies the equation

$$L(f_1+f_2) = L(f_1)\cdot L(f_2) \tag{4.10}$$

The separated condition may be expressed as the support condition for the Fourier transform of moments

$$\kappa(f_1,\dots,f_n) = \tilde{\kappa}(\tilde{f}_1,\dots,\tilde{f}_n) \tag{4.11}$$

$$\tilde{f}(p) = \int e^{ipx} f(x)\, d^4x \in J_p$$

We say that A(f) satisfies the spectral condition if $\tilde{\kappa}(p_1,\ldots,p_n)$ vanishes outside of the domain

$$p_1 \varepsilon \bar{V}_+ \;, \; p_1+p_2 \varepsilon \bar{V}_+,\ldots,p_1+\ldots+p_{n-1} \varepsilon \bar{V}_+ \;,$$

$$p_1+\ldots+p_n = 0 \tag{4.12}$$

Finally, the completeness of A(f) may be expressed as follows. We say that A(f) is a complete stochastic field, if the relation

$$Q[A(f).X] = Q[A(f)].Q[X] \quad \forall \; f \varepsilon J_X \tag{4.13}$$

implies that the variable X is constant.
This means that there are no stochastically independent variables outside of the set

$$\{A(f) \; ; \; f \; \varepsilon \; J_X\} \tag{4.14}$$

besides the trivial one.

Thus we have expressed the physical contents of local relativistic quantum field theory in the language of the generalized quantum stochastic fields. One may hope that it will be possible to find non-trivial solutions (besides the quantum Gaussian processes corresponding to the free field case) fulfilling above axioms, or at least to answer the existence question. This hope is based on the analogy with the classical generalized stochastic processes theory where the existence problem has been solved for various cases and strong technical means were developed.

The author is thankful to Professor K. Urbanik for introducing into recent advances in the stochastic processes theory.

REFERENCES

1. W. GARCZYNSKI, Quantum Mechanics as Quantum Markovian Process, Acta Phys. Polon. (in Press).
2. W. GARCZYNSKI, Quantum Markovian Process for One Nonrelativistic Particle, Bull. Acad. Polon. Sci., (in press).
3. W. GARCZYNSKI, Quantum Markovian Process for Nonrelativistic Particle With Spin, Bull Acad. Polon.Sci., (in press).
4. W. GARCZYNSKI, Quantum Markovian Process With Denumerable Set of States, Bull. Acad. Polon. Sci., (in press).
5. W. GARCZYNSKI, Dynamical Equations for Spinless, Neutral Bosons System. Part I, On Mass-shell Formulation, Acta Phys. Polon., (in press).
6. W. GARCZYNSKI, Dynamical Equations for Spinless, Neutral Bosons System, Part II, Off Mass-shell Formulation, Acta Phys. Polon., (in press).
7. W. GARCZYNSKI, On Quantum Stochastic Processes, Bull. Acad . Polon. Sci., (in press) .
8. W. GARCZYNSKI, Lecture at the VI-th Winterschool for Theoretical Physics in Karpacz, February 1969, Proceedings.
9. A. N. KOLMOGOROV, Math. Ann., 104, 415 (1931).
10. A. N. KOLMOGOROV, Uspekhi Mat. Nauk., 33, 567 (1932).
11. W. FELLER , An Introduction to Probability Theory and Its Applications, Vol. II., John Wiley and Sons, Inc. 1966.
12. M. LOÈVE, Probability Theory, Sec. Ed. D. Van Nostrand Co. Inc. 1960.
13. J. L. DOOB, Stochastic Processes, New York, 1953 .
14. E. B. Dynkin, Osnovanija teorii markovskikh processov (in Russian), Moscow 1959.
15. E. B. DYNKIN, Markovskije Processy, (in Russian),

Moscow 1963.

16. K. ITÔ, Stochastic Processes, I and II (in Japanese), Russian translation, Moscow 1960.
17. K. ITÔ and H. D. McKean Jr., Diffusion Processes, and their Sample Paths, Springer-Verlag Berlin 1965.
18. M. SMOLUCHOWSKI, Bull. Int. Acad. Polonaise, Sci. Lett., A, 418 (1913).
19. P. A. M. DIRAC, The Principles of Quantum Mechanics, 4-th ed., Oxford 1958.
20. R. P. FEYNMAN, Rev. Mod. Phys. 20, 367 (1948).
21. R. P. FEYNMAN AND A. R. HIBBS, Quantum Mechanics and Path Integrals, New York 1965.
22. L. STREIT, Acta Phys. Austriaca, Suppl. II, 2 (1965).
23. E. W. MONTROLL, Comm. Pure and Applied Math. 5,415 (1952).
24. J. M. GELFAND and A. M. YAGLOM, Usp. Mat. Nauk, Vol. XI, 77 (1956).
25. M. KAC, Probability and Related Topics in Physical Science, Inc. London 1959.
26. A. S. WIGHTMAN, Phys. Rev. 101 (1956), 860.
27. W. HEISENBERG, Z. Physik., 120 (1943), 513 and 673.
28. G. F. CHEW, S-Matrix Theory of Strong Interactions, Inc. New York 1962.
29. J. RZEWUSKI, Report at the VI-th Winterschool in Karpacz, February 1967, Proceedings, Vol. I, 1-42.
30. N. N. BOGOLIUBOV and D. V. SHIRKOV, Introduction to the Theory of Quantized Fields (in Russian), Moscow 1957.
31. J. RZEWUSKI, Field Theory, Part II (to appear).
32. R. JOST, The General Theory of Quantized Fields, Am. Math. Soc., Providence 1965.
33. J. M. GELFAND and J. J. VILENKIN, Generalized Functions, Vol. 4, (in Russian), Moscow 1961.
34. K. URBANIK, Teoria verojatnostej i jejo prim. 1, 146 (1956).

EXTENSION OF THE CONVENTIONAL FRAMEWORK OF LOCAL QUANTUM FIELD THEORY AND THE DESCRIPTION OF RESONANCES†

By

Jerzy LUKIERSKI

Institute of Theoretical Physics, University of Wroclaw,
Wroclaw, Poland

1. INTRODUCTION

In last ten years many authors have attempted to formulate a consistent relativistic quantum field theory with unstable particles [1]. At present one can state only that the role of unstable particles was clarified in some soluble models: the lowest Lee model sectors [2], Zachariasen model [3]. The main difficulties in the construction of general relativistic theory with unstable particles appeared to be

1. to incorporate correctly the instability property into the theory,
2. to get the consistency with an experimental fact that in asymptotic space - time regions there are only observed free stable particles.

The aim of this lecture is to discuss the applicability to the description of strongly unstable objects the field operator $\phi(x;s)$, with <u>continuous</u> spectrum of asymptotic masses.

† Seminar given at the VIII. Internationale Universitätswochen für Kernphysik, Schladming, Feb. 24 - March 8,1969.

Such field operator were introduced by Licht [4], who also give suitable generalization of the LSZ asymptotic condition. Our purpose is to show that the field operator $\phi(x;s)$ $(s \geq s_0 > o)$ can be used in the construction of self-contained quantum field theory with unstable elementary objects.

Such formalism represents an alternative to the conventional formulation of quantum field theory, where it is assumed that every intermediate state occurring in the collision process is described solely in terms of free stable particles, with definite spins and masses. In conventional formulation any unstable state is expressed in terms of multiparticle states, with suitably peaked total mass spectrum. The dynamical appearance of these mass peaks cannot be explained, however, in terms of finite sums of Feynman diagrams, generated by renormalizable local interaction of stable particles. The ordinary perturbation theory becomes useless, and the only way is to separate out the terms, which are not approximated by perturbation series. Such procedure was proposed in nonrelativistic scattering theory by Weinberg [5] who introduced so-called quasi-particles. In the field-theoretic formulation one can describe these additional elementary objects by separate field operators. If we consider unstable quasi-particles (resonances), the corresponding field operator cannot be characterized by definite value of the mass parameter, and we are forced to introduce the continuous mass parameter [6],[7].

All further considerations will be presented for spin zero stable particles and the unstable elementary objects representing S-wave two-particle resonances. All the results have, however, rather obvious generalization to the case of a physical system of interacting stable particles with spin and arbitrary multiparticle resonances [8].

2. FIELD THEORY WITH CONTINUOUS ASYMPTOTIC MASS SPECTRUM AND THE DESCRIPTION OF ASYMPTOTICALLY INDEPENDENT PARTIAL WAVES

In usual formulation of relativistic QFT [9] one introduces as a basic object the field operator $\phi(x)$, which defines by means of the asymptotic condition the full set of asymptotic fields [10]. These asymptotic fields describe free stable particles, which are observed in the scattering experiments at infinite past and infinite future. The asymptotic creation and annihilation operators $a_{\substack{in\\out}}(\vec{p})$, $a^{+}_{\substack{in\\out}}(\vec{p})$, satisfying the commutation relations

$$[a^{+}_{\substack{in\\out}}(\vec{p}) \, , \, a_{\substack{in\\out}}(\vec{p}\,')] = \delta(\vec{p}-\vec{p}\,') \tag{2.1}$$

define by means of the relation

$$|\vec{p}_1,\ldots,\vec{p}_n>_{\substack{in\\out}} = \prod_{i=1}^{n} a_{\substack{in\\out}}(\vec{p}_i)|0> \tag{2.2}$$

the basic state vectors, which span the asymptotic spaces of states. The $\hat{S}$-matrix relates these states

$$\hat{S}|\vec{p}_1,\ldots,\vec{p}_n>_{in} = |\vec{p}_1,\ldots,\vec{p}_n>_{out} \tag{2.3}$$

and defines the conventional scattering amplitudes.

Let us imagine, however, the world in which we observe asymptotically <u>only</u> correlated multiparticle states. For simplicity we shall assume that the asymptotic measurements register only the two-particle S-wave states, with its invariant masses $s \geq s_0 > 0$. Further we make a <u>dynamical assumption</u>, that for sufficiently large times these pairs can be treated as independent ones. In order to describe suchp<u>physical</u> situation, with asymptotically free and independent S-wave pairs, one can use the operators $a_{\substack{in\\out}}(\vec{p};s)$, $a^{+}_{\substack{in\\out}}(\vec{p};s)$, where s is a continuous parameter, satisfying the relation $s \geq s_o > 0$, and

$$[a^+_{\substack{in\\out}}(\vec{p};s),\ a_{\substack{in\\out}}(\vec{p}\,';s)] = \delta(\vec{p}-\vec{p}\,')\ \delta(s-s') \tag{2.4}$$

The one-particle asymptotic states

$$|\vec{p};s\rangle_{\substack{in\\out}} = a_{\substack{in\\out}}(\vec{p};s)\,|0\rangle \tag{2.5}$$

describe the S-wave pair with total four-momentum $p_\mu =$ $= (\vec{p}, \sqrt{\vec{p}^2+s}\,)$, which can be treated as a primary elementary object [11]. If we wish to introduce n independent asymptotic S-wave pairs, with four-momenta $p_{\mu i} = (\vec{p}_i, \sqrt{\vec{p}_i^2+s_i})$, one can generalize (2.5) as follows:

$$|\vec{p}_1,\ldots,\vec{p}_n;s_1,\ldots,s_n\rangle_{\substack{in\\out}} = \prod_{i=1}^{n} a_{\substack{in\\out}}(\vec{p}_i;s_i)\,|0\rangle \tag{2.6}$$

Let us consider underlying quantum field theory, with basic field operator $\phi(x;s)$, which leads to the asymptotic states (2.6). In such field theory the asymptotic fields $\phi_{\substack{in\\out}}(x;s)$ satisfy the following relations:

$$(\Box - s)\ \phi_{\substack{in\\out}}(x;s) = 0 \tag{2.7}$$

$$[\phi_{\substack{in\\out}}(x;s),\ \phi_{\substack{in\\out}}(x';s')] = i\Delta(x-x';s)\,\delta(s-s') \tag{2.8}$$

One assumes for the field operator $\phi(x;s)$ the conventional basic postulates of relativistic QFT

1. Poincaré Invariance
2. Locality
3. Asymptotic Condition

The postulate of Poincaré invariance is obvious. The locality condition can be expressed as follows:

$$[\phi(x;s),\ \phi(x';s')] = 0 \qquad \begin{matrix} x \sim x' \\ s,s' \text{ arbitrary} \end{matrix} \tag{2.9}$$

The generalized LSZ asymptotic condition [12]

$$\lim_{t\to\mp\infty} \phi(x;s) = \phi_{\substack{in\\out}}(x;s) \tag{2.10}$$

should be understood as a weak convergence of suitably smeared out field operator on the dense set of states. If one introduces instead of the axioms 1. - 3. , which are of LSZ type, the axioms of Wightman type, suitably adjusted for the field $\phi(x;s)$, the generalized LSZ asymptotic condition can be proved along the lines of Hepp's argumentation [10b].

The conventional theory, with basic field operator $\phi(x)$, describes only interacting stable particles, and the theory with the basic field operator $\phi(x;s)$ only interacting elementary objects with continuous mass spectrum. The most interesting from physical point of view is the case when the theory describes the interaction between stable particles and unstable objects, which can be formed out of the elementary objects with continuous mass spectrum. Such theory is defined by <u>two</u> field operators $\phi(x)$ and $\phi(x;s)$, and occurs for example if the field operator $\phi(x)$ describes the interacting S-wave bound state, and the field operator $\phi(x;s)$ represents the rest of interacting S-wave, with the continuous mass spectrum. In such theory the asymptotic space is the tensor product of two Fock spaces, spanned by the state vectors (2.2) and (2.6).

It is interesting to mention that it is possible to construct the operators $a(\vec{p};s)$, satisfying (2.4), in the Fock space generated from the vacuum by usual creation and annihilation operators. We have

$$a(\vec{p};s) = \frac{1}{\rho^{1/2}(s)(2p_0)^{1/2}} \int d^3k \; \frac{\delta[p_0-(\omega(\frac{\vec{p}}{2}+\vec{k})+\omega(\frac{\vec{p}}{2}-\vec{k}))]}{(2\omega(\frac{\vec{p}}{2}+\vec{k}))^{1/2}(2\omega(\frac{\vec{p}}{2}-\vec{k}))^{1/2}} \times$$

$$\times\; a(\frac{\vec{p}}{2}+\vec{k})\; a(\frac{\vec{p}}{2}-\vec{k})\,|0\rangle\langle 0| + \ldots \tag{2.11}$$

where $\omega(\vec{k}) = (\vec{k}^2+m^2)^{1/2}$, $p_o = (\vec{p}^2+s)^{1/2}$ and $\rho(s)$ is the two-particle phase space factor. The proof that the general formula exists has been given by Streit [13] and studied in the zero-dimensional case by Brandt and Greenberg [14].

3. THE LAGRANGEAN FORMALISM AND GENERALIZED FEYNMAN DIAGRAMS

In this Section we are going to introduce for the field $\phi(x;s)$ the Lagrangean formalism just in order to get the rules how to construct the corresponding generalized Feynman diagrams.

If we wish to write down the set of Feynman diagrams, representing causal propagators in conventional theory, we should know:

a) the masses and spins of stable particles - defined by the free Lagrangean density $L_o(x)$*

b) the vertices, determining local interaction of stable particles - defined by the interaction Lagrangean density L_{int}.

The $\hat{S}$-matrix operator is given by the following expression

$$\hat{S} = T \exp\{i\int L^o_{int}(x)d^4x\} \tag{3.1}$$

where $L^o_{int}(x)$ denotes the interaction Lagrangean density with substituted free field operators.

Such scheme can be repeated for the theory with the field operator $\phi(x;s)$. The notion of mass of a free particle is replaced by the notion of range M of masses of free elementary object with continuous mass spectrum. The free Lagrangian density, describing such object, can

* One can include the mass renormalization terms into L_o, what leads to the physical values of masses.

be written as follows:

$$L_o(x) = -\frac{1}{2}\int_M ds\{\partial_\mu\phi_o(x;s)\partial^\mu\phi_o(x;s) + s\phi_o(x;s)\phi_o(x;s)\} \tag{3.2}$$

where $M \subset R_1^+$ and the causal propagator which represents virtual lines in generalized Feynman diagrams, has the form:

$$\Delta^c(p^2;s,s') = \delta(s-s')\Delta^c(p^2;s) = \frac{\delta(s-s')}{p^2-s+i\,\varepsilon} \tag{3.3}$$

The presence of continuous mass parameter causes the generalization of conventional notion of local vertex. In usual relativistic local QFT the number of interaction Lagrangeans, leading to renormalizable theory, is very limited. In the theory with the field operator $\phi(x;s)$ one can construct much more general vertices, and in general the interaction Lagrangean density depends on some arbitrary s-dependent vertex factor . For example the conventional four-leg vertex can be generalized as follows:

$$\lambda\phi^4(x) \to \int_M ds_1 \;\ldots\; \int_M ds_4\; \Lambda(s_1 \ldots s_4)\phi(x;s_1) \ldots \times \\ \times \ldots \phi(x;s_4) \tag{3.4}$$

Having chosen the interaction Lagrangean L_{int} one can get, using the formula (3.1), the set of rules, determining generalized Feynman diagrams. These graphs differ from the usual ones because in every vertex we integrate the variable masses of incoming particles with the vertex factor. The theory, however, becomes simplified, if the vertex factors are separable, i.e., if for example in eq. (3.4) we assume

$$\Lambda(s_1 \ldots s_4) = \lambda\Lambda(s_1) \ldots \Lambda(s_4) \tag{3.5}$$

If it is the case, the replacement (3.4) can be written

as follows

$$\lambda\phi^4(x) \rightarrow \lambda\Phi^4(x) \tag{3.6}$$

where the field operator

$$\Phi(x) = \int_M \Lambda(s)\ \phi(x;s)\ ds \tag{3.7}$$

denotes the interacting field operator for an elementary unstable object [6]. The theory with separable vertex factors becomes four-dimensional one. In the interaction representation, characterized by means of the replacement $\phi(x;s) \rightarrow \phi_o(x;s)$ the field operator

$$\Phi_o(s) = \int_M \Lambda(s)\ \phi_o(x;s) ds \tag{3.8}$$

satisfies the following commutation relations

$$[\Phi_o(x),\ \Phi_o(x')] = i \int_M \Lambda^2(s)\ \Delta(x-x';s)\ ds \tag{3.9}$$

and represents generalized free field [15]. According to our terminology, such generalized free field describes <u>free elementary unstable object</u> [16].

The theory, where the local interaction Lagrangean density couples the four-dimensional field operators of type (3.7) leads to the set of Feynman diagrams with the vertices described by numerical coupling constants, but the lines representing propagation of generalized free fields. Such virtual lines, with suitable choice of the spectral function $\Lambda^2(s)$, were considered as appropriate for the description of virtual unstable particles by Matthews and Salam [1a] and Schwinger [1c] . As shown by Jacob and Sachs [17] such description is in agreement with the experimental situation. The definition of an unstable free object as characterized by a continuous Källén-Lehmann spectral function was also justified on the ground of group-theoretical analysis by Lurcat [9]. We see, that our formulation provides field-theoretic back-ground for these already more than ten years old ideas, and provides a physical interpretation for the generalized free fields.

4. TWO EXAMPLES

We will discuss two examples of the Lagrangean theories describing interaction of stable particles with unstable objects.

a) The Bilinear Interaction Model

We consider first the known example of Zachariasen model [18] describing the S-wave scattering obtained by the summation of the bubbles in direct channel (see Fig. 1).

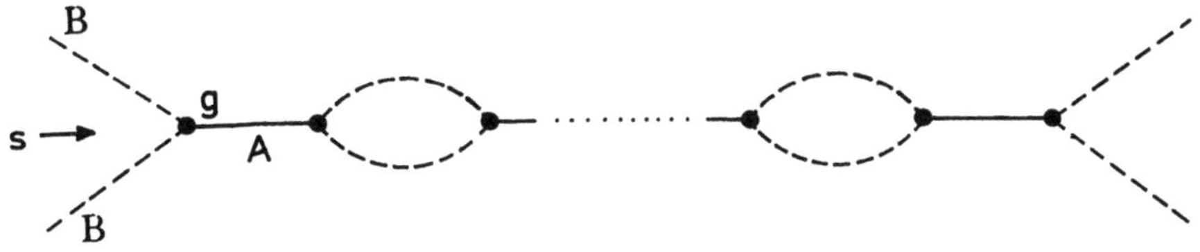

Figure 1 - The S-wave scattering diagram in Zachariasen model.

It has been shown by Thirring [19] already in 1962 that it is easy to describe such scattering by means of the Lagrangean model, provided that we introduce the field $\phi(x;s)$, describing the S-wave BB-pair with the total mass square by means of the following operator*

$$B(x) = \int_4^\infty \rho^{1/2}(s)\ \phi(x;s)\ ds \tag{4.1}$$

where $\rho(s)$ is the two-particle phase space factor. The propagator of the field $B(x)$ in the interaction representation is given by the formula

* We assume for simplicity that $m_B=1$.

$$\Delta_B(p^2) = \int_4^\infty \frac{\rho(\kappa^2)\, d\kappa^2}{p^2-\kappa^2+ i\varepsilon} \tag{4.2}$$

and represents the bubble diagram. One can express the scattering diagram on Fig. 1 as follows:

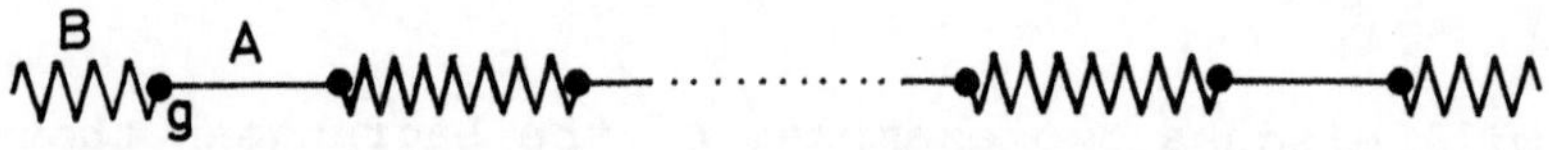

Figure 2 - The S-wave scattering diagram in the Lagrangean formulation of Zachariasen model.

The diagram on Fig. 2 one gets if

$$L_{int}(x) = g\, A(x)\, B(x) \tag{4.3}$$

The S-wave scattering amplitude T(s) is given by the sum of diagrams on Fig. 2 with the external B-lines extracted. One gets

$$T(s) = \frac{g^2}{s-m_A^2} + \frac{g^2}{s-m_A^2}\,\Delta_B(s)\,\frac{g^2}{s-m_A^2} + \ldots = g^2(s-m_A^2- - g^2\Delta_B(s))^{-1} \tag{4.4}$$

Introducing the S-wave phase shift $\eta(s)$ one can write the $\hat{S}$-matrix element for the S-wave scattering as follows:

$$e^{2i\eta(s)} = \frac{s-m_A^2-g^2\Delta_B^*(s)}{s-m_A^2-g^2\Delta_B(s)} \tag{4.5}$$

In Thirring-Zachariasen model the field operator $\phi(x;s)$ can be calculated explicitly, and commutes to the c-number. In general case if

$$[\phi(x;s),\phi(x';s')] = i\int\rho(s,s';\kappa^2)\,\Delta(x-x';\kappa^2)\,d\kappa^2 \tag{4.6}$$

one gets the phase-factor scattering of one-particle states

$$|\vec{p};s\rangle_{out} = e^{2i\eta(s)}|\vec{p};s\rangle_{in} \tag{4.7}$$

or the formula

$$\phi^{(+)}_{out}(x;s) = e^{2i\eta(s)}\phi^{(+)}_{in}(x;s) \tag{4.8}$$

where $\eta(s)\neq 0$ <u>if the spectral function $\rho(s,s';\kappa^2)$ has the singularities proportional to $(s-\kappa^2)^{-1}.(s'-\kappa^2)^{-1}$</u> . Particularly if

$$\rho(s,s';\kappa^2) = K(s;\kappa^2)\ K(s';\kappa^2) \tag{4.9}$$

and

$$K(s;\kappa^2) = A(s)\delta(s-\kappa^2) + B(s;\kappa^2)\ P\frac{1}{s-\kappa^2} + $$
$$+ \text{ terms continuous for } s \approx \kappa^2 \tag{4.10}$$

the free field operator (4.7) describes the elastic S-wave scattering with the following phase shift [20]

$$e^{2i\eta(s)} = \frac{A(s) - i\pi B(s;s)}{A(s) + i\pi B(s;s)} \tag{4.11}$$

We see, therefore, that one can get description of an arbitrary elastic resonance. Generalizing, one can conclude that <u>every</u> decoupled scattering channel can be described by a free field operator $\phi_\alpha(x;s)$, provided that the indices α describe the parameters, characterizing all angular momenta which determine this channel [21] .

b) The Decay Vertex of Yukawa Type

Let us consider the conventional Yukawa vertex

$$L_{int}(x) = g\ A^2(x)B(x) \tag{4.12}$$

describing in the case if $m_B > 2m_A = 2$ the decay of B-particle. Because the B-particles are unstable, we do not observe them at infinite times, and the zero-order approximation in conventional perturbation theory - the stable free B-particles with mass m_B - differs qualitatively from the physical B-particles. In conventional formalism one describes formally the instability by means of complex mass renormalization [4h],[6b]. It is possible to include this renormalization effects in the zero order approximation, if we resign from the notion of stable virtual B - particles. We come to the replacement of the interaction Lagrangean (4.12) by the following one:

$$L_{int}(x) = g\, A^2(x)\, B(x) \tag{4.13}$$

where

$$B(x) = \int_4^\infty \Lambda(s)\phi(x;s)\, ds \tag{4.14}$$

and $\Lambda(s)$ represents the covariant vertex factor.

The scattering amplitudes for the theory with the interaction Lagrangean (4.13) are the same as for the Yukawa coupling (4.12), with the only change that the B-lines describe the propagation of unstable elementary object, characterized by means of the spectral function $\Lambda^2(s)$, or we replace

$$\Delta_B(p^2) = \frac{1}{p^2-m_B^2+i\varepsilon} \rightarrow \Delta_B(p^2) = \int_4^\infty \frac{\Lambda^2(\kappa^2)d\kappa^2}{p^2-\kappa^2+i\varepsilon} \tag{4.15}$$

For example the lowest order diagrams, describing the contribution to the AA → AA amplitude, are depicted on Fig.3

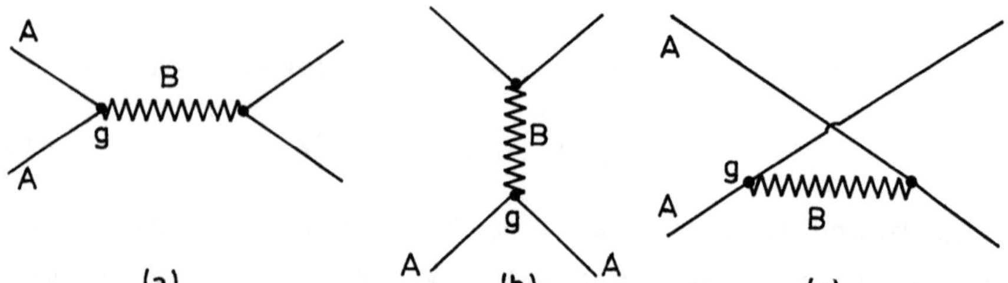

Figure 3 - The lowest order graphs with virtual unstable B-object, exchanged in s-channel, (3a), t-channel (3b) and u-channel (3c).

and give the following contribution to the scattering amplitude:

$$T^{(2)}(s,t,u) = g^2\Delta_B(s) + g^2\Delta_B(t) + g^2\Delta_B(u) \qquad (4.16)$$

The formula (4.16) is identical with the crossing-symmetric one-dimensional spectral representation of Cini and Fubini [22].

The vertex factor $\Lambda(s)$ determines all higher corrections to the elastic scattering amplitude. Particularly the lowest order diagrams which contribute to the double spectral function are given by the box diagram and correspondingly the crossed ones. The next step would be to consider the sum of ladder diagrams (see Fig. 4)

Figure 4 - The ladder diagram , contributing also to the elastic region.

The asymptotic estimates of the sum of ladders on Fig. 4 were studied by Lee and Sawyer [23], and for small momentum transfer it has been shown that one gets also for the <u>continuous mass spectrum</u> of the exchanged objects in the limit $s \to \infty$ the Regge behaviour [24].

The vertex factor $\Lambda(s)$ determines also all other scattering, production and decay processes. One of the consequences of such approach is the fact that the simple decay $B \to AA$ is always accompanied by the decays $B \to 2nA$ provided that support of $\Lambda(s)$ extends above the threshold for the production of 2n A - particles.

Further one can also discuss the renormalization program.

5. FINAL REMARKS

It should be stressed that the notion of elementarity of a given physical object depends crucially on interaction. If the object interacts always as a whole entity, and there is not present an interaction which singles out its constituents, one can say that such object is elementary. In this lecture we presented a field-theoretic scheme in the case where the elementary objects have continuous mass spectrum. Such objects are necessary if one wishes to introduce primary unstable objects.

In our considerations follows that the coupling constants are replaced by vertex factors, which are arbitrary. In order to introduce some restrictions on these vertex factors we could a) assume, that the spectral function $\sum(\kappa^2) = g^2\Lambda^2(s)$ is parametrized by the formula of Breit - Wigner type, b) impose on the Born term, representing lowest order scattering amplitude (see Fig. 3a), the elastic unitarity condition. In such way one gets

$$\sum(\kappa^2) = \frac{\rho(\kappa^2)}{\pi} [(\kappa^2-M_0^2+A(\kappa^2))^2+\rho^2(\kappa^2)]^{-1} \qquad (5.1)$$

where M_0 and $A(\kappa^2)$ are arbitrary and parametrized different choices of unitarized spectral function;
c) impose the equation of bootstrap type. For example the off-shell bootstrap vertex equation (Fig. 5) discussed extensively in [25], leads to the exponential damping of the vertex factor $\Lambda(s_1,s_2,s_3)$ for large values of s_i.

The theory, presented here, can be treated as an al-

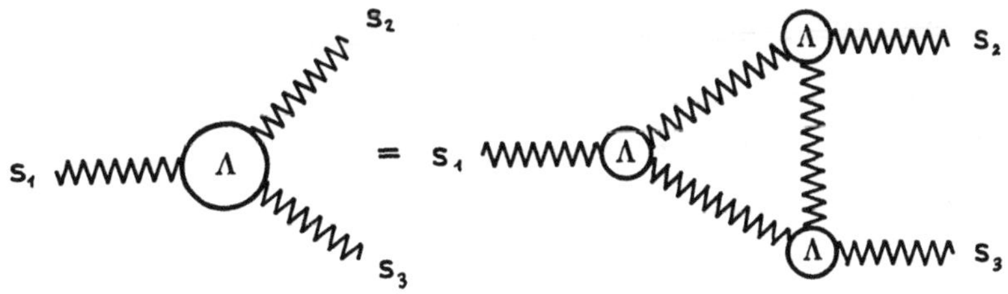

Figure 5 - The bootstrap vertex equation .

gebraization of the Faddeev approach to the multiparticle scattering [26]. In the Faddeev-Lovelace formalism one replaces the usual scattering amplitudes by means of the scattering amplitudes for composite 2-particle subsystems. In such formalism the 2→2 scattering amplitude should be treated as an input and its choice determine arbitrary in our formalism vertex factors. Full algebraization of the Faddeev approach needs, however, the introduction of infinite-component field $\phi_{\ell}(x;s)$ where $\ell=0,1,\ldots$ and describes correspondingly the S,P,... partial waves. Such formalism will be discussed in further publication.

REFERENCES

1. The unstable particles were mainly studied in the framework of $\hat{S}$-matrix theory, where the Peierl's concept of complex pole on second Rieman sheet of the propagator has been used. Below we list some papers, from "classical" period 1958 - 1963, where there were made many attempts to find the place of unstable particles in general framework of relativistic local QFT:

a) P. T. MATTHEWS, A. SALAM, Phys. Rev. 112, 283 (1958); 115, 1079 (1959).

b) M. LEVY, Nuovo Cim. 13, 115 (1959); 14, 612 (1959).

c) J. SCHWINGER, Ann.Phys. 9, 169 (1960).

d) M. IDA, Progr. Theor. Phys. 24, 1135 (1960).

e) M. HAMA, S. TANAKA, Progr. Theor. Phys. 26, 829 (1961).

f) P. J. PEEBLES, Phys. Rev. 128, 1412 (1962).

g) J. McEVAN, Phys. Rev. 132,2353 (1963).

h) M. VELTMAN, Physica 29, 186 (1963).

As basic paper for field-theoretic understanding of the properties of unstable state should be quoted N. S. KRYLOV and B. A. FOCK, JETF 17, 93 (1947). These ideas were elaborated further by L. A. KHALFIN in his papers (see DAN SSSR and JETF; 1956-1969).

2. See, for example:

a) V. GLASER and G. KÄLLÉN, Nucl. Phys. 2, 706 (1957).

b) T. OKABAYASHI and SATO, S., Progr. Theor. Phys. 17, 30 (1957).

c) T. GOTO, Progr. Theor. Phys. 21, 1 (1959).

d) W. JAUS, Helv. Phys. Acta 39, 617 (1966).

3. See, for example:

a) M. GELL-MANN and ZACHARIASEN, Phys. Rev. 124, 953 (1961).

b) C. R. HAGEN, Nuovo Cim. 50A, 545 (1967).

4. a) A. L. LICHT, Maryland Thesis 1963.

b) A. L. LICHT, Ann. of Phys. (N.Y.), 34, 161 (1965).

c) See also A. S. WIGHTMAN, Cargese Summer School Lectures, Corsica, 1964.

5. a) S. WEINBERG, Phys. Rev. 130, 776 (1963); 131, 440 (1963).

b) M. SCADRON and S. WEINBERG, Phys. Rev. 133B, 1589 (1964).

c) M. SCADRON and J. WRIGHT, Phys. Rev. 135B, 202 (1964).

6. Free objects with continuous mass spectrum were discussed from a group-theoretical point of view by F. LURCAT, Phys. Rev. 173, 1461 (1968).
7. See also J. LUKIERSKI, "Field Theory Describing Interacting Two-Particle Subsystems, I. General Formalism", Nuovo Cim., in press.
8. LUKIERSKI, J., Proceedings of Balaton Seminar on Hadrons, September 1968, publ. Acta Phys. Hungarica, 1969.
9. See, for example:
   a) R. JOST, "General Theory of Quantized Fields", Providence, R. I. 1965.
   b) R. F. STREATER and A. S. WIGHTMAN, "PCT, Spin and Statistics, and All That", Bejamin Inc., New York 1965.
10. a) H. LEHMANN, K. SYMANZIK and W. ZIMMERMANN, Nuovo Cim. 1, 205 (1955).
    b) K. HEPP, Comm. Math. Phys., 1, 95 (1965).
11. See [6]: also H. JOOS, Fortschr. d. Phys. 10, 65-146 (1962); A. J. McFARLANE, Rev. Mod. Phys. 34, 41 (1962) and Journ. Math. Phys. 4, 490 (1963).
12. See [4a-b] and [7]. The proof of generalized LSZ asymptotic condition has been given by W. KARWOWSKI and N. SZNAJDER, Acta Phys. Polon., in press: see also W. KARWOWSKI, J. LUKIERSKI and N. SZNAJDER, Nuovo Cim., to be published.
13. L. STREIT, Helv. Phys. Acta 39, 65 (1965).
14. R. A. BRANDT and O. W. GREENBERG, University of Maryland, preprint No. 817, June 1968.
15. The notion of generalized free field was introduced by O. W. GREENBERG, Ann. Phys. 16, 158 (1961). See also G. F. DELL'ANTONIO, Journ. Math. Phys. 2, 759 (1961).
16. See J. LUKIERSKI, Bull. Acad. Sci. Polon. 16, 343 (1968).

17. R. JACOB and R. G. SACHS, Phys. Rev. 121, 350 (1961).
18. F. ZACHARIASEN, Phys. Rev. 121, 1851 (1961).
19. W. THIRRING, Phys. Rev. 126, 1209 (1962).
20. See [12], the second reference.
21. See, for example, F. R. HALPERN, Phys. Rev. 137, B1587 (1965).
22. M. CINI and S. FUBINI, Ann. Phys. 10, 362 (1960).
23. B. W. LEE and R. F. SAWYER, Phys. Rev. 127, 2266 (1962).
24. It should be mentioned, that in order to get the Regge behaviour the spectral function $\Lambda^2(s)$ should tend to zero for large values of $s \to \infty$, but can be non-integrable.
25. See, for example, J. HARTE, Phys. Rev. 171, 1825 (1968), and the references given therein.
26. a) L. D. FADDEEV, Mathematical Problems of Quantum Theory of Scattering of a Three-Particle State, publ. Steklov Institute, No. 39 (1963).
b) C. LOVELACE, Phys. Rev. 135B, 1225 (1964).

SUMMARY - FIRST WEEK*

By

S. GASIOROWICZ

University of Minnesota, Minneapolis and
Deutsches Elektronen Synchrotron (DESY), Hamburg

When I look at the program for the first week, I am struck by the fact that the subjects discussed were not quite so new, so that the process in which answers are separated from assumptions has had a chance to work. What characterized most of the first week's proceedings was a certain conservatism, and that was a good preparation for the fireworks of the second week.

I would like to say a little about each of the lecture series. I shall not discuss the seminars, since these tended to be very specialized. I am not sure that I am capable, in many cases, to give a technical critique, and it would go against the general spirit of tolerance that has enveloped this difficult field, to stand on the outside and criticize this or that assumption as unfounded .

Let me start chronologically and alphabetically with BARUT's lectures, in which a variety of topics, only loosely connected, were discussed. Barut first discussed massless particles in the S-matrix kinematics. There is an obvious difference between particles with and without mass, as is evident from a helicity count. This difference is taken care of in the description of the transformation properties of the one-particle states under the

---

* given at the VIII. Internationale Universitätswochen für Kernphysik, Schladming, Feb. 24 - March 8, 1969 .

Poincaré group. If, however, one requires that the scattering matrix elements, have the proper kinematical structure so that the $m \to 0$ limit can be taken, and if one requires that poles of the type $(p_i \cdot k)^{-1}$ appear - these correspond to the massless particle of momentum k being absorbed by the external particle of momentum p_i - then Barut shows that (a) for J = 1 the residues for these poles, F_i , satisfy the condition $\sum F_i=0$ (charge conservation); (b) for J=2 the residues satisfy the condition $F_i=F$ (universal coupling), and (c) for $J \geq 3$, all the $F_i=0$. The last result was surprising to me, but contrary to first impressions it does not mean that massless spin 3 particles cannot exist; it only means that they do not couple in the low frequency limit; for example, there cannot be any classical fields corresponding to $J \geq 3$.

Next Barut considered an interaction involving two massless particles, e.g. $e+\nu \to \nu+e$. The kinematics here dictate a V-A form (this is perhaps even more obvious with the conventional notation than with that employed by Barut), but the usual coupling constant may be replaced by G(s,t), for which a kind of strip approximation without subtractions was written down. This led to an integral equation that looked like

$$G(s,t) = \int ds' \frac{s'-m_e^2}{s'(s'-s)} |G(s',t)|^2 + \ldots$$

and Barut expressed the hope that this would have a unique solution, and that G(0,0) would correspond to the weak coupling constant. In electron mass units one would expect $G \simeq 35$. In dealing with the $k \to 0$ limit for J=0 particles, Barut tried to formulate the condition $\lim_{k\to 0} M(k)=0$ (equivalent to the "gauge invariance" $\phi \to \phi+\text{const}$) in terms of the invariance of the matrix element under dilatation of the (say) pion momentum alone. I must say that I had trouble with this notion, since after taking energy conservation into account there is an ambiguity in whether you express things in terms of k (for which you require invariance under $k \to \lambda k$) or in terms of other variables which are presumably not changed. I may not have understood the argument, but it is clear that there is no "invariance" justification for the procedure, even though it does, from $M(k) = \lambda^{-1} M(\lambda k)$ yield $M(0) = 0$.

I have a similar uneasy feeling about the use of O(4,2). Here again the group is not a symmetry of a system. Rather, the idea is that the particle spectrum, which should finally emerge out of the Grand Hamiltonian, the Saturation of the Current Commutation Relations or the Ultimate Bootstrap, can be guessed at before all of these problems are solved. For this one used infinite component wave equations. There are some successes - for the scoffers it should be pointed out that Barut gets $(t-a)^{-2}$ for the electromagnetic form factors, a simple, approximately correct result which neither current commutators, nor Regge poles, nor anything else give us the slightest clue about. There are also troubles, "ghosts" of imaginary mass, and I do not believe that Barut's interpretation of them as u-channel singularities will work, since there are amplitudes that are even in $s \leftrightarrow u$ for which this interpretation will not work.

Next FLAMM presented a detailed survey of the nonleptonic hyperon decays, starting with a phenomenological description of what can be measured, and what has been measured. As far as theory is concerned, one must distinguish two appraoches. One is more "fundamental" since it starts with the current-current interaction, and takes for the currents the usual Cabibbo weak currents. With this approach the $\Delta T = 1/2$ rule is not understood. If one assumes it, by taking $H_W^{(27)} = 0$ then the soft pion techniques have had some success with the s-wave decays, but not with the p-wave decays. Flamm briefly described a calculation by Chan where the same assumption + 6 parameters give a reasonable description of the phenomena. This is not one of the great successes of theory! A less fundamental approach is to treat the weak Hamiltonian as a i-spin 1/2 "spurion". Here the connection with the rest of the weak interactions is dropped, and one looks for simplicity in the correlation of various nonleptonic decays with each other. Flamm described his calculations using the quark model. As frequently happens with the quark model, some things emerge immediately and very nicely, e.g. $S(\Sigma_{+}^{+})=0$, which agrees with experiment. For other things, e.g. the p-wave decays one must make more definite models, and Flamm described his work on a model that, with 3 parameters, fits the data. The parameters are reasonable, but I am not an expert on quarks, and would not know whether the particular quark wave functions agree with other ones constructed to fit other low energy phenomena.

I had mentioned soft pion techniques. The validity of extrapolating from threshold with $m_\pi^2=0$ to the physical region is a dynamical question, and this was the subject of a very beautiful series of talks by FURLAN. Furlan spoke about the saturation of commutation relations, e.g.

$$\langle p_2|[Q_\beta^A,Q_\alpha^A]|p_1\rangle = i\, e_{\beta\alpha\gamma}\langle p_2|Q_\gamma|p_1\rangle$$

Intermediate state contributions can be classified as in the figure shown below

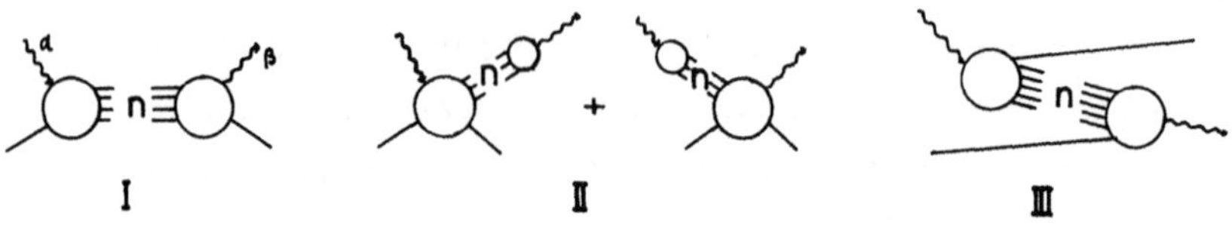

and in different frames different contributions are emphasized. The commutation relation may be rewritten in the form

$$1 = (1-\frac{m^2}{E^2})g_A^2 + \frac{1}{\pi}\int\frac{d\nu}{\nu^2}\int du\,\delta(u-\frac{\nu^2}{E^2})D^{(-)}(\nu,u)$$

where $D^{(-)}$ is the antisymmetric absorptive part for the scattering of the "divergence of the axial current" by a nucleon in the frame in which the nucleon has energy E. The path of integration in the (ν,u) plane is a parabola, and depending on the choice of parabola one gets different contributions to $D^{(-)}$. Thus Fubini and Furlan were able to sharpen the Weisberger derivation of the Adler-Weisberger sum-rule, show that E=M gives the Weinberg-Tomozawa form of the sum-rule, $E=\infty$ the Adler-Weisberger form, and in general discuss under what circumstances soft pion techniques are good. The goodness of extrapolation depends on (i) the smallness of m_π^2 compared with the mass difference between the nucleon and the effective position of the $\frac{1}{2}^-$ contribution, when a frame is chosen which emphasizes I; (ii) the smallness of m_π^2 compared with the mass of the effective pion isobar (i.e. $\gtrsim (3m_\pi)^2$) in the cases that II contributions are emphasized. Through this analysis of Fubini and Furlan, one can say that effectively not much more can be said within the present framework of our poor knowledge of higher mass states and our inability to play the unitarity game.

This framework may be changing through some developments which were discussed by HORN and in the second week by JACOB. Horn gave a very interesting and conservative discussion of finite energy sum rules (FESR). On the basis of a very few assumptions, analyticity (or even less) and Regge asymptotics for an amplitude F one gets a class of relations of the form

$$\frac{1}{N}\int_0^N d\nu\ \nu^n\ \mathrm{Im}\ F(\nu,t) = \sum_i \beta_i(t)\ \frac{N^{\alpha_i(t)+n}}{\alpha_i(t)+n}$$

where N is in the "asymptotic" region and α_i, β_i are Regge pole parameters. The usefulness of the relations stems from the fact that one apparently can choose N low enough, in a region in which one can still describe the amplitudes as sums of resonances (N $\gtrsim$ 1 GeV for amplitudes in which resonances tend to add up with the same signs; N $\gtrsim$ 3 GeV for the other ones). Thus one can relate Regge pole parameters to resonance parameters. Horn described his calculations for the pion-nucleon system, and the successes in fitting the ρ trajectory parameters, e. g. the zero of $\alpha_\rho(t)$ near $t = -0.6\ (\mathrm{GeV})^2$. In addition Horn discussed the extrapolation of these ideas to the suggestion that this overlap of the resonance and asymptotic region is not a lucky accident, but a reflection of a "duality" of descriptions in resonance language or asymptotics language of the whole energy region. Economy of parameters leads to a preference for the first at low energies and the second at high energies,but these are related. This duality emerges from using FESR as bootstrap equations, and it has an interesting byproduct, the prediction of exchange degeneracy among trajectories as a result of the absence of "exotic" resonances. Horn completed his discussion with an introduction to the Veneziano model which contains all of these notions and which was discussed in detail by Jacob.

KLAUDER spoke about Hamiltonian quantum field theory. Specifically he discussed operator theory, the significance of some of the assumptions that are usually made, and some recent work of Glimm and Jaffe on the $\lambda\phi^4$ theory in two dimensions. I am very much of a nonexpert in this field: I can understand the words but cannot carry the tune, so that the best I can do is refer you to the lecture notes for details. The important result described there is the existence of a nontrivial quantum field theory. This does not, as yet, have much to do with the physics of particles, but one may hope that out of these studies will emerge some new, powerful approximation techniques. It is only with the help of these that we will be able to decide whether local field theory is as good for strong interactions as it is for electromagnetic ones. When these new methods come, we will at least know who to go to explain them to us.

In conclusion, on behalf of those who had to leave early, and those who were fortunate to be able to stay until today, I would like to thank Professor Urban and his staff for a very wonderful two weeks.

SUMMARY - SECOND WEEK*

By

Herbert PIETSCHMANN

Institute for Theoretical Physics,
University of Vienna

1. INTRODUCTION

Another of the very fine Schladming winterschool has come to an end. We have heard a lot and learned a great deal.

If we try to look into physics in general and high energy physics in particular as it stands in our days we find that physics has changed a lot over the few hundred years of its existence.In the old times it was,for instance, the prediction of new planets by the distortion they caused to the orbits of inner ones that turned out to be one of the highlights of the history of human investigation of nature. When these planets were found at the predicted location, this was a proof of the success of mathematical methods in science.

In the first half of this century it was Yukawa's prediction of the meson or Dirac's prediction of the antiparticles that paralleled the old tradition. Even in recent years Gell-Mann's prediction of the Ω^- was one of the corner stones for the foundation of SU(3).

It seems that so far a good indication for the value of a theory has been the fact that its predictions were met by experience. Today, however, this gradually ceases to be so. Needless to say that it is a very regretable

---

* given at the VIII. Internationale Universitätswochen für Kernphysik, Schladming, February 24 - March 8, 1969.

fact. It seems that today a theory can still be valued by means of its predictions. But it is not the agreement with experiment, rather whether or not the theory makes definite predictions at all. (By "definite prediction" we mean a prediction which can possibly be checked within the present generation of physicists, say.)*

In this light it seems worth while to go back to last years Schladming proceedings and look upon some of the predictions. It turns out that there were many. Most of them are more or less forgotten. So let me just take the most spectacular one: S. L. Glashow predicted the so-called "U-matter". Among other things it should concentrate on deep sea bugs and he said: "Great scientific and technological vistas would be revealed, should U-matter really exist!"**

After this introduction it seems natural to order the five lecture series of the second week according to their immediate link to experience. We will therefore start with

2. G. CONFORTO:

Experimental Tests of CPT, CP, T and C Invariance

Confortos lectures form a very nice and self-contained representation of the material. Hence there are no remarks to the actual content. However, it is worth while to contemplate the origin of CP violation a little bit. It was found in 1964 by accident. We should always keep this in mind when we are looking for arguments in favour of a new larger accelerator or the like. Stretching the point to the extrem we can say that whenever we know exactly what

* In our days physicists are so busy to write down their ideas that they have no time to think them up.

** After this revelation the mere fact that no high energy physicist has gone into a "sea lab" so far is very strange. However, it may be the reason, why so many physicists go into biology these days.

the interesting outcome of a new machine will be, the money is almost certainly waisted. It is the new unknown and unexpected discovery we are out for when we design a new experiment.

As far as the theoretical side of CP violation is concerned, we can only state the fact that we have no theory as yet. Before we can even attack this problem, two more fundamental questions have to be answered which are still open: a) What kind of interaction violates CP (There was a seminar by O. Nachtmann on this subject to which we will come back later); b) What experiments besides the K^o-system will show the effect.

All we know is that CPT is still good. There was a comment from the floor, saying: "If quantum field theory is extended, it is easier to do without CPT than with CPT"; the comment came from an axiomatician and is therefore certainly correct. However, it seems to me that in theories which aim at being used to calculate numbers, CPT is still the most fundamental concept.

3. M. GOURDIN:

e^+e^- Collisions in Storage Rings

Gourdin's lectures were extremely nice from two points of view. Firstly, they were close to experiments in the sense that they dealt with calculations which are needed for experimental progress. Secondly they were also ment for younger physicists and this is certainly one of the important goals of this winterschool*

Out of the many facts and derivations he presented very neetly and to which I have no additions let me just

---

* Gourdin called his lectures apologetically "Physics with Integrals" which is symptomatic for the "AA" (Algebraic Age).

emphasize one thing. If one defines

$$\langle 0 | j^{em} | V \rangle \propto \frac{m_V^2}{f_V^2}$$

he can try essentially 3+1 models of SU(3) breaking. One can apply the SU(3) relation to $1/f_V$, m_V^2/f_V or m_V/f_V. In the last case one has the two alternatives of using the mixing angle of the m^2 or of the $1/m^2$ mass formula. With exactly three experimental numbers, Gourdin killed three out of four models.*

The three numbers used were

$$\Gamma(\omega \to e^+e^-)$$

$$\Gamma(\phi \to e^+e^-)$$

$$tg^2\Theta_Y$$

I cannot help being very satisfied by the reduction of models whose number seems to outgrow even the worst population explosion.

4. M. SUGAWARA:
Off Mass Shell Approach to Baryon Decays

Sugawara dealt with processes of the kind $B \to B' + \pi$. He wrote down the matrix element

$$\langle B' | j_\pi | B \rangle$$

took out one of the baryons, used current algebra and soft pion techniques and obtained self-consistency conditions. In this way he arrived at a number of very nice explana-

* For this achievment we should award him the name "Model Sleigher" which should be considered one of the highest awards in high energy physics.

tions of the baryon mass spectrum. For example: for decays with change of spin the final particles must form a quartet. It follows that baryons with spin 1/2 and 3/2 have to form quartets which they actually do.

In order to facilitate the task of next years "Summerer"* I would like to list some predictions so that they can easily be dugged out

Particle	Spin Parity	Mass(MeV)	Comment
Σ	$\frac{7}{2}^-$	1924	
Δ	$\frac{1}{2}^-$	1900	broad, highly inelastic
Δ	$\frac{3}{2}^-$	1700 - 1800	not definite
Σ	$\frac{1}{2}^-$	. 1700	
Λ	$\frac{1}{2}^+$	2030	very broad
Σ	$\frac{3}{2}^-$	> 1700	
Λ	$\frac{3}{2}^+$	1600	extremely narrow

I would like to mention that in the sense of the introduction Sugawara has produced a very good theory, since it can certainly be tested. A word of caution may be in order about the relative F-D coupling because these conclusions were reached by means of soft K-mesons.

* Not to be confused with "Winterer"

5. M. JACOB and K. DIETZ:

The Veneziano Model

After the simple minded Regge model, SU(3), higher symmetries and current algebra it is now the Veneziano model that occupies the vast majority (including the "back ground") of high energy theoreticians. Therefore, it naturally took two of the main lectures. The importance of this recent development need not be emphasized since it is already clear from the percentage of publications dedicated to this field in the current literature.

Let me therefore start with two comments which I cannot help recalling when it comes to this field. First of all there is the philologically nice nomenclature[*] . It reminds me to a quotation of Ludwig Boltzmann who said about philosophy: " Die Philosophie ist der ständige Mißbrauch einer eigens dazu erfundenen Terminologie" .[**]

Secondly, the fact that a theoretical model can really predict so many details although it does not fulfill all the axioms, is very interesting[***](we should be familiar with this situation from unrenormalizable theories like weak interactions or even from quantum electrodynamics although the situation is somewhat different there). Once these two little grains of salt are swallowed, the rest is extremely nice and a lot of open problems can be understood.

---

* "Daughters with poles that violate crossing of ancestors and a lot more Freudian stuff"

** Today he would probably say the same thing but not necessarily restricted to philosophy.

*** Once upon a time there was a Swami in India who claimed that he could walk on water. I tried to understand this with a theoretical model and to my great surprise it was easily possible. There is the surface tension which keeps him from diving in (In this model one has to neglect gravity of course).

Jacob presented duality and showed the impressive power of explanation. We have to be grateful to him in particular for the nice and intelligible presentation of his material. K. Dietz dealt with a bootstrap of rising trajectories.* The question whether Regge trajectories are indefinitely rising or will eventually bend down again is certainly of great importance. I would like to point out that in the case of indefinitely rising trajectories considerations may come into play which are usually neglected in high energy physics. This is for the simple reason that infinity is such a terribly large number. When the mass of a particle increases its Compton wave lengths decreases and will eventually reach Planck's length derived from the gravitational coupling constant. Hence a particle of rest mass larger than 10^{19} BeV will be unstable against gravitational collaps and therefore this mass is a ceiling which no trajectory can break through.

A very decisive test of the model is whether or not the daughter particles $\varepsilon^{0}, \kappa, \rho'$ and so on will actually be found. It seems to me that there is certainly something fundamental within this approach. We should not forget that the object of this investigation is perhaps the most complicated one: Dynamics of strong interactions.

6. SEMINARS

There where six seminars in the second week.

J. Lukierski: Description of Variable-Mass States in Quantum Field Theory

W. Garczynski: Stochastic Approach to Quantum Mechanics and Quantum Field Theory

W. Zakrzewski: Regge Pole Couplings

---

* Bootstraping was originally invented by Münchhausen but he did not know that one can use it also to produce daughters.

W. Sandhas: Three-Body Problem

K. Dietz: Nonlinear Representation of SU(3)⊗SU(3)

O. Nachtmann: CP Violation and Cosmological Fields

Out of these I would like to discuss two. W. Sandhas talked about what is called the"AGS" (Alt-Grassberger-Sandhas) approach to Faddeev's equations. I would like to point out the amazing agreement of his calculations which have no free parameters. We can congratulate him to his results for n-d scattering and break up as well as

$$d + d \rightarrow p + H^3$$

O. Nachtmann talked about the possible cause of CP violation by cosmological fields. It turns out that such a field of scalar type is not jet ruled out if its coupling constant is about 10^{-7} - 10^{-13} times the usual gravitational coupling constant. I would like to reiterate that it is one of the main questions of high energy physics to find out where the CP violating interaction fits into the hierarchy of interactions.

At the end of each winter school in Schladming it is the honour and the pleasure of the last speaker to express the sincere thanks of all participants to Professor Dr. Paul Urban and his organizing committee.

This year I have the outstanding satisfaction to announce to you that Professor Urban has been awarded the "Großes Silbernes Ehrenzeichen für Verdienste um die Republik Österreich" for his extremely valuable contributions to high energy physics by the organization of this winter school. Let us thank him and his staff again very cordially.

Errata

Page 102: In Table 3 in the last line it should be: $\langle B_f | O_B{}^S | B_i \rangle = 0$ and $\overline{S} = 1.225$.
For $P(\Xi_-^-)$ a graph has been omitted. A corrected and extended version of this model which essentially gives the same results can be found in: Refined Quark Model of Weak Hadron Decays, D. FLAMM and W. MAJEROTTO, preprint (to be published in Nuclear Physics).

Page 103: On the top of the page the following sentence is missing: .... right sign for the trouble maker $P(\Xi_-^-)$. To calculate the decay amplitudes we have to express the graphs in Fig. 2 and Fig. 3 by spin unitary spin operators O_A and O_B, respectively, which then are sandwiched between the baryon quark wave functions [15].

GPSR Compliance

The European Union's (EU) General Product Safety Regulation (GPSR) is a set of rules that requires consumer products to be safe and our obligations to ensure this.

If you have any concerns about our products, you can contact us on ProductSafety@springernature.com

In case Publisher is established outside the EU, the EU authorized representative is:

Springer Nature Customer Service Center GmbH
Europaplatz 3
69115 Heidelberg, Germany

Batch number: 10726298

Printed by Printforce, the Netherlands